T0191924

Quantum Computers

Belal Ehsan Baaquie · Leong-Chuan Kwek

Quantum Computers

Theory and Algorithms

 Springer

Belal Ehsan Baaquie
Helixtap Technologies
Singapore, Singapore

Leong-Chuan Kwek
Centre for Quantum Technologies
National University of Singapore
Singapore, Singapore

ISBN 978-981-19-7519-6   ISBN 978-981-19-7517-2   (eBook)
https://doi.org/10.1007/978-981-19-7517-2

This Springer imprint is published by the registered company Springer Nature Singapore Pte Ltd.
The registered company address is: 152 Beach Road, #21-01/04 Gateway East, Singapore 189721,
Singapore

Preface

Quantum theory introduces a fundamentally new framework for thinking about nature and entails a radical break from the paradigm of classical physics. In spite of the fact that the *shift of paradigm* from classical to quantum mechanics has been ongoing for more than a century, a complete conceptual grasp of quantum mechanics has till today proved elusive. According to leading quantum theorist Richard Feynman, *'It is safe to say that no one understands quantum mechanics'* [1].

The idea of quantum computers was pioneered by Yuri Manin (1980) and Richard Feynman (1981) when they discussed simulations that could not be carried out by a classical computer—but only by a quantum computer. Feynman discussed the ideas of quantum gates and quantum computers in [2]. In 1983, David Deutsch made a crucial advance by proposing that the quantum qubit is the natural generalization of the bit of classical computers [3]. The field over the last two decades has seen phenomenal growth, and the development of a practical quantum computer is poised to be game-changing technology of the twenty-first century.

This book is divided into the following four parts. The introduction in Part I gives a more detailed discussion of these four parts.

1. The topics in Part I have been chosen to introduce non-specialists and newcomers to this field.
2. Part II discusses in great detail some of the important special cases that illustrate the workings of a quantum algorithm.
3. Part III discusses relatively more advanced topics as well as some important applications of quantum algorithms.
4. The last part is a discussions of some of the general features of quantum algorithms.

Hopefully, after reading this book, one can have a rudimentary understanding of quantum algorithms.

Singapore

Belal Ehsan Baaquie
Leong-Chuan Kwek

References

1. Feynman RP (2007) The character of physical law. Penguin Books, USA
2. Feynman RP (2018) Lectures on computation. CRC Press, USA
3. Deutsch D (1985) Quantum theory, the church–turing principle and the universal quantum computer. Proceedings of the Royal Society

Acknowledgments

We would like to acknowledge and express our thanks to many outstanding researchers in the field of quantum computing.

We would like to thank our respective families for their unflagging support and encouragement, without which this book could not have been written.

One of us (BEB) would like to thank George Zweig, Gerald 'Gerry' Neugerbauer, Clifford M. Will, Jon Mathews, Jeffrey E. Mandula, David Mermin, David Ashcroft and Kurt Gottfried for shaping my views on quantum mechanics. He would also like to thank M. Mahmudul Karim, Behzad Mansouri and Frederick H. Willeboordse for many stimulating discussions on computation and algorithms. He had the privilege of having enlightening discussions on quantum mechanics with peerless scholars and visionaries Richard P. Feynman and Kenneth G. Wilson, for which he is eternally indebted to them.

The other author (KLC) would like to thank Luigi Amico, Berge Englert, Jiangfeng Du, Thomas Durt, Artur Ekert, Rosario Fazio, Fan Heng, David Hutchinson, Myungshik Kim, Jose Ignacio Latorre, Aiqun Liu, Christian Miniatura, Bill Munro, C. H. Oh, Xiangbin Wang, Vlatko Vedral, Marek Zukowski and many others for their constant encouragement and advice. He has also benefited a lot from numerous stimulating discussions with many of his current and past students, post-doctoral fellows and group members (A partial list follows: Davit Aghamalyan, Joonwoo Bae, Kishor Bharti, Shaohen Chew, Andy Chia, Keng Wah Choo, Tobias Haug, Masahito Hayashi, Hermanni Heimonen, Mingxia Huo, Alexander Hue, Kelvin Koor, Thiha Kyaw, Elica Kyoseva, Chee Kong Lee, Ying Li, Cao Lin, Jonathan Lau, Wei Luo, Kian Hwee Lim, David A. Herrera-Martí, Dariel Mok, Ewan Munro, Tiong Eng Ng, Wei Nie, Yuan Shen, Harshank Shrotriya, Sivakumar M. Maniam, Sai Vinjanampathy, Lingxiao Wan, Gong Zhang, Hui Zhang, Zisheng Wang, Janus H. Wesenberg and so forth).

Contents

About the Authors

Belal Ehsan Baaquie is an adviser to Helixtap Technologies, Singapore. He is a former professor of Finance at INCEIF University, Malaysia, and a former professor of Physics at the National University of Singapore. He has a Bachelor of Science degree from the California Institute of Technology and a PhD from Cornell University in theoretical physics, specializing in quantum field theory. He later developed an interest in finance and economics and started applying quantum mathematics to these fields. He has written four books on quantum finance: *Quantum Finance* (Cambridge, 2007), *Interest Rates and Coupon Bonds in Quantum Finance* (Cambridge, 2009), *Quantum Field Theory for Economics and Finance* (Cambridge, 2018) and *Mathematical Methods and Quantum Mathematics for Economics and Finance* (Springer, 2020). In addition, he has written several books focusing on topics from quantum mechanics and mathematics to books on the leading ideas in science, including *The Theoretical Foundations of Quantum Mechanics* (Springer, 2013), with his most recent coauthored book being *Quantum Computers: Theory and Algorithms* (Springer, 2023). (Amazon author page at https://www.amazon.com/-/e/ B07GCC7WQ4).

Leong-Chuan Kwek is a principal investigator at the Center for Quantum Technologies, NUS, the Co-director of the Quantum Science and Engineering Center at NTU and a faculty of the National Institute of Education, Singapore. He is the Immediate Past President of the Institute of Physics, Singapore and a Council Member of the Association of Asia Pacific Physical Societies. He is also an elected Fellow of the American Association for the Advancement of Science (AAAS) and the Institute of Physics, UK (IOP).

Chapter 1
Introduction

Quantum mechanics describes and explains the workings of nature—and is by far the most successful theory of science. Although quantum mechanics has qualitatively changed our view of nature, a satisfactory understanding of it is still far from complete. The 'enigmatic' aspects of quantum mechanics—specially the pivotal and non-classical role that measurements plays in connecting the mathematical formalism with observations—are discussed in the context of quantum computers in the concluding chapters. In fact, the workings of quantum computers throw new light on the mysteries and conundrums of quantum mechanics.

Similar to the application of quantum mechanics for the understanding of nature, quantum mechanics embodies a radical *shift of paradigm* from classical information science to quantum information science. The book focuses on bringing this shift of paradigm to the forefront.

According to quantum mechanics, the state vector encodes all the *information* about a physical system. For example, if one asks: 'What is a hydrogen atom?'—then a response based on quantum mechanics is that it is a specific state vector. In general, all physical entities are represented by state vectors.

The rather unexpected and counterintuitive result of quantum information science is that the state vector can be *designed*, using the laws of quantum physics, to encode information that originates from non-physical sources—in particular from algorithms that arise in information and computer science. Furthermore, unitary (reversible) transformations can be carried out on an initial state vector, again using the laws of quantum physics, that execute (quantum) algorithms designed and based on the requirements of computer science. What is noteworthy is that due to the enigmatic 'quantum parallelism' of quantum mechanics, the evolution of the state vector executes algorithms much faster that any classical computer—and in some cases even exponentially faster.

The focus of this introductory book is on the theoretical and algorithmic aspects of quantum computers. A few algorithms are employed to illustrate the main ideas in quantum algorithms. The main aim of this introductory text is to clarify precisely why a quantum computer is so much more powerful than any possible classical computer.

The ideas that underpin a quantum computer are explained with a minimum recourse to advanced mathematics, and wherever possible, quantum algorithm are stated in a non-mathematical language. The physics of how to create a quantum computer, the hardware, has been eschewed as it is not essential for understanding the logic of quantum algorithms. This is similar to using a laptop without knowing anything about electrical engineering.

The field of quantum computers is an emerging technology, and quantum algorithms are codes that can vastly speed up information processing, far outpacing currently used classical computers. Classical computers refer to computers that are constructed with electronics circuits and are based on the existing paradigm of computer science. Quantum computers have potential applications in a vast number of fields, such as cryptography, manufacturing, information and communications technology, finance, autonomous vehicles, pharmaceuticals, energy, logistics, artificial intelligence and so on. In recent years, there is hardly a day that goes by without some articles touching on the potential importance of quantum computers, as well as on the intense competition among various companies and countries that are trying to take the lead in developing a practical quantum computing device.

What has really caught the imagination of the public at large is the term 'quantum advantage'. This is a project to develop programmable quantum codes that can solve problems that a classical computer, including the fastest of the supercomputers, can never beat or solve—even in principle, since for a classical computer, to solve these computational problems would need a time longer than the age of the Universe. On October 23, 2019, the Associated Press reported that 'Google's leaked paper showed that its quantum processor, Sycamore, finished a calculation in three minutes and 20 s—and that it would take the world's fastest supercomputer 10,000 years to do the same thing'.[1] Although this claim was contested by IBM and other groups, what is clear is that it is just a matter of time that some form of quantum advantage (or quantum supremacy) is established. On October 28, 2021, the mantle of quantum advantage was claimed by Pan Jianwei and his group, with their quantum computing system of 'Jiuzhang' carrying out large-scale photonic Gaussian Boson Sampling (a model for quantum computing) machine 100 trillion times faster than the world's fastest existing supercomputer.[2] Quantum advantage will probably be claimed by other groups in the future as quantum computing is a rapidly evolving field.

All computing devices are based on the principles of quantum mechanics; in particular, semiconductors run on the principles of quantum mechanics. The term quantum computer does not refer to the hardware of a computer. Instead, it refers

---

[1] https://www.marketwatch.com/story/google-touts-quantum-computing-milestone-2019-10-23. The result has since been published [1].

[2] https://english.cas.cn/newsroom/cas_media/202110/t20211028_289325.shtml#::text=Dec%2005%2C%202020,tackle%20within%20a%20reas....

to *information processing quantum algorithms* based directly on the principles of quantum mechanics: algorithms that carry out computational operations that are not allowed in a classical computer are at the core of a quantum computer.

It will be seen that the algorithms in quantum computers are expressed in terms of quantum gates that refer to specific quantum circuits (hardware) that encode these quantum gates; the algorithms are directly based on the underlying quantum circuits that realize these algorithms. This is the reason that a high-level programming language like C++ has not yet emerged for quantum computers because these are largely based on intuition and thinking derived from everyday life—and which does not refer to any of the underlying gates and circuits. The current level of understanding of quantum algorithms has a logical structure that is quite counterintuitive and not similar to the algorithms employed in classical computers.

The hardware required to build a quantum computer is not discussed in this book as it belongs to the domain of quantum physics and engineering—whereas the focus of this book is on information theory and algorithms that emerges out of quantum mechanics. There are still many challenges in hardware implementation of quantum algorithms, and hardware continues to be the bottle-neck in constructing a working quantum computer. A scalable quantum computing device poses a number of formidable, but not insurmountable, technological problems. Since the primary focus of this introduction is to understand the logic of a quantum computer, there is no need for a detailed discussion on how to build such a device. This is similar to studying classical computers, where mastering the logic of the codes does not require any knowledge of semiconductors and electronics that run these codes.

In this book, we provide a brief introduction to quantum computation. All the necessary concepts are derived from first principles, and this book is self-contained in its presentation.

Part I of the book is focused on the foundations of the subject. The concept of classical computers and classical circuits and gates is reviewed, as classical algorithms provide a platform for introducing quantum algorithms. A brief review of quantum mechanics is provided as it forms the necessary foundation for understanding both quantum algorithms as well as the hardware that is required for these algorithms; however, as mentioned above, we will not discuss the hardware aspect of quantum computers. The concepts of quantum superposition (also called quantum parallelism) and quantum entanglement are discussed in order to understand the key advantages of a quantum over a classical computer. A detailed discussion of quantum gates is given as well as a derivation of phase estimation and quantum Fourier transforms, as these form the bedrock for many of the algorithms that we discuss later.

Part II discusses some well-known quantum algorithms, including the Deutsch and the Deutsch–Jozsa algorithm. The Shor and Grover algorithms, considered to be major milestones of the subject, are discussed in detail to exemplify the techniques used in quantum algorithms.

Part III discusses some of the possible applications of quantum algorithms. Option pricing is discussed in some detail as one expects that the first major application of quantum algorithms are going to be in finance. Solving linear equations using the HHL algorithm is discussed as well as quantum error correction and hybrid algorithms.

Part IV is a summary of the subject, with a general analysis of some of the key features and novel aspects of a quantum computer.

There are many references on quantum computers, both advanced and introductory, and the following are a few of them [2–9].

1.1   Interview on Quantum Computers

One of us (BEB) spoke to Future of Finance cofounder Dominic Hobson in 2021 about the near as well as the long-term prospects of quantum computers.[3] It is recommended that the interview be re-read after having gone through this book, as the technical aspects of the interview, which are not spelt out, should then be clearer. The interview has been modified for inclusion in this book (Fig. 1.1).

Hobson: How long must we wait for quantum computers to be adopted and applied?

Baaquie: 30 years have passed since the beginning of quantum computers in 1982, without, even now, any practical application. A linear progression would place applications in the distant future. However, progress in science happens through sudden discontinuous advances and hence a breakthrough can be expected. It is difficult to put a time scale on it.

Hobson: What engineering problems remain to be solved in quantum computing?

Baaquie: The fundamental problem is that quantum qubits are very fragile, with the smallest disturbance breaking what is the called quantum superposition that is the basis of quantum parallelism—and which is responsible for the exponential speeding up of a quantum algorithm. The breaking of quantum superposition is also called decoherence. This is a hardware problem that needs basic understanding of quantum physics to obtain a solution.

Hobson: Do we have enough engineers with that basic understanding of quantum physics?

Baaquie: The semiconductor industry (which had US$430 billion sales in 2019) requires quantum mechanics. The design of all semiconductor devices is based on quantum effects in materials. Semiconductor engineers could foreseeably master the physics of quantum computers if such an undertaking becomes sufficiently rewarding.

Hobson: Who—governments, venture capitalists, technology vendors and corporates—do you see investing in quantum computing at the moment and why?

---

[3] https://futureoffinance.biz/2021/07/13/where-quantum-computers-are-now-and-where-they-are-going-next/.

Where Quantum Computers Are Now and Where They Are Going Next

◆ Quantum Computing

Fig. 1.1 Interview on quantum computer's prospects. Published with permission of © Belal E. Baaquie and L. C. Kwek. All Rights Reserved

Baaquie: The USA has invested US$1.2 billion in 2019 over five years, and China is not far behind. The Singapore government has invested US$70 million in 2020 on a one-year horizon. Venture capital has invested US$300 million in 2020 compared to US$4 million in 2005: PsiQuantum has received US$215 million in investments and Rigetti has received US$190 million. Why? Governments are probably driven by geostrategic considerations, whereas venture capital investments show that the market thinks a breakthrough in quantum computing applications is close.

Hobson: Quantum computers do not always give the right answer. Is that a fundamental problem or an engineering problem?

Baaquie: The error in a quantum algorithm is a fundamental engineering problem that is linked to noise. For example, error correction can be achieved for a classical computer by sending three bits of 0 and 1 instead of a single bit; on receiving three bits, the average of the three bits is taken to correct for possible errors. There is also a phase (sign) that is carried by qubits when they are superposed and could flip its sign and needs to be corrected. So a quantum qubit would need $3 \times 3 = 9$ qubits for error correction. If one adds the quantum decoherence of the fragile quantum-superposed state, it is estimated that one needs 50 qubits to obtain a single error free qubit. This is the formidable problem—posed by errors in transmission and by quantum decoherence—for a quantum computer.

Hobson: What does quantum computing cost at the moment—measured by, say, an hour of computing time—and will that cost fall over time?

Baaquie: Rigetti apparently charges US$1000 per hour for a 36-qubit system. One expects the cost of quantum computing to fall, but again predictions are difficult to make as to when breakthrough technologies will occur that lower the cost.

Hobson: Is a hybrid of quantum and digital computing the immediate solution or the long-term path forward?

Baaquie: Hybrid computing is currently one of the favored approaches, with parameters provided by a classical algorithm and the simulation being carried out on a quantum computer. In the long run, there is no need for hybrid computing since classical computers are a special case of the quantum computer, and its algorithms can be factored into the quantum algorithm. Since qubits are much more expensive than classical bits, hybrid algorithms are a matter of cost efficiency.

Hobson: Are quantum computers indifferent to the physical substrate or device they run on?

Baaquie: Most quantum computers are devices that can support qubit states. Currently, the preferred device for a qubit is the Josephson junction. This is a two-state system based on a superconductor and hence requires low temperature. Other devices can run on photons and with the advantage of functioning at room temperature. Photonic quantum computers are based on chip-integrated silicon photonic devices and claim to be scalable and error-resistant. The most efficient device for a quantum computer at present is difficult to discern.

Hobson: What can quantum computers do now that digital computers cannot?

Baaquie: There are two famous quantum algorithms. The first is the Grover algorithm that searches for a particular entry in a database of size N—and takes $\sqrt{N}$ steps compared to N steps of a classical computer. For example, to find an entry in a ledger of 1 million entries, Grover's algorithm would take only 1000 instead of a million steps. The second is the Shor's algorithm, which can factorize a number into its primes (used in all encryption and secure communications) exponentially faster than a classical computer. The problem is that, for any practical problem, Grover's algorithm can only efficiently search databases with $N = 7$ and Shor's algorithm, as of now, can only factorize $21 = 3 \times 7$. This is because of the currently limited number of qubits and the noise problem.

Hobson: How do you write code on a quantum computer?

Baaquie: A classical computer's code is written according to one's day-to-day thinking. For instance, if one wants to add two numbers, one takes the first number and then adds to it the second number. A quantum computer code is counterintuitive, being based on the laws of quantum mechanics. To add two numbers, one has first to write the numbers in terms of the qubits. All operations (or gates, to use the classical computing term) in a quantum algorithm are special matrices called unitary transformations and so one has to decide how the operation of addition is represented by unitary matrices. For example, to add $1 + 1$ using a quantum algorithm, as discussed in Chap. 7, one would need to employ a $16 \, times \, 16$ matrix.

Quantum algorithms are at the rudimentary level of machine language with codes written in terms of gates; it is hoped that high-level languages will be developed for quantum computers that are independent of how the quantum computer realizes the algorithms, with the high-level language being task-oriented and not being tied to the hardware. When a high-level language for quantum computers is developed, it will make quantum algorithms easily accessible for applications to professionals working in many fields.

Hobson: What impact will quantum computing have on the current state of artificial intelligence (AI) and machine learning (ML)?

Baaquie: An application of machine learning is, for example, taking the present day value of stocks and applying an algorithm to generate a prediction for tomorrow's prices. Quantum algorithms are being developed to address machine learning. Similarly for AI and deep learning.

Hobson: How can firms actually access quantum computing services? For example, are they accessible from the Cloud?

Baaquie: IBM's Qiskit and Amazon's Braket provide a few qubits free of charge to the public.

Hobson: What practical applications of quantum computing are taking place in financial services today?

Baaquie: It is the view of many experts that the breakthrough for quantum algorithms is going to take place in finance. Unlike in devices such as solar panels where the algorithmic content is negligible compared to the hardware component, in finance, efficient algorithms can have a far greater impact. The reason is that even the slightest improvement in the performance of a quantum algorithm over the classical algorithms will lead to substantial gains in finance, given the massive volume of the debt and capital markets, with global debt reaching $226 trillion by the end of 2020.[4]

Furthermore, due to the financial markets constantly providing up-to-date data, quantum algorithms can be calibrated and tested far more efficiently than for hardware-oriented devices.

Hobson: In what areas of financial services will quantum computing have an effect and why?

Baaquie: There is an almost universal consensus amongst experts that two areas of finance hold the best promise for applying quantum algorithms. The first is portfolio optimization. A large hedge fund usually has a collection of 10,000 to 20,000 instruments in its investment universe from which to form its portfolio. Typically, the fund managers would like to run their models over a 30-year period.

The optimization is to maximize returns, subtracting the dispersion of the portfolio as well as transaction costs, and with a budget constraint. Classical computers are quite inadequate to exhaustively study all the different possibilities, including generating the outliers that are so important to fund managers. A quantum computer can simulate outliers since its powerful algorithms can scan all possibilities, even those

---

[4] https://www.visualcapitalist.com/global-debt-to-gdp-ratio/.

events that for a portfolio are highly unlikely. The quantum computer can compute the probability of the occurrence of these outliers—a computation that is beyond the computing power of classical computers. Outliers are not important for arbitrage but rather for risk management. Outliers can be useful for detecting financial fraud as well, since the occurrence of a fraud would most likely violate the computed likelihood of outliers.

The second application is arbitrage opportunities, based on the development of advanced pricing models that can spot arbitrage opportunities. The options market, in 2021, had a notional value of over US$600 trillion (with actual value of US$12 trillion) and to scan this market in real-time needs the computing power of a quantum computer. There are many other fields of finance where quantum algorithms can make a difference, such as the credit rating of customers, risk management and so on.

Hobson: Why is business so slow to invest in quantum computing?

Baaquie: The horizon of when one can get returns is not clear. Once there is a winning application, it will open the floodgates for benefits to end-users and businesses—and will lead to massive investment in this game-changing technology.

Hobson: Quantum computers can complete calculations that would otherwise be impossible. What are the implications of that for current cryptographic methods?

Baaquie: The Sycamore (IBM 54 qubits) performed a calculation in 200 s that would have taken the best supercomputers 10,000 years. So, clearly, the quantum computer can perform computations that no classical computer can ever perform. A quantum computer would need about 20 million qubits and eight hours (rather than requiring 1 billion qubits as previously theorized) to break the 2048 RSA code, so it is still quite a distant possibility. Shor's algorithm has the potential to completely overpower all classical encryption, but maybe not in the immediate future.

Hobson: Quantum computing has a flavor of the Manhattan Project about it. What geopolitical concerns does it foster?

Baaquie: Quantum computers will impact all military technology, cybersecurity, secure communications and so on. So leading nations in the world probably consider quantum computers to be a strategic asset that needs to be mastered at any economic cost.

Hobson: Does quantum computing create a risk that humanity is overwhelmed by a Superintelligence?

Baaquie: Yes, surprisingly enough, quantum computers do pose a serious risk to humanity for the following reason. Artificial intelligence (AI) is premised on analyzing patterns in big data and then making predictions and taking decisions. Technologies such as autonomous vehicles, cyberwarfare and so on will be greatly enhanced by combining AI with quantum computers.

There are two kinds of AI algorithms. The first is artificial narrow intelligence (ANI), which can carry out a task assigned to it such as driving an autonomous vehicle. The second is artificial general intelligence (AGI), which can do a wide range of tasks and can also decide by itself what task it will carry out. For example, consider an AGI-empowered robotic soldier. This machine will decide on who lives and who dies on the battlefield and maybe even off the battlefield. At present, AGI is comparable to human intelligence, so human beings can create devices to manage and control an autonomous AGI. But if AGI is enhanced and empowered by a quantum computer, it will have algorithms and computing power that no human being can match or fathom and can give rise to artificial superintelligence (ASI).

The quantum computer will give ASI access to unknown algorithms, patterns and predictions that no human being can understand, let alone control. The danger is that we may (inadvertently) mishandle a quantum computer-empowered ASI, since we have no clue about what it is doing. An out-of-control quantum computer-empowered ASI can wreak havoc on human society and on the environment. It is superintelligence, a quantum computer-empowered ASI, that carries the greatest threat to humanity. The development of ASI should definitely be watched closely, especially when it is combined with a quantum computer.

Hobson: What are quantum computers teaching us about our quantum mechanical universe?

Baaquie: This question is answered in the concluding Sect. 19, as one can use the formalism developed in this book to give an accurate answer.

References

1. Frank Arute, Kunal Arya, Ryan Babbush, Dave Bacon, Joseph C Bardin, Rami Barends, Rupak Biswas, Sergio Boixo, Fernando GSL Brandao, David A Buell, et al. Quantum supremacy using a programmable superconducting processor. *Nature*, 574(7779):505–510, 2019
2. Le Bellac Michel (2006) A short introduction to quantum information and computation. Cambridge University Press, UK
3. Bernhardt Chris (2019) Quantum Computing for Everyone. MIT Press, USA
4. Kaye Phillip, Laflamme Raymond, Mosca Michele (2007) Introduction to Quantum Computing. Cambridge University Press, USA
5. McMahon David (2008) Quantum computing explained. MIT Press, USA
6. N. David Mermin. *Quantum Computer Science*. Cambridge University Press, UK, 2007
7. Michael A (2012) Nielsen and Isaac L. Chuang. Quantum Computation and Quantum Information. Cambridge University Press, UK
8. Rieffel Eleanor G, Polak Wolfgang H (2011) Quantum Computing: A Gentle Introduction. MIT Press, USA
9. Zygelman Bernard (2018) A First Introduction to Quantum Computing and Information. Springer, UK

Part I
Fundamentals

Chapter 2
Binary Numbers, Vectors, Matrices and Tensor Products

In this part of the book, we briefly reviewed the general mathematical framework behind gates, bits and qubits. These topics form a recurring theme, and they are employed in all the derivations of this book.

A fundamental bit of classical computers is a binary set $\{0, 1\}$, called a binary number, and it is represented by two-dimensional vectors belonging to a linear vector space; the n-bits are represented by vectors that are elements of an n-dimensional vector space, which is isomorphic to the n-dimensional Euclidean space E_n. With this description, a quantum state built from n-bits (see discussion in Chap. 6) is an element of a complex linear vector space, with vectors having a unit norm and is called a Hilbert space.

All gates, both classical and quantum are in general $N \times N$ square matrices, with the matrices having some special features. Tenor product of matrices and vectors is a way of combining gates and bits and is discussed with especial emphasis on square matrices. Tenor product of vectors of single bits can be used to form a system with many bits.

A vector in this book is usually denoted by bold face notation $\mathbf{v}$, as is the case in most linear algebra textbooks. In quantum mechanics and for quantum computers, the Dirac notation is used in which $\mathbf{v} \equiv |v\rangle$. The Dirac notation for vectors and matrices is discussed to foreground the use of this notation in the rest of the chapters. The results of linear algebra are briefly reviewed using the conventional vector notation based on $\mathbf{v}$. A detailed mapping from $\mathbf{v}$ to $|v\rangle$ is then discussed so that readers who not familiar with the Dirac notation have clear idea of the mapping.

This chapter reviews the properties of binary numbers, vector space, matrices and tensor products. The derivations are based on [1].

B. E. Baaquie and L.-C. Kwek, *Quantum Computers*,
https://doi.org/10.1007/978-981-19-7517-2_2

2.1 Binary Representation

Binary notation is based on representing all real numbers using the base 2 instead of the decimal expansion that uses the base 10. If effect, all real numbers can be represented by a series of 0's and 1's with an expansion similar to the decimal expansion. As will become clear in this chapter, an arbitrary real number can always be represented by

$$\cdots 1000111010.001010101101111 \cdots$$

The input and output of an algorithm are strings of symbols. All symbols can be mapped to the real numbers, and all real numbers are then represented by strings of binary number 0's and 1's.

Every number has a binary expansion, similar to the usual expansion that has the base 10. For example, consider the number 203; its base 10 expansion is given by

$$203 = 200 + 0 + 3 = 2 \times 10^2 + 0 \times 10^1 + 3 \times 10^0$$

The binary expansion is given by

$$203 = 128 + 64 + 8 + 2 + 1 = 1 \times 2^7 + 1 \times 2^6 + 0 \times 2^5$$
$$+ 0 \times 2^4 + 1 \times 2^3 + 0 \times 2^2 + 1 \times 2^1 + 1 \times 2^0$$
$$\rightarrow 11001011 \quad : \quad \text{Binary representation} \tag{2.1}$$

Hence, in the binary representation, the number 203 is equal to 11001011. For a general integer x that needs n binary bits, the leading (largest) binary term is x_1, and the last term is x_n. Hence, one has the following binary representation

$$x = x_1 2^{n-1} + x_2 2^{n-2} \cdots + x_i 2^{n-i} + \cdots + x_{n-1} 2^1 + x_n 2^0$$
$$= \sum_{j=1}^{n} x_j 2^{n-j} \; ; \; x_j = 0, 1 \tag{2.2}$$

The integer x has the binary representation

$$x = x_1 x_2 x_3 \cdots x_j \cdots x_{n-1} x_n$$

The minimum is for all $x_i = 0$ since

$$x_i = 0 \; \Rightarrow x = 0$$

The maximum is for all $x_i = 1$ since

$$x_i = 1$$

$$\Rightarrow x = \sum_{j=1}^{n} 2^{n-j} = 2^{n-1} + 2^{n-2} \cdots + 2^{n-i} + \cdots + 2^1 + 2^0$$

$$= 2^n - 1$$

Hence, the range of the integer x that can represent n-bits is given by

$$0 \le x \le 2^n - 1 \tag{2.3}$$

The 2^n distinct values for the integer x is given by

$$x = 0, 1, 2, \cdots, 2^n - 1$$

Hence, one can label the different binary strings by an integer x with range

$$0 \le x \le 2^n - 1$$

For n bits, the number of different *binary strings* of length n is given by 2^n; a typical n-bits binary strings are represented by Dirac's ket vectors

$$|x\rangle = |x_1 x_2 x_3 \cdots x_j \cdots x_{n-1} x_n\rangle$$

For example,

$$|0\rangle = |00000\cdots 00\rangle \;; \; |1\rangle = |00000\cdots 01\rangle \;; \; |2\rangle = |00000\cdots 10\rangle$$
$$|3\rangle = |00000\cdots 11\rangle; \cdots ; \; |2^n - 1\rangle = |11111\cdots 11\rangle \tag{2.4}$$

The finite binary expansion for number $x < 1$

$$x = x_1 2^{-1} + x_2 2^{-2} \cdots + x_i 2^{-i} \cdots + x_n 2^{-n}$$
$$\equiv 0 \cdot x_1 x_2 \cdots x_i \cdots x_n$$
$$= \sum_{j=1}^{n} x_j 2^{-j} \;; \; x_j = 0, 1 \tag{2.5}$$

For example

$$0 \cdot 1011 = 1 \cdot \frac{1}{2} + 0 \cdot \frac{1}{2^2} + 1 \cdot \frac{1}{2^3} + 1 \cdot \frac{1}{2^4} = 0.5 + 0.125 + 0.0625 = 0.6875$$

Note that

$$2^m x = \text{integer} + 0 \cdot x_{m+1} \cdots x_n$$

Consider the binary expansion of $x < 1$

$$x = 0 \cdot x_1 x_2 \cdots x_m \cdots$$

Multiplying the binary expansion of $x < 1$ by 2^k yields $(m > k)$

$$
\begin{aligned}
2^k x &= x_1 x_2 \ldots x_k . x_{k+1} x_{k+2} \ldots x_m \ldots \\
&= x_1 x_2 \ldots x_k + 0. x_{k+1} x_{k+2} \ldots x_m \ldots \ldots \\
&= \text{integer} + 0. x_{k+1} x_{k+2} \ldots x_m \ldots
\end{aligned}
\tag{2.6}
$$

Hence, for example

$$
\begin{aligned}
\exp(2\pi i \cdot 2^k x) &= \exp\left(2\pi i (x_1 x_2 \ldots x_k + 0. x_{k+1} x_{k+2} \ldots x_m \ldots)\right) \\
&= \exp\left(2\pi i (0. x_{k+1} x_{k+2} \ldots x_m \ldots)\right)
\end{aligned}
\tag{2.7}
$$

The result above is required for phase estimation and quantum Fourier Transform, and is discussed in Chap. 8.

An arbitrary real number has the binary decimal expansion given by

$$x_1 x_2 \cdots x_k . x_{k+1} x_{k+2} \cdots x_n$$

Using the results given above, one has the following binary decimal expansion of 203.6875

$$203 = 11001011 \; ; \; 0.6875 = 0.1011 \;\; \Rightarrow \;\; 203.6875 = 11001011.1011$$

Binary numbers can be added, and all other arithmetic operations—including subtraction, division and multiplication—can be reduced to the procedure of addition. Binary addition, denoted by $\oplus$, is ordinary addition mod 2, and there is carry over, as is the case for ordinary addition. The basic binary addition, that can be used recursively to add any two binary numbers, is given by the following.

$$
\begin{array}{cccc}
 & 0 & 0 & 1 & 1 \\
\oplus \; 0 & 1 & 0 & 1 \\
\hline
 & 0 & 1 & 1 & 10
\end{array}
$$

Equivalently

$$0 \oplus 0 = 0 \; ; \; 0 \oplus 1 = 01 ; \; 1 \oplus 0 = 01 ; \; 1 \oplus 1 = 10 \; ;$$

Note there is a carry of 1 for the addition of $1 \oplus 1 = 10$, similar to ordinary arithmetic with base 10. Using the above rules, we obtain the following binary addition

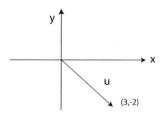

$$7 + 1 = 8 \quad \Rightarrow \quad 111 \oplus 01 = 111 \oplus 001 = 1000$$

Binary addition is discussed in great detail in Sect. 3.8 and its quantum generalization in Sect. 7.4 since the arithmetic operation reveals a lot of general features of classical and quantum algorithms.

2.2 Linear Vector Space

A vector is a collection of numbers; an example of a two-dimensional vector is the position of a particle on a plane and denoted by 3, -2. as in Fig. 2.1. A vector can be geometrically thought of as an arrow in space. An arrow has both length and direction. How does one mathematically represent the arrow? An arrow is a straight line from its base to its tip.

For simplicity, consider a two-dimensional space, denoted by E_2, and consider an xy-grid, with sides of unit length. Let **u** be a vector that can point in any direction, and is shown in Fig. 2.1. A vector, such as **u**, will be denoted by a boldface letter to differentiate it from a single number.

There two rudimentary and fundamental operations that one can perform with vectors, which are the following.

- **Vector addition**: From two vectors **u**, **v**, one can form a third vector which is their vector sum and given by

$$\mathbf{w} = \mathbf{u} + \mathbf{v}$$

One can think of the vector **w** as the result of starting with vector **u**, with its base at the origin, and then placing the vector **v** on the tip of vector **u**; the resulting vector **w** starts at the origin, which recall is the origin of vector **u**, and ends at the tip of vector **v**. In component notation, let

$$\mathbf{u} = \begin{bmatrix} a \\ b \end{bmatrix} \; ; \; \mathbf{v} = \begin{bmatrix} c \\ d \end{bmatrix}$$

Vector addition, which was described geometrically, is expressed by adding the coordinates in the x and y-directions separately and yields

$$\mathbf{w} = \mathbf{u} + \mathbf{v} = \begin{bmatrix} a+c \\ b+d \end{bmatrix} \tag{2.8}$$

Vector addition is associative in the sense that

$$(\mathbf{u} + \mathbf{v}) + \mathbf{w} = \mathbf{u} + (\mathbf{v} + \mathbf{w}) = \mathbf{u} + \mathbf{v} + \mathbf{w}$$

- **Scalar multiplication**: for a real or complex number λ, one has a new vector $\mathbf{w}$ given by

$$\mathbf{w} = \lambda \mathbf{v}$$

By scalar multiplication, the length of a given vector is changed. If $\lambda = 2$, the vector $\mathbf{w}$ has twice the length of $\mathbf{v}$ and points in the same direction and for $\lambda = -2$, the vector $\mathbf{w}$ has twice the length of $\mathbf{v}$ but points in the *opposite direction*. In general, λ can be a complex number.

- **Scalar product**: To define the scalar product one needs to define Hermitian conjugation, denoted by †. In general, the components of a vector can be complex numbers. The Hermitian conjugation of a complex vector is given by taking its transpose complex conjugating its components and yields

$$\mathbf{u}^\dagger = \begin{bmatrix} a^* & b^* \end{bmatrix} \; ; \; \mathbf{v}^\dagger = \begin{bmatrix} c^* & d^* \end{bmatrix}$$

Two vectors $\mathbf{u}$, $\mathbf{v}$ having the same number of components can be combined using the procedure of a scalar product; for the two-dimensional vectors, we have

$$\mathbf{u}^\dagger \cdot \mathbf{v} = a^* c + b^* d = \left(\mathbf{v}^\dagger \cdot \mathbf{u} \right)^*$$

For $\mathbf{u}$, $\mathbf{v}$ having n-components having complex values, we have

$$\mathbf{u}^\dagger \cdot \mathbf{v} = \sum_{i=1}^{n} u_i^* v_i = \left(\mathbf{v}^\dagger \cdot \mathbf{u} \right)^*$$

The norm of a n-dimensional vector is defined by the following

$$|\mathbf{u}|^2 = \mathbf{u}^\dagger \cdot \mathbf{u} = \sum_{i=1}^{n} u_i^* u_i > 0 \tag{2.9}$$

2.3 N-Dimensional Complex Linear Vector Space

The discussion for two-dimensional vector space E_2 generalizes to N-dimensions. Let $\mathbf{e}_i$, $i = 1, 2 \ldots, N$ be N orthogonal basis states with unit norm. Then

$$\mathbf{e}_i^\dagger \cdot \mathbf{e}_j = 0 : i \neq j \; ; \; \mathbf{e}_i^\dagger \cdot \mathbf{e}_i = 1 \tag{2.10}$$

A realization of the basis vectors is given by

$$\mathbf{e}_n = \begin{bmatrix} 0 \\ \vdots \\ 1 \\ 0 \\ \vdots \\ 0 \end{bmatrix} \; ; \; \mathbf{e}_n^\dagger = \underbrace{\begin{bmatrix} 0 \ldots 1 \, 0 \ldots \end{bmatrix}}_{n\text{-th position}} \tag{2.11}$$

A compact way of writing Eq. 2.10 is the following

$$\mathbf{e}_m^\dagger \cdot \mathbf{e}_n = \delta_{n-m} \equiv \begin{cases} 1 \; n = m \\ 0 \; n \neq m \end{cases} : m, n = 1, \ldots, N \tag{2.12}$$

Note δ_{n-m} is the Kronecker delta function and has only integer arguments.

The basis states are linearly independent since

$$\sum_{i=1}^{N} a_i \mathbf{e}_i = 0 \; \Rightarrow \; a_i = 0 \text{ for all } i$$

The linear span of $\mathbf{e}_i$ is all linear combinations of the basis vectors and covers the entire space E_N. To represent an arbitrary vector $\mathbf{x}$ in E_N, the following notation is used

$$\mathbf{x} = \begin{bmatrix} x(1) \\ \vdots \\ x(i) \\ \vdots \\ x(N) \end{bmatrix} \; ; \; \mathbf{e}_i^\dagger \cdot \mathbf{x} = x(i) \tag{2.13}$$

In terms of the basis vectors, $\mathbf{x}$ can be represented by

$$\mathbf{x} = \sum_{i=1}^{N} x(i)\mathbf{e}_i \; ; \; \mathbf{x}^\dagger = \sum_{i=1}^{N} x^*(i)\mathbf{e}_i^\dagger$$

The scalar product for vectors x, y, using Eq. 2.10, is given by

$$\mathbf{x}^\dagger \cdot \mathbf{y} = \sum_{i,j=1}^{N} x^*(i)y(j)(\mathbf{e}_i^\dagger \cdot \mathbf{e}_j) = \sum_{i,j=1}^{N} x^*(i)y(j)\delta_{i-j} = \sum_{i=1}^{N} x^*(i)y(i)$$

The angle θ, which in principle can be complex, between two vectors $\mathbf{x}, \mathbf{y}$ in the N-dimensional complex vector space is defined, as one would expect, by the following

$$\mathbf{x}^\dagger \cdot \mathbf{y} = |\mathbf{x}||\mathbf{y}|\cos(\theta) \; ; \; |\mathbf{x}| = \sqrt{\sum_i x^*(i)x(i)} \; ; \; |\mathbf{y}| = \sqrt{\sum_i y^*(i)y(i)}$$

A Hilbert space for N-dimensional linear vector space is defined to be the collection of N-dimensional vectors that have unit norm, namely

$$|\mathbf{u}|^2 = \mathbf{u}^\dagger \cdot \mathbf{u} = \sum_{i=1}^{N} u_i^* u_i = 1 \; ; \; \mathbf{u} \in \text{Hilbert space} \tag{2.14}$$

The Hilbert space consists of vectors that lie of the N-dimensional sphere S^N. For $N = 2$, Hilbert space is the circle S^1, and for $N = 3$, it is the two-dimensional surface of a sphere S^2.

2.4 Matrices

In general, a matrix is an array that consists of M-rows, that are arranged horizontally, and N columns, arranged vertically. A typical $M \times N$ matrix A is given below.

$$A = \begin{bmatrix} a_{11} & a_{12} & \cdots & a_{1N} \\ a_{21} & a_{22} & \cdots & a_{2N} \\ \cdots & & & \\ a_{M1} & a_{M2} & \cdots & a_{MN} \end{bmatrix}$$

The individual entries of the matrix, denoted by a_{mn}, are the elements of the matrix. The elements can be real or complex numbers. For matrix A, there are in total MN elements.

A particular row for matrix A consists of N elements; for example, the second row is given by

$$\begin{bmatrix} a_{21} & a_{22} & \cdots & a_{2N} \end{bmatrix}$$

One can think of a row as the transpose of a vector. Similarly, each column is given by M elements and can be thought of as vector of length M. The second column for matrix A is given by

$$\begin{bmatrix} a_{12} \\ a_{22} \\ \cdots \\ a_{M2} \end{bmatrix}$$

A matrix that is $1 \times M$, and denoted by the column vector $\mathbf{v}$, is given by

$$\mathbf{v} = \begin{bmatrix} v(1) \\ v(2) \\ \dots \\ v(M) \end{bmatrix}$$

One can see that $\mathbf{v}$ is equivalent to a M-dimensional vector and all vectors, in fact, are special cases of matrices.

The following are some of the properties of matrices. Let A, B be $M \times N$ matrices. Then

●

$$(A + B)\mathbf{v} = A\mathbf{v} + B\mathbf{v}$$

In components

$$[(A + B)\mathbf{v}]_i = \sum_{j=1}^{N}(A + B)_{ij}v(j) = \sum_{j=1}^{N}A_{ij}v(j) + \sum_{j=1}^{N}B_{ij}v(j)$$

Since $\mathbf{v}$ is arbitrary, matrix addition is done element by element

$$(A + B) = A + B \quad \Rightarrow \quad (A + B)_{ij} = A_{ij} + B_{ij}$$

One has the null matrix

$$(A - A) = A - A = 0 = \begin{bmatrix} 0 & 0 & \dots & 0 \\ 0 & 0 & \dots & 0 \\ & \ddots & & \\ 0 & 0 & \dots & 0 \end{bmatrix}$$

● The result of multiplying A by a scalar λ results in λA, which in components is given by

$$(\lambda A)_{ij} = \lambda A_{ij}$$

The rule for matrix multiplication is that each row on the left, one by one multiplies, each column on the right, one by one. This is the same rule that was obtained earlier in multiplying a matrix into a vector. The result depends on the dimensions of the matrices. The rules obtained for the 2×2 matrices generalize to the case of multiplying any two matrices.

Let A be a square $M \times M$ matrix and B be a $M \times L$ matrix, as given below

$$A = \begin{bmatrix} a_{11} & a_{12} & \ldots & a_{1M} \\ a_{21} & a_{22} & \ldots & a_{2M} \\ & \ldots & & \\ a_{M1} & a_{M2} & \ldots & a_{MM} \end{bmatrix} \quad ; \quad B = \begin{bmatrix} b_{11} & b_{12} & \ldots & b_{1L} \\ b_{21} & b_{22} & \ldots & b_{2L} \\ & \ldots & & \\ b_{M1} & b_{N2} & \ldots & b_{ML} \end{bmatrix}$$

Matrix multiplication requires the number of columns in A be equal to the number of rows in B. The general result for AB yields a $M \times L$ matrix and is given by

$$AB = \begin{bmatrix} \sum a_{1i}b_{i1} & \sum a_{1i}b_{i2} & \ldots & \sum a_{1i}b_{iL} \\ \sum a_{2i}b_{i1} & \sum a_{2i}b_{i1} & \ldots & \sum a_{2i}b_{iL} \\ & \ldots & & \\ \sum a_{Mi}b_{i1} & \sum a_{2i}b_{i1} & \ldots & \sum a_{Mi}b_{iL} \end{bmatrix} \quad ; \quad \sum \equiv \sum_{i=1}^{M}$$

The unit matrix $\mathbb{I}$ leaves the matrix unchanged on multiplication

$$A\mathbb{I} = \mathbb{I}A \quad ; \quad \mathbb{I} = \begin{bmatrix} 1 & 0 & \ldots & 0 \\ 0 & 1 & \ldots & 0 \\ & \ldots & & \\ 0 & 0 & \ldots & 1 \end{bmatrix} \tag{2.15}$$

Consider the $M \times L$ matrix B; it acts on a $L \times 1$ (vector) $\mathbf{v}$ and yields a matrix (vector) $M \times 1$, given by $\mathbf{v}$. In matrix notation, the action of the matrix B on $\mathbf{v}$ is represented as follows ($\sum \equiv \sum_{i=1}^{L}$)

$$B \begin{bmatrix} v(1) \\ v(2) \\ \ldots \\ v(L) \end{bmatrix} = \begin{bmatrix} b_{11} & b_{12} & \ldots & b_{1L} \\ b_{21} & b_{22} & \ldots & b_{2L} \\ & \ldots & & \\ b_{M1} & b_{N2} & \ldots & b_{ML} \end{bmatrix} \begin{bmatrix} v(1) \\ v(2) \\ \ldots \\ v(L) \end{bmatrix} = \begin{bmatrix} \sum b_{1i}v(i) \\ \sum b_{2i}v(i) \\ \ldots \\ \sum b_{Mi}v(i) \end{bmatrix}$$

In vector notation

$$B\mathbf{v} = \mathbf{w} \implies w(i) = \sum_{j=1}^{L} b_{ij}v(j) \quad ; \quad i = 1, \ldots, M$$

In other words, B maps a L dimensional vector $\mathbf{v}$ to a M dimensional vector $\mathbf{w}$. In terms of complex vector space, B maps linear vector space $\mathbb{C}_L$ to vector space $\mathbb{C}_M$, and yields

$$B : \mathbb{C}_L \to \mathbb{C}_M$$

2.5 Properties of $N \times N$ Matrices

The rest of the discussion will treat only $N \times N$ *square matrices*, as these matrices play a specially important role in many problems. Let A, B, C be $N \times N$ square matrices and let $\mathbf{v}$ be an N-dimensional vector.

- Matrix multiplication is *not commutative* and is given by

$$(AB)_{ij} = \sum_k A_{ik} B_{kj} \quad \Rightarrow \quad AB \neq BA$$

- Matrix multiplication is associative with

$$(AB)C = A(BC) = ABC$$

To illustrate the non-commutativity of matrices, consider two matrices

$$A = \begin{bmatrix} 1 & -3 \\ 2 & 5 \end{bmatrix} \; ; \; B = \begin{bmatrix} 1 & 4 \\ -2 & 3 \end{bmatrix}$$

Then

$$AB = \begin{bmatrix} 7 & -5 \\ -8 & 23 \end{bmatrix} \neq BA = \begin{bmatrix} 9 & 17 \\ 4 & 21 \end{bmatrix} \quad \Rightarrow \quad AB - BA \neq 0$$

The non-commutativity of matrices has far-reaching consequences.

2.5.1 Hermitian Conjugation

The Hermitian conjugation of a vector is a special case of the Hermitian conjugation of a matrix. Similar to the case of a vector, where Hermitian conjugation entails mapping a column vector into a row vector and complex conjugating, in case of a matrix, all the columns and rows are interchanged and complex conjugated. In symbols, the Hermitian conjugation of a matrix A and vector $\mathbf{v}$ is defined by

$$(A^\dagger)_{ij} = (A)^*_{ji} \; : \; \text{Transpose and complex conjugate} \tag{2.16}$$
$$(A + B)^\dagger = A^\dagger + B^\dagger$$
$$(ABC)^\dagger = C^\dagger B^\dagger A^\dagger \tag{2.17}$$
$$((A^\dagger)^\dagger) = A$$
$$(\mathbf{v})^\dagger = \mathbf{v}^\dagger \; ; \; ((\mathbf{v})^\dagger)^\dagger = \mathbf{v}$$

Transforming the basis states can be obtained by analyzing the effect of a linear transformation on the dual basis states $\mathbf{e}^\dagger$. The change of volume is identical since both the vector and its dual provide an equivalent description of the linear transformation. The dual basis vectors are transformed by the Hermitian conjugate of the linear transformation given by

$$A^\dagger : \mathbb{C}_2 \rightarrow \mathbb{C}_2$$

where $\mathbb{C}_2$ is a two-dimensional complex manifold. Hence the determinant has the following property

$$\det(A)^* = \det(A^\dagger) \tag{2.18}$$

2.6   Tensor (Outer) Product

Consider the following set of xy-basis states

$$\mathbf{e}_1 = \begin{bmatrix} 1 \\ 0 \end{bmatrix} \; ; \; \mathbf{e}_2 = \begin{bmatrix} 0 \\ 1 \end{bmatrix}$$

To express a matrix in terms of the basis states, one needs to evaluate the tensor product of the basis states. Recall the scalar product of two vectors has been defined by $\mathbf{v}^\dagger \mathbf{w}$. One can also define the *tensor product* or *outer product* of two vectors (and of matrices in general). The outer product is defined by the symbol $\mathbf{w} \otimes \mathbf{v}^\dagger$, with the explicit component wise representation given by

$$\mathbf{w} \otimes \mathbf{v}^\dagger = \begin{bmatrix} w_1 \\ w_2 \\ \ldots \\ w_N \end{bmatrix} \otimes \begin{bmatrix} v_1 & v_2 & \ldots & v_N \end{bmatrix} \equiv \begin{bmatrix} w_1 v_1 & w_1 v_2 & \ldots & w_1 v_N \\ \ldots & & & \\ w_i v_1 & w_i v_2 & \ldots & w_i v_N \\ \ldots & & & \\ w_N v_1 & w_N v_2 & \ldots & w_N v_N \end{bmatrix}$$

In general, for N-component vectors w_i, v_j, the outer product, denoted by A, is given by

$$A = \mathbf{w} \otimes \mathbf{v}^\dagger \; ; \; A_{ij} = w_i v_j \; ; \; i, j = 1, \ldots N$$

For xy-basis states, one has the following four outer products

$$\mathbf{e}_1 \otimes \mathbf{e}_1^\dagger = \begin{bmatrix} 1 \\ 0 \end{bmatrix} \otimes \begin{bmatrix} 1 & 0 \end{bmatrix} = \begin{bmatrix} 1 & 0 \\ 0 & 0 \end{bmatrix} \; ; \; \mathbf{e}_1 \otimes \mathbf{e}_2^\dagger = \begin{bmatrix} 1 \\ 0 \end{bmatrix} \otimes \begin{bmatrix} 0 & 1 \end{bmatrix} = \begin{bmatrix} 0 & 1 \\ 0 & 0 \end{bmatrix}$$

and

$$\mathbf{e}_2 \otimes \mathbf{e}_1^\dagger = \begin{bmatrix} 0 \\ 1 \end{bmatrix} \otimes \begin{bmatrix} 1 & 0 \end{bmatrix} = \begin{bmatrix} 0 & 0 \\ 1 & 0 \end{bmatrix} \; ; \; \mathbf{e}_2 \otimes \mathbf{e}_2^\dagger = \begin{bmatrix} 0 \\ 1 \end{bmatrix} \otimes \begin{bmatrix} 0 & 1 \end{bmatrix} = \begin{bmatrix} 0 & 0 \\ 0 & 1 \end{bmatrix}$$

The tensor product of the basis states yields a basis for the 2×2 matrix since each position in the matrix corresponds to one matrix with a nonzero entry only at that position. For the N-dimensional case of E_N, define the tensor product of any two basis vectors by

$$\mathbf{e}_i \otimes \mathbf{e}_j^{\dagger} \ ; \ \ i, j = 1, \ldots, N$$

Note that

$$\mathbf{e}_i \otimes \mathbf{e}_j^{\dagger} \neq \mathbf{e}_j^{\dagger} \otimes \mathbf{e}_i \ ; \ \ i, j = 1, \ldots, N$$

The fact that the tensor product of the basis vectors is sufficient to represent any vector of of E_N requires that the basis states must satisfy the *completeness equation* given by

$$\sum_{i=1}^{N} \mathbf{e}_i \otimes \mathbf{e}_i^{\dagger} = \mathbb{I}_{N \times N} \tag{2.19}$$

For the 2×2 matrices, one has

$$\mathbf{e}_1 \otimes \mathbf{e}_1^{\dagger} + \mathbf{e}_2 \otimes \mathbf{e}_2^{\dagger} = \begin{bmatrix} 1 & 0 \\ 0 & 0 \end{bmatrix} + \begin{bmatrix} 0 & 0 \\ 0 & 1 \end{bmatrix} = \begin{bmatrix} 1 & 0 \\ 0 & 1 \end{bmatrix} = \mathbb{I}$$

Completeness equation Eq. 2.19 yields the following

$$\mathbf{v} = \mathbb{I}\mathbf{v} = \sum_{i=1}^{N} \mathbf{e}_i \otimes (\mathbf{e}_i^{\dagger} \cdot \mathbf{v}) = \sum_{i=1}^{N} v_i \mathbf{e}_i \tag{2.20}$$

where

$$v_i = \mathbf{e}_i^{\dagger} \cdot \mathbf{v} \ : \ \text{components of } \mathbf{v}$$

Due to the completeness equation, for an arbitrary matrix[1]

$$A = \sum_{ij} A_{ij} \mathbf{e}_i \otimes \mathbf{e}_j^{\dagger} = \sum_{ij} A_{ij} \mathbf{e}_i \mathbf{e}_j^{\dagger} \tag{2.21}$$

Note that

$$\mathbf{e}_I^{\dagger} A \mathbf{e}_J = \sum_{ij} A_{ij} (\mathbf{e}_I^{\dagger} \cdot \mathbf{e}_i) \otimes (\mathbf{e}_j^{\dagger} \cdot \mathbf{e}_J) = \sum_{ij} A_{ij} \delta_{I-i} \delta_{J-j} = A_{IJ}$$

In the notation of tensor product, the action of A on a general vector is given by

---

[1] Where the symbol $\otimes$ is sometimes dropped if the expression has no ambiguity.

$$A \begin{bmatrix} v_1 \\ v_2 \end{bmatrix} = \sum_{ij} A_{ij} \mathbf{e}_i \otimes \left(\mathbf{e}_j^\dagger \cdot \sum_k v(k) \mathbf{e}_k \right) = \sum_{ijk} A_{ij} \mathbf{e}_i v_k \delta_{j-k}$$

$$= \sum_{ij} A_{ij} v_j \mathbf{e}_i = \sum_i w_i \mathbf{e}_i \qquad (2.22)$$

where

$$\mathbf{w} = \sum_i w_i \mathbf{e}_i = \begin{bmatrix} w_1 \\ w_2 \end{bmatrix} = A\mathbf{v} = \begin{bmatrix} a_{11} & a_{12} \\ a_{21} & a_{22} \end{bmatrix} \begin{bmatrix} v_1 \\ v_2 \end{bmatrix} \qquad (2.23)$$

Then, from Eq. 2.22

$$w_I = \mathbf{e}_I^\dagger \cdot \mathbf{w} = \sum_{ij} A_{ij} v_j (\mathbf{e}_I^\dagger \cdot \mathbf{e}_i) = \sum_{ij} A_{ij} v_j \delta_{i-I}$$

$$\Rightarrow w_I = \sum_j A_{Ij} v_j \qquad (2.24)$$

From the definition of Hermitian conjugation

$$A^\dagger = \left(\sum_{ij} A_{ij}^* \mathbf{e}_i \otimes \mathbf{e}_j^\dagger \right)^\dagger = \sum_{ij} A_{ij}^* (\mathbf{e}_i \otimes \mathbf{e}_j^\dagger)^\dagger$$

$$= \sum_{ij} A_{ij}^* \mathbf{e}_j \otimes \mathbf{e}_i^\dagger = \sum_{ij} A_{ji}^\dagger \mathbf{e}_j \otimes \mathbf{e}_i^\dagger \; ; \; A_{ji}^\dagger \equiv A_{ij}^* \qquad (2.25)$$

The tenor product of two matrices is defined the following manner

$$A = \begin{bmatrix} a_{11} & a_{12} \\ a_{21} & a_{22} \end{bmatrix} \; ; \; B = \begin{bmatrix} b_{11} & b_{12} \\ b_{21} & b_{22} \end{bmatrix} \Rightarrow A \otimes B = \begin{bmatrix} a_{11}B & a_{12}B \\ a_{21}B & a_{22}B \end{bmatrix}$$

$$\Rightarrow A \otimes B = \begin{bmatrix} a_{11}b_{11} & a_{11}b_{12} & a_{12}b_{11} & a_{12}b_{12} \\ a_{11}b_{21} & a_{11}b_{22} & a_{12}b_{21} & a_{12}b_{22} \\ a_{21}b_{11} & a_{21}b_{12} & a_{22}b_{11} & a_{22}b_{12} \\ a_{21}b_{21} & a_{21}b_{22} & a_{22}b_{21} & a_{22}b_{22} \end{bmatrix}$$

One also has

$$B \otimes A = \begin{bmatrix} b_{11} & b_{12} \\ b_{21} & b_{22} \end{bmatrix} \otimes \begin{bmatrix} a_{11} & a_{12} \\ a_{21} & a_{22} \end{bmatrix} = \begin{bmatrix} b_{11}A & b_{12}A \\ b_{21}A & b_{22}A \end{bmatrix} \neq A \otimes B$$

Note that unlike the case of the ordinary product of two matrices, the order of the tensor product of two matrices is not reversed under Hermitian conjugation and is given by the following

$$(A \otimes B)^\dagger = A^\dagger \otimes B^\dagger$$

Table 2.1 Matrix A and related matrices

Matrix	Symbol	Components	Example
	A	A_{ij}	$\begin{bmatrix} 1 & 2+i \\ -i & 2 \end{bmatrix}$
Transpose	A^T	A^T_{ij}	$\begin{bmatrix} 1 & -i \\ 2+i & 2 \end{bmatrix}$
Complex conjugate	A^*	A^*_{ij}	$\begin{bmatrix} 1 & 2-i \\ +i & 2 \end{bmatrix}$
Hermitian conjugate	$A^\dagger = (A^*)^T$	$A^\dagger_{ij}$	$\begin{bmatrix} 1 & i \\ 2-i & 2 \end{bmatrix}$
Inverse	A^{-1}	A^{-1}_{ij}	$\frac{1}{1+2i}\begin{bmatrix} 2 & -(2+i) \\ i & 1 \end{bmatrix}$

2.7 Square Matrices

A number of properties of square matrices are given in Table 2.1. Based on these properties, special square matrices can be defined that occur in many applications in economics and finance.

Let A be a complex valued $N \times N$ square matrix. Suppose A is an invertible matrix. The following are some of the matrices related to A, and given in the Table 2.1.

- Transpose: $A^T_{ij} = A_{ji}$
- Complex conjugate: $A^*_{ij} = (A_{ji})^*$
- Hermitian conjugate: $A^\dagger_{ij} = (A^*_{ij})^T = A^*_{ji}$
- Inverse:

$$A^{-1}_{ij} = \left(\frac{1}{A}\right)_{ij}$$

The following are some special $N \times N$ square matrices that have far-reaching applications.

- Symmetric matrix: $S^T = S \;\Rightarrow\; S^T_{ij} = S_{ji} = S_{ij}$
- Hermitian matrix: $H^\dagger = H \;\Rightarrow\; H^\dagger_{ij} = H^*_{ji} = H_{ij}$
- Orthogonal matrix: $O^T O = \mathbb{I} \;\Rightarrow\; O^{-1} = O^T$
- Unitary matrix: $U^\dagger U = \mathbb{I} \;\Rightarrow\; U^{-1} = U^\dagger$

All quantum gates are unitary matrices, and hence, only these are of relevance for quantum algorithms. The inverse of a unitary matrix U is $U^{-1} = U^\dagger$, and hence finding the inverse, which requires major computational procedures for general matrices, is trivial and almost available automatically. In drawing quantum circuits, the matrix $U^\dagger$ is sometimes replaced by its equivalent U^{-1} for notational convenience.

2.8   Dirac Bracket: Vector Notation

Euclidean space E_N is an N-dimensional linear vector space. Vectors are elements of E_N and matrices are linear transformations from E_N to E_M. The synthesis of linear algebra with calculus is most easily carried out by expressing vectors and matrices in Dirac's notation. Although the notation can also be applied to the general case, the discussion is confined to square matrices that are linear transformations from E_N to E_N since all matrices in quantum algorithms are square matrices.

The basic ingredient of the Dirac **bracket** notation is the following

$$\text{ket} : \ |..\rangle = \text{ vector } ; \quad \text{bra} : \ \langle..| = \text{ dual vector}$$

Taken together they form the complete bracket given by

$$\text{bracket} : \ \langle..|..\rangle = \text{scalar product} = \text{complex number}$$

Consider, for generality, a complex valued vectors $\mathbf{v}$, $\mathbf{w}_i$ that are represented by the ket vector

$$\mathbf{v} = |v\rangle \ ; \quad \mathbf{w}_i = |w_i\rangle$$

There is no need to make the symbol v, w_i boldface inside the ket since the notation makes it clear that it is a vector. Hermitian conjugation yields the dual bra vector that is given by

$$\mathbf{v}^\dagger = \langle v|$$

The canonical basis and its dual vectors are given by

$$\mathbf{e}_i = |i\rangle \ ; \quad \mathbf{e}_i^\dagger = \langle i|$$

Linear superposition is written as expected

$$\mathbf{u} = a\mathbf{v} + b\mathbf{w} \ \Rightarrow \ |u\rangle = a|v\rangle + b|w\rangle$$

with the dual expression given by

$$\mathbf{u}^\dagger = a^*\mathbf{v}^\dagger + b^*\mathbf{w}^\dagger \ \Rightarrow \ \langle u| = a^*\langle v| + b^*\langle w|$$

The expansion of a vector into its components is

$$\mathbf{v} = \sum_i v(i)\mathbf{e}_i \ \Rightarrow \ |v\rangle = \sum_i v(i)|i\rangle$$

where

$$v(i) = \langle i|v\rangle$$

is defined more precisely later in Eq. 2.27.

The two possible values $\{0, 1\}$ of the classical bit, in Dirac's notation, can be represented by the two-dimensional column vectors. The two possible values of a bit are represented by what are called **ket vectors** and their transpose, called **bra vectors**, in the following manner:

$$|0\rangle = \begin{bmatrix} 1 \\ 0 \end{bmatrix} \; ; \; |1\rangle = \begin{bmatrix} 0 \\ 1 \end{bmatrix} \; : \; \langle 0| = [1 \; 0] \; ; \; \langle 1| = [0 \; 1] \tag{2.26}$$

Note the important fact that each string is *determinate*, with each position in the ket and bra vector having a definite value of either 0 or 1. Note that arithmetic operations on bits is not the addition of the vectors that are used to represent binary numbers.

The representation of binary numbers in Eq. 2.26 is by a two-dimensional vector that belongs to a two-dimensional linear vector space, which is the two-dimensional Euclidean space E_2. The n-binary numbers can be represented by an n-dimensional vector, which is an element on the n-dimensional Euclidean space E_2.

In Dirac's notation, the scalar product is given by

$$\mathbf{v}^\dagger \cdot \mathbf{w} = ((v|)(|w\rangle) = \langle v||w\rangle = \langle v|w\rangle$$

Furthermore

$$\mathbf{w}^\dagger \cdot \mathbf{v} = (\mathbf{v}^\dagger \cdot \mathbf{w})^* = \langle v|w\rangle^* \Rightarrow \mathbf{w}^\dagger \cdot \mathbf{v} = \langle v|w\rangle^* = \langle w|v\rangle$$

For the orthonormal basis states

$$\mathbf{e}_i^\dagger \cdot \mathbf{e}_j = \langle i|j\rangle = \delta_{i-j} = \begin{cases} 0 \; i \neq j \\ 1 \; i = j \end{cases}$$

Hence, the component of a vector is given by

$$v(i) = \mathbf{e}_i^\dagger \cdot \mathbf{v} = \langle i|v\rangle \tag{2.27}$$

The outer product is another notation for the tensor product and given by

$$\mathbf{u} \otimes \mathbf{v}^\dagger = |u\rangle\langle v|$$

A matrix has the following expression

$$A = \sum_{ij} A_{ij} \mathbf{e}_i \otimes \mathbf{e}_j^\dagger = \sum_{ij} A_{ij} |i\rangle\langle j| \tag{2.28}$$

where, using $\langle j|k\rangle = \delta_{j-k}$, yields

$$A_{ij} = \langle i|A|j\rangle \tag{2.29}$$

Note that the Dirac notation is a **major innovation** for the field of linear algebra [2]. The ith component of the vector $\mathbf{v}$ is denoted by a subscript in ordinary vector notation by $v(i)$ and the ijth component of the matrix A is denoted by a subscript in ordinary matrix notation by A_{ij}. In Dirac's notation, the component label i and ij of a vector and a matrix, respectively, are elevated to be on par with the vector and the matrix itself, and are represented by bra and ket vectors. In particular

$$\langle i|v\rangle = v(i) \;\; ; \;\; \langle i|A|j\rangle = A_{ij}$$

This notation brings much greater clarity on the use of indices since these are being combined and transformed in most linear algebra calculations.

Matrix multiplication, using $\langle j|k\rangle = \delta_{j-k}$, is given by

$$AB = \sum_{ij}\sum_{kl} A_{ij}B_{kl}|i\rangle\langle j|k\rangle\langle l| = \sum_{ij}\sum_{kl} A_{ij}B_{kl}|i\rangle\delta_{j-k}\langle l|$$

$$= \sum_{ijk} A_{ij}B_{jk}|i\rangle\langle k| = \sum_{ik} C_{ik}|i\rangle\langle k| \;\; ; \;\; C_{ik} = \sum_{j} A_{ij}B_{jk} \qquad (2.30)$$

$$\Rightarrow \;\; AB = C$$

The completeness equation for the canonical basis is given by

$$\mathbb{I} = \sum_{I=1}^{N} \mathbf{e}_I \otimes \mathbf{e}_I^\dagger = \sum_{I=1}^{N} |I\rangle\langle I| \qquad (2.31)$$

Any complete basis states $\mathbf{v}_I$ yield

$$\mathbb{I} = \sum_{I=1}^{N} \mathbf{v}_I \otimes \mathbf{v}_I^\dagger = \sum_{I=1}^{N} |v_I\rangle\langle v_I| \qquad (2.32)$$

Suppose $\mathbf{v}_I$ are eigenfunctions of the matrix S; the eigenfunction equation is given by

$$S\mathbf{v}_I = \lambda_I\mathbf{v}_I \;\; \Rightarrow \;\; S|v_I\rangle = \lambda_I|v_I\rangle \;\; : I = 1, \ldots N$$

The spectral decomposition of matrix S is given by

$$S = \sum_{I=1}^{N} \lambda_I\mathbf{v}_I \otimes \mathbf{v}_I^\dagger = \sum_{I=1}^{N} \lambda_I|v_I\rangle\langle v_I| \qquad (2.33)$$

Unitary matrix $\mathcal{U}$ yields

$$|e_I\rangle = \mathcal{U}^\dagger|v_I\rangle \;\; ; \;\; \langle e_I| = \langle v_I|\mathcal{U} \;\; \Rightarrow \;\; \mathcal{U}|e_I\rangle = |v_I\rangle \;\; ; \;\; \langle e_I|\mathcal{U}^\dagger = \langle v_I| \qquad (2.34)$$

A Hermitian matrix S is diagonalized by a unitary matrix $\mathcal{U}$ matrix such that [1]

$$\mathcal{U}^\dagger S \mathcal{U} = \sum_{I=1}^{N} \lambda_I \mathbf{e}_I \otimes \mathbf{e}_I^\dagger = \sum_{I=1}^{N} \lambda_I |I\rangle\langle I|$$

Conversely, Eq. 2.34 yields the spectral decomposition

$$S = \sum_{I=1}^{N} \lambda_I \, \mathcal{U}|e_I\rangle\langle e_I|\mathcal{U}^\dagger = \sum_{I=1}^{N} \lambda_I \, |v_I\rangle\langle v_I| \tag{2.35}$$

Consider the expression

$$S^2 = \sum_{I,J=1}^{N} \lambda_I \lambda_J |v_I\rangle\langle v_I|v_J\rangle\langle v_J| = \sum_{I=1}^{N} \lambda_I^2 |v_I\rangle\langle v_I| \tag{2.36}$$

Hence, one has in Dirac's notation

$$S^n = \sum_{I=1}^{N} \lambda_I^n |v_I\rangle\langle v_I| \tag{2.37}$$

In general, for any function $F(S)$, one has

$$F(S) = \sum_{I=1}^{N} F(\lambda_I)|v_I\rangle\langle v_I| \tag{2.38}$$

In particular

$$\exp(S) = \sum_{I=1}^{N} \exp(\lambda_I)|v_I\rangle\langle v_I| \tag{2.39}$$

2.9 Tensor and Outer Product: Strings and Gates

Starting from the single bit and gates acting on a single bit, the mathematics of tensor and outer products allow us to construct n-bit strings and gates acting on these multiple bit strings.

Consider a single bit; using Dirac's notation, the bit is given by

$$|0\rangle = \begin{bmatrix} 1 \\ 0 \end{bmatrix} \; ; \; |1\rangle = \begin{bmatrix} 0 \\ 1 \end{bmatrix}$$

The NOT gate, represented by X, is given by the following mapping

$$X|0\rangle = |1\rangle \; ; \;\; X|1\rangle|0\rangle$$

In matrix notation, the NOT gate is given by

$$X = \begin{bmatrix} 0 & 1 \\ 1 & 0 \end{bmatrix}$$

One can verify that, as expected

$$\begin{bmatrix} 0 & 1 \\ 1 & 0 \end{bmatrix}\begin{bmatrix} 1 \\ 0 \end{bmatrix} = \begin{bmatrix} 0 \\ 1 \end{bmatrix} \; ; \; \begin{bmatrix} 0 & 1 \\ 1 & 0 \end{bmatrix}\begin{bmatrix} 0 \\ 1 \end{bmatrix} = \begin{bmatrix} 1 \\ 0 \end{bmatrix}$$

Multiple bits are represented by the tensor product of the classical single bit. Tenor products are a way of combining vectors and matrices to form other matrices.

For example, a 2-bit string is represented by the tensor product of two vectors and is given by the following

$$\begin{bmatrix} a \\ b \end{bmatrix} \otimes \begin{bmatrix} c \\ d \end{bmatrix} = \begin{bmatrix} ac \\ ad \\ bc \\ bd \end{bmatrix}$$

Using Dirac's notation, the tensor product of two ket vectors is given by

$$|u\rangle = \begin{bmatrix} a \\ b \end{bmatrix} \; ; \; |v\rangle = \begin{bmatrix} c \\ d \end{bmatrix} \;\; \Rightarrow \;\; |u\rangle \otimes |v\rangle = \begin{bmatrix} ac \\ ad \\ bc \\ bd \end{bmatrix}$$

Hence, 2-bit strings have a representation in terms of two 1-bits given as follows

$$|00\rangle = \begin{bmatrix} 1 \\ 0 \end{bmatrix} \otimes \begin{bmatrix} 1 \\ 0 \end{bmatrix} = \begin{bmatrix} 1 \\ 0 \\ 0 \\ 0 \end{bmatrix}$$

$$|01\rangle = \begin{bmatrix} 0 \\ 1 \\ 0 \\ 0 \end{bmatrix} \; ; \; |10\rangle = \begin{bmatrix} 0 \\ 0 \\ 1 \\ 0 \end{bmatrix} \; ; \; |11\rangle = \begin{bmatrix} 0 \\ 0 \\ 0 \\ 1 \end{bmatrix} \tag{2.40}$$

Gates that act on a 2-bits string input are represented by the tensor product of the matrices in the following manner. The NOT gate X acting on the n-bit string classical string $|c\rangle^{\otimes n}$ is given by a n-fold tensor product of matrices $X^{\otimes n}$ and given by

$$X^{\otimes n} \equiv \underbrace{X \otimes X \cdots \otimes X}_{n \text{ times}}$$

For the case of $n = 2$, the NOT gate acting on a 2-bit string is given as follows

$$X^{\otimes 2} = X \otimes X = \begin{bmatrix} 0 & 1 \\ 1 & 0 \end{bmatrix} \otimes \begin{bmatrix} 0 & 1 \\ 1 & 0 \end{bmatrix} = \begin{bmatrix} 0 & 0 & 0 & 1 \\ 0 & 0 & 1 & 0 \\ 0 & 1 & 0 & 0 \\ 1 & 0 & 0 & 0 \end{bmatrix} \tag{2.41}$$

To check the matrix notation and our intuition about the gates, one can check whether the gate $X^{\otimes 2}$ does indeed reverse the value of the binary bits. The NOT gate on a 2-bit string has the following action

$$|00\rangle \to |11\rangle \; ; \; |01\rangle \to |10\rangle \; ; \; |10\rangle \to |01\rangle \; ; \; |11\rangle \to |00\rangle$$

Hence, one can check that

$$X^{\otimes 2}|00\rangle = X^{\otimes 2} \begin{bmatrix} 1 \\ 0 \\ 0 \\ 0 \end{bmatrix} = \begin{bmatrix} 0 \\ 0 \\ 0 \\ 1 \end{bmatrix} = |11\rangle$$

$$X^{\otimes 2}|01\rangle = X^{\otimes 2} \begin{bmatrix} 0 \\ 1 \\ 0 \\ 0 \end{bmatrix} = \begin{bmatrix} 0 \\ 0 \\ 1 \\ 0 \end{bmatrix} = |10\rangle$$

$$X^{\otimes 2}|10\rangle = X^{\otimes 2} \begin{bmatrix} 0 \\ 0 \\ 1 \\ 0 \end{bmatrix} = \begin{bmatrix} 0 \\ 1 \\ 0 \\ 0 \end{bmatrix} = |01\rangle$$

$$X^{\otimes 2}|11\rangle = X^{\otimes 2} \begin{bmatrix} 0 \\ 0 \\ 0 \\ 1 \end{bmatrix} = \begin{bmatrix} 1 \\ 0 \\ 0 \\ 0 \end{bmatrix} = |00\rangle \tag{2.42}$$

Let $|c\rangle$, $c = 0, 1$ represent one classical bit $\{0, 1\}$. The n-bits strings $\{0, 1\}^n$ are given by the n-fold tensor product and yields

$$|c\rangle^{\otimes n} \equiv \underbrace{|c_1\rangle \otimes |c_2\rangle \cdots \otimes |c_n\rangle}_{n \text{ bits}} \equiv \{0, 1\}^{\otimes n}$$

The tensor product is written as follows

$$|c_1\rangle \otimes |c_2\rangle \cdots \otimes |c_n\rangle = |c_1 c_2 \cdots c_n\rangle \; ; \; c_i = 0, 1$$

Writing out the $n-$bits strings yields the following

$$|000\ldots000\rangle,\ |000\ldots001\rangle,\ |000\ldots010\rangle,\ |000\ldots011\rangle,\ldots,\ |111\ldots111\rangle$$

There are 2^n number of distinct n-bits strings, and which are written in compact notation as follows

$$|x\rangle\ ;\quad x = 0, 1, 2, \ldots, 2^n - 1, \tag{2.43}$$

For example

$$|x = 0\rangle = |0\rangle = |000\ldots000\rangle, \ldots, |x = 2^n - 1\rangle = |2^n - 1\rangle = |111\ldots111\rangle$$

Recall that vectors are a special case of matrices, and the outer product of two vector is the tensor product of two vectors. Consider two vectors

$$|v\rangle = \begin{bmatrix} v_1 \\ v_2 \\ \cdots \\ v_n \end{bmatrix} \ ;\quad |w\rangle = \begin{bmatrix} w_1 \\ w_2 \\ \cdots \\ w_k \end{bmatrix}$$

The scalar product of two vectors is defined only if $k = n$ and is given by

$$\langle w|v\rangle = \begin{bmatrix} w_1^* & w_2^* & \cdots & w_n^* \end{bmatrix} \cdot \begin{bmatrix} v_1 \\ v_2 \\ \cdots \\ v_n \end{bmatrix} = \sum_{i=1}^n w_i^* v_i$$

The outer product is given by the following

$$|v\rangle\langle w| = \begin{bmatrix} v_1 \\ v_2 \\ \cdots \\ v_n \end{bmatrix} \otimes \begin{bmatrix} w_1^* & w_2^* & \cdots & w_k^* \end{bmatrix}$$

$$= \begin{bmatrix} v_1 w_1^* & v_1 w_2^* & \cdots & v_n w_k^* \\ v_2 w_1^* & v_2 w_2^* & \cdots & v_2 w_k^* \\ \cdots \\ v_n w_1^* & v_n w_2^* & \cdots & v_n w_k^* \end{bmatrix}_{n \times k} \tag{2.44}$$

For the bits, the scalar product is given by

$$\langle 1|1\rangle = \begin{bmatrix} 0 & 1 \end{bmatrix} \cdot \begin{bmatrix} 0 \\ 1 \end{bmatrix} = 1$$

whereas the outer product of a bit is given by

$$|1\rangle\langle 1| = \begin{bmatrix} 0 \\ 1 \end{bmatrix} \otimes [0\ 1] = \begin{bmatrix} 0 & 0 \\ 0 & 1 \end{bmatrix}$$

The identity matrix and the NOT gate are given by

$$\mathbb{I} = |0\rangle\langle 0| + |1\rangle\langle 1| \ ; \quad X = |1\rangle\langle 0| + |0\rangle\langle 1|$$

Similarly, for 2-bits string, the outer product for example is given by

$$|00\rangle\langle 01| = \begin{bmatrix} 1 \\ 0 \\ 0 \\ 0 \end{bmatrix} \otimes [0\ 1\ 0\ 0] = \begin{bmatrix} 0 & 1 & 0 & 0 \\ 0 & 0 & 0 & 0 \\ 0 & 0 & 0 & 0 \\ 0 & 0 & 0 & 0 \end{bmatrix} \tag{2.45}$$

Equation 2.45 above will be used in deriving the AND gate for a 2-bits string.

2.9.1 3-Bits String

Three bits are organized as follow:

$$000\ ;\ 001\ ;\ 010\ ;\ 011\ ;\ 100\ ;\ 101\ ;\ 110\ ;\ 111 \tag{2.46}$$

corresponding to

$$0\ ;\ 1\ ;\ 2\ ;\ 3\ ;\ 4\ ;\ 5\ ;\ 6\ ;\ 7$$

respectively. In terms of Dirac's notation for the state vectors for the 3-bits strings, one has

$$|ABC\rangle \equiv |A\rangle \otimes |B\rangle \otimes |C\rangle \ : \quad A, B, C = 0, 1$$

For example

$$|011\rangle \equiv |0\rangle \otimes |1\rangle \otimes |1\rangle = \begin{bmatrix} 1 \\ 0 \end{bmatrix} \otimes \begin{bmatrix} 0 \\ 1 \end{bmatrix} \otimes \begin{bmatrix} 0 \\ 1 \end{bmatrix} = \begin{bmatrix} 1 \\ 0 \end{bmatrix} \otimes \begin{bmatrix} 0 \\ 0 \\ 0 \\ 1 \end{bmatrix}$$

$$= [0\ 0\ 0\ 1\ 0\ 0\ 0\ 0]^T \tag{2.47}$$

where T stands for transpose, since for transpose is equal to Hermitian conjugation for vectors with real components.

References

1. Baaquie BE (2020) Mathematical methods and quantum mathematics for economics and finance. Springer, Singapore
2. Mermin ND (2007) Quantum computer science. Cambridge University Press, UK

Chapter 3
Classical Gates and Algorithms

An algorithm is defined to be a well-defined finite set of instructions that are carried out systematically in a given number of steps for solving a well-defined problem. An algorithm, in particular, can have the purpose of carrying out a specific information processing task. The algorithm is applied to an *input* (initial data) and the result of applying the computational steps constituting an algorithm yields the *output* (result). In effect, an *algorithm* is a function that produces an output from a given input.

A computer is a physical device designed for carrying out a set of computational instructions encoded in an algorithm. The question that naturally arises is: What is the collection of functions that an algorithm can, in principle, utilize in carrying out its computation? To answer this question, one needs an idealized model of what constitutes a computer. There are two aspects to the idealized model of a computer, the first—which stems from the point of view of computer science—is to define what constitutes a collection of *computable functions*; the second aspect is the relationship of the physical device that carries out this computation to physics, since all real computing devices must be based on the laws of physics.

The idealized computer that emerges from information science is called the Turing machine, named after its inventor Alan Turing. It is this idealized computer that is the basis of designing complex algorithms and protocols for information processing tasks and computing various functions. So far, all algorithms encountered in information science and Nature belong to the collection of functions defined by the Church–Turing theorem and they can all be used for computing by a Turing machine. There may appear, in the future, non-Turing-computable-functions, which would then require an extension of computable functions to a larger class of functions not covered by the Church–Turing theorem.

Classical algorithms are based on logical truth tables that are in turn realized by gates. A rather detailed analysis is carried out of classical gates as these are the necessary precursors to understanding quantum gates and quantum algorithms.

© The Author(s), under exclusive license to Springer Nature Singapore Pte Ltd. 2023   37
B. E. Baaquie and L.-C. Kwek, *Quantum Computers*,
https://doi.org/10.1007/978-981-19-7517-2_3

All quantum gates are matrices, similar to classical gates, with one significant difference, that all quantum gates are unitary. However, the practice gained from studying classical gates carries over completely to quantum gates since the same mathematics of linear algebra and tensor products are required in the study of both classical and quantum gates. **Hence, as a preparation for studying quantum qubits and gates, this chapter goes into great detail in studying classical bits and gates.**

The effect of classical gates on incoming bits is in general *irreversible*, in contrast to quantum gates that in principle are reversible. By introducing auxiliary incoming bits, the action of classical gates are extended to *reversible* gates, since these in turn can be generalized to quantum gates.

The algorithm for adding two binary numbers is studied in great detail as it is an exemplar of a classical algorithm that is realized by a classical circuit. Both the half-adder and full-adder are analyzed and sets the stage for the quantum generalization of the full-adder to its quantum algorithm, which adds two quantum bits (qubits).

The classical full-adder gates carry over unchanged from the classical addition of two bits to the quantum addition of two qubits. What changes in going from the classical to quantum addition is that the input and output states are qubits instead of classical bits: the classical gates of the full-adder are unchanged in going from the classical to the quantum case.

Quantum algorithms have no advantage over classical algorithms in arithmetic operations. The reason that the classical problem is worth studying is because one encounters all the mathematical ingredients that are required for defining quantum algorithms in a simpler and more intuitively obvious context [1].

3.1   Classical Algorithm

The input and output strings of a computer are binary numbers of the form $100110101011001\ldots$. Note in the binary representation of a string, each position in the string can be either 0 or 1. This is called a *bit* and is equal to the set $\{0, 1\}$. An input depending on m bits is denoted by

$$\underbrace{\{0, 1\} \otimes \{0, 1\} \cdots \otimes \{0, 1\}}_{m \text{ bits}} \equiv \{0, 1\}^{\otimes m}$$

We can now define a classical algorithm by the following: a computer is a physical device that transforms an input string of length m into an output string of length n using an algorithm. In other words

Computer algorithm :

$$\cdots \underbrace{100011101010000111}_{\text{input string of length } m} \cdots \rightarrow \text{algorithm} \rightarrow \cdots \underbrace{111101001101000100}_{\text{output string of length } n} \cdots$$

A classical computer, and this includes all computers today, is based on a definition of strings being *determinate*, and all algorithms transforming strings at every
stage of the computation are based on their determinate nature: at each step of the
computational process for a classical computer, the value of the string changes from
one determinate value to another determinate value. At any intermediate stage of
a classical algorithm, the state of the binary string can be observed. It is precisely
in the nature of the intermediate stage of the algorithm that a classical and a quantum algorithm differ: only the initial and final state of the quantum algorithm can
be observed: any measurement carried out to determine the intermediate state of
the string, between the input and output state, will result in the termination of the
quantum algorithm.

3.2 Classical Gates

A computer is essentially built from logical gates and circuits, which are physical
devices. A circuit is a collection of physical conducting wires with currents flowing
in the wires—and with the devices (gates) placed appropriately in the circuit. For
example, a binary circuit is one in which the current flowing in the circuit has only
two values, for instance the presence and the absence of a current. A gate is a physical
device that acts on the current to produce a result, which is a new current. Bits are
represented by lines and devices by various symbols. A circuit can have many bits as
inputs and outputs as well as many gates carrying out the operations of the algorithm.

For the purpose of discussing the logic of a computer, both classical and quantum,
the physical aspect of the device need not be discussed and the mathematical representation of the device is sufficient for describing the algorithms being executed by
a computer.

A classical gate is a device that acts on one or more binary inputs and produces
a single or more binary outputs. A single binary bit has two possible states, a 0 (for
the absence of a current) and a 1 (for the presence of a current).

We study classical gates in great detail firstly, to recast classical gates in the formalism of quantum mechanics and secondly, and secondly to re-express all classical
gates as reversible gates. The reason being that classical reversible gates are identical
to similar quantum gates, so the classical gates foreground the discussion on quantum
gates. It should be noted that there are many quantum gates that have no counterpart
in classical gates.

As mentioned above, the real difference between classical and quantum algorithms stem from the fact that the classical bit is replaced by the quantum qubit.
Classical reversible gates are augmented by many other quantum gates, and it is the
combination of qubits and quantum gates that allow computations that are beyond
the scope of classical computers.

Combinations of gates are used for carrying out the computer algorithm. With
a collection of gates, having one or more binary inputs, one can perform various
arithmetic operations on binary inputs. For a classical computation, a set of gates

is universal if, for any positive integers n, m and function $f : \{0, 1\}^m \rightarrow \{0, 1\}^n$, a circuit can be constructed for computing f using only the gates from the universal set. A universal gate is defined to be capable of realizing any computer gate. This definition of a universal gate is based on the Church–Turing theorem [2].

There are seven basic logic gates, six of which are shown in Fig. 3.1, that are sufficient for constructing a circuit for a classical computer; these gates are discussed in books on classical computers. Combination of gates forms a circuit, with two examples given in Fig. 3.2.

A Turing machine is realized in a classical computer by a combination of gates composing a circuit, with the circuit being designed to execute a particular algorithm. Two examples of a circuit are shown in Fig. 3.2.

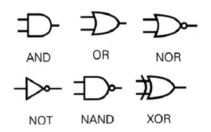

Fig. 3.1   Logic gates used in a classical computer. Published with permission of © Belal E. Baaquie and L. C. Kwek. All Rights Reserved

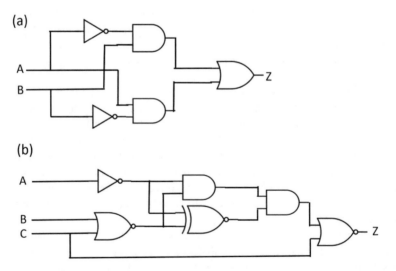

Fig. 3.2   A circuit with **a** two and **b** three input bits and one output bit, with many gates between the input and output. Published with permission of © Belal E. Baaquie and L. C. Kwek. All Rights Reserved

Fig. 3.3 Identity and NOT gate. Published with permission of © Belal E. Baaquie and L. C. Kwek. All Rights Reserved

Table 3.1 Identity and NOT gate truth tables

Input	Output
A	Z
0	0
1	1
Input	Output
A	Z
0	1
1	0

Two of the gates, namely the Identity and NOT gates, are shown in Fig. 3.3. Their truth tables determine how the binary inputs are combined to give a single output denoted by Z and are given by the following rules

- The identity gate makes no change in the input bit.
- The NOT gate acts on a single bit denoted by A, and the value of the output bit Z is reversed.

The symbols for the Identity and NOT gates are given in Fig. 3.3, and the truth table is given in Table 3.1; this provides a complete description of the Identity and NOT gates—and is sufficient for building and programming a classical computer. For a quantum computer, there are additional requirements linked primarily to making the gates reversible. To foreground the discussion on qubits (quantum bits), the bits and gates are represented in the framework of quantum mechanics.

The results obtained earlier in Sect. 2.9, in the context of the mathematical discussion of vectors and matrices, are repeated here for completeness. Since a bit has two possible values, 0 and 1, denote the two possible values by $|0\rangle$ and $|1\rangle$, respectively; the bits can be represented by a two-dimensional vector and given by

$$|0\rangle = \begin{bmatrix} 1 \\ 0 \end{bmatrix} \; ; \; |1\rangle = \begin{bmatrix} 0 \\ 1 \end{bmatrix}$$

Let X represent the NAND gate. The truth table for the NAND gate is given in Table 3.1 and is represented by and yields

$$X|0\rangle = |1\rangle \; ; \; X|1\rangle = |0\rangle$$

The logic gates are represented by matrices acting on the vector representation of the bits. The NOT gate can be represented by X: a 2×2 matrix given by the following

$$X = \begin{bmatrix} 0 & 1 \\ 1 & 0 \end{bmatrix} \quad \Rightarrow \quad X \begin{bmatrix} 1 \\ 0 \end{bmatrix} = \begin{bmatrix} 0 \\ 1 \end{bmatrix} \; ; \; X \begin{bmatrix} 0 \\ 1 \end{bmatrix} = \begin{bmatrix} 1 \\ 0 \end{bmatrix}$$

The NOT is reversible since one bit goes into one bit and yields $X^2 = X$; hence

$$X^{-1} = X$$

3.3 2-Bits String Gates

The truth table for the 2-bits strings is given in Fig. 3.4. The AND, NAND and XOR gates are analyzed to illustrate how all gates can be represented by matrices.

Lines in the logic gates are incoming and outgoing bits, and the gate itself is shown by various symbols. Note that all gates shown in Fig. 3.4 have two incoming bits, but only one outgoing bit. All quantum gates have equal numbers of incoming lines and outgoing lines. This is required for having unitary gates. To foreground the discussion on quantum gates, the classical logic gates will be re-represented by **two incoming bits**, which is the usual input but now with an auxiliary target bit as another input bit; the effect of the gate is to produce **two outgoing bits**, with the input bit being unchanged and the target bit being transformed into the output bit.

Since the input bit remained unchanged, it was not necessary to include it in the truth table given in Fig. 3.4. However, quantum gates require equal number of incoming and outgoing bits, and the input bit—although it is unchanged by the gate— is also included in the outgoing lines for making the generalization of classical to quantum gates more transparent.

Name	NOT	AND	NAND	OR	NOR	XOR	XNOR
Algebraic Expression	$\bar{A}$	AB	$\overline{AB}$	$A + B$	$\overline{A + B}$	$A \oplus B$	$\overline{A \oplus B}$
Symbol							

A	Z		A	B	Z		A	B	Z		A	B	Z		A	B	Z		A	B	Z		A	B	Z		A	B	Z
0	1		0	0	0		0	0	1		0	0	0		0	0	1		0	0	0		0	0	1
1	0		0	1	0		0	1	1		0	1	1		0	1	0		0	1	1		0	1	0
			1	0	0		1	0	1		1	0	1		1	0	0		1	0	1		1	0	0
			1	1	1		1	1	0		1	1	1		1	1	0		1	1	0		1	1	1

Fig. 3.4 Truth tables for the logic gates. Published with permission of © Belal E. Baaquie and L. C. Kwek. All Rights Reserved

Table 3.2 AND truth table

Input	Target	Input	Output
A	B	A	Z_a
0	0	0	0
0	1	0	0
1	0	1	0
1	1	1	1

We rewrite the truth table for the 2-bits string AND gate in Fig. 3.2. The gate acts on two input bits that are denoted by A and the auxiliary input bit B; the gate changes the value of B to the output bit, denoted by Z. Note that the bit A is unchanged and is called the control bit, since it does not change, and the bit B is called the target bit—as it is the 'target' of the gate. The truth table given in Fig. 3.4 is rewritten for the AND gate, in the notation of A, B, Z, in Table 3.2 (and later for the NAND gate in Table 3.3). This notation is used for the truth tables of all (classical) gates.

In Dirac's bracket notation, the initial state of a circuit starts with the 2-bits string state $|A\rangle \otimes |B\rangle = |AB\rangle$. Due to the action of a classical gate $\mathcal{G}$ on two 2-bits string $|AB\rangle$, to indicate which bit the gate is acting upon, one uses the notation

$$\mathcal{G} : \mathcal{G}(|A\rangle \otimes |B\rangle) = \mathcal{G}|AB\rangle$$

For the AND gate $\mathcal{G} = \mathcal{A}$, the output 1 if both input bits have a 1 and is 0 otherwise. An AND gate gives a nonzero current only if both the input and target 1-bits have currents flowing in them. The AND gate $\mathcal{A}$ acts on a 2-bits string state and requires a 4×4 matrix; the classical AND $\mathcal{A}$ gate has the truth table given in Table 3.2. The result can be given in condensed vector notation

$$\mathcal{A}|AB\rangle = |A, Z_a = AB\rangle = |AZ_a\rangle \tag{3.1}$$

where $AB = Z_a$ is binary multiplication.

In terms of the 2-bits string, the action of $\mathcal{A}$ given in the truth table, Table 3.2, by the following

$$|00\rangle \rightarrow |00\rangle$$
$$|01\rangle \rightarrow |00\rangle$$
$$|10\rangle \rightarrow |10\rangle$$
$$|11\rangle \rightarrow |11\rangle$$

Using the notation of outer product for writing the AND gate as the sum of terms expressed as $|output\rangle\langle input|$ and the result given in Eq. 2.45 yields

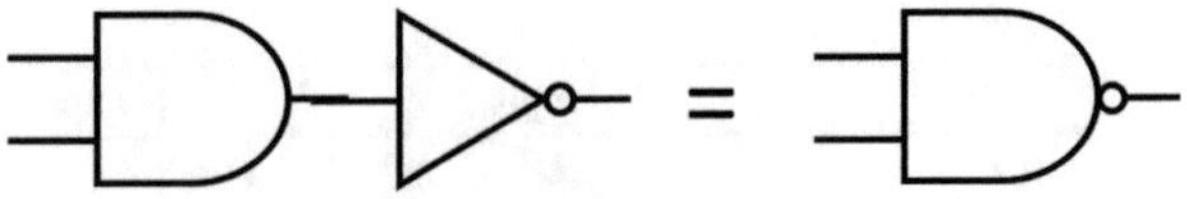

Fig. 3.5 NAND gate as the composition of the AND and NOT gates. Published with permission of © Belal E. Baaquie and L. C. Kwek. All Rights Reserved

$$\mathcal{A} = |00\rangle\langle 00| + |00\rangle\langle 01| + |10\rangle\langle 10| + |11\rangle\langle 11|$$

$$= \begin{bmatrix} 1 & 0 & 0 & 0 \\ 0 & 0 & 0 & 0 \\ 0 & 0 & 0 & 0 \\ 0 & 0 & 0 & 0 \end{bmatrix} + \begin{bmatrix} 0 & 1 & 0 & 0 \\ 0 & 0 & 0 & 0 \\ 0 & 0 & 0 & 0 \\ 0 & 0 & 0 & 0 \end{bmatrix} + \begin{bmatrix} 0 & 0 & 0 & 0 \\ 0 & 0 & 0 & 0 \\ 0 & 0 & 1 & 0 \\ 0 & 0 & 0 & 0 \end{bmatrix} + \begin{bmatrix} 0 & 0 & 0 & 0 \\ 0 & 0 & 0 & 0 \\ 0 & 0 & 0 & 0 \\ 0 & 0 & 0 & 1 \end{bmatrix} \tag{3.2}$$

Hence, from Eq. 3.2

$$\mathcal{A} = \begin{bmatrix} 1 & 1 & 0 & 0 \\ 0 & 0 & 0 & 0 \\ 0 & 0 & 1 & 0 \\ 0 & 0 & 0 & 1 \end{bmatrix} : \text{ AND gate} \tag{3.3}$$

Using the representation of the 2-bits string given in Eq. 2.40 it can be verified that $\mathcal{A}$ produces the result given in truth table, Table 3.2. The action of the gate in Eq. 3.1 in the notation of linear algebra has the following realization given in Eq. 3.3. For example, using Eq. 3.3

$$\mathcal{A}|00\rangle = \mathcal{A} \begin{bmatrix} 1 \\ 0 \\ 0 \\ 0 \end{bmatrix} = \begin{bmatrix} 1 \\ 0 \\ 0 \\ 0 \end{bmatrix} = |00\rangle$$

Similarly

$$\mathcal{A}|01\rangle = \mathcal{A} \begin{bmatrix} 0 \\ 1 \\ 0 \\ 0 \end{bmatrix} = \begin{bmatrix} 1 \\ 0 \\ 0 \\ 0 \end{bmatrix} = |00\rangle$$

and so on. Note that $\mathcal{A}$ is not invertible.

The logic gates of a classical computer are in general irreversible because, except for the NOT and XOR gates, they correspond to a transformation of the target bits with an equal number of 0's and 1's into the output bits with an unequal number of 0's and 1's. Hence, one cannot obtain the target bits state from the final 1-bit output bits, even if one retains one of the input bits. Note from Fig. 3.4, the XOR and XNOR gates have an output with equal number of 0's and 1's and hence they are invertible if we retain one of the input bits.

Table 3.3 NAND truth table

Input	Target	Input	Output
A	B	A	Z_n
0	0	0	1
0	1	0	1
1	0	1	1
1	1	1	0

For example, the final state of the AND gate (given in Eq. 3.3), with truth table given in Table 3.2, does not have enough information in the output state to obtain the values of the target 1-bits; the output bits Z_a with value 0 correspond to three different target bits and hence cannot be inverted to yield the target bit. This is reflected in the fact that the AND gate $\mathcal{A}$, considered as a matrix, is not invertible.

The NAND gate is combination of an AND and a NOT gate as shown in Fig. 3.5; in truth table given in Table 3.3, for the output column denoted by Z_n of the NAND gate has interchanged all the 0's and 1's of the AND gate, which is given in Table 3.2.

One can represent the NAND gate by the transformation $\mathcal{N}$ given by

$$\mathcal{N} = \begin{bmatrix} 0 & 0 & 0 & 0 \\ 1 & 1 & 0 & 0 \\ 0 & 0 & 0 & 1 \\ 0 & 0 & 1 & 0 \end{bmatrix} : \text{ NAND gate} \tag{3.4}$$

One can verify that the NAND is correct by directly applying the expression for the NAND gate given in Eq. 3.4 on the 2-bits strings and obtain the result given in the truth table Table 3.3. The transformation $\mathcal{N}$ yields the NAND truth table given in Table 3.3 is given by

$$\mathcal{N}|AB\rangle = |AZ_n\rangle \tag{3.5}$$

Similar to the AND gate, the NAND gate does not have an inverse and hence is not reversible. The decomposition of the NAND gate given in Fig. 3.5 shows that the AND and NOT gates—being applied sequentially—yield the NAND gate. In other words

$$|AB\rangle \rightarrow \mathcal{A}|AB\rangle \rightarrow (\mathbb{I} \otimes X)(\mathcal{A}|AB\rangle) \Rightarrow |AZ_n\rangle = (\mathbb{I} \otimes X)\mathcal{A}|AB\rangle \tag{3.6}$$

The order of the gates is *reversed* compared to the way one thinks of applying the NOT gate to the output of the AND gate as given in Eq. 3.6 [2]. From Eq. 3.1, one has for the NAND truth table given in Table 3.3

$$\mathcal{N}|AB\rangle = (\mathbb{I} \otimes X)\mathcal{A}\Big(|AB\rangle\Big) = (\mathbb{I} \otimes X)\Big(|AZ_a\rangle\Big) = |AZ\rangle \tag{3.7}$$

Fig. 3.6 XOR Gate. Published with permission of © Belal E. Baaquie and L. C. Kwek. All Rights Reserved

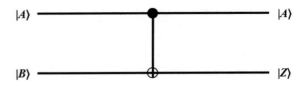

The gate $\mathbb{I} \otimes X$ acts as a NOT gate only on the output bit Z_a and is given by

$$X = \begin{bmatrix} 0 & 1 \\ 1 & 0 \end{bmatrix}$$

$$X_2 \equiv \mathbb{I} \otimes X = \begin{bmatrix} 1 & 0 \\ 0 & 1 \end{bmatrix} \otimes \begin{bmatrix} 0 & 1 \\ 1 & 0 \end{bmatrix} = \begin{bmatrix} X & 0 \\ 0 & X \end{bmatrix} = \begin{bmatrix} 0 & 1 & 0 & 0 \\ 1 & 0 & 0 & 0 \\ 0 & 0 & 0 & 1 \\ 0 & 0 & 1 & 0 \end{bmatrix} \qquad (3.8)$$

Is the sequential application of the gates mathematically realized by the matrix multiplication of the two individual gates? The answer is a Yes and shows the utility of using vectors and matrices to represent bits and gates. Hence, the realization of the circuit for the NAND gate given in Eq. 3.6 is the following matrix product of two gates

$$\mathcal{N} = (\mathbb{I} \otimes X)\mathcal{A} = X_2\mathcal{A} \qquad (3.9)$$

and yields the result given in Eq. 3.4.

3.4   XOR Reversible Gate

To foreground the representation of quantum algorithms by quantum circuits, we represent a classical circuit using the same framework as the one we will use for quantum circuits. The notation for the circuit diagrams, as shown in Fig. 3.6, is to represent bits by horizontal lines; the gates are represented by vertical lines. The bits on which the gate acts are marked, as in Fig. 3.6 by a heavy dot for the bit that is unchanged (conditions the gate) and an empty circle with a crossing for the bit that is transformed. This notation is used consistently for both classical and quantum gates for all the circuits in this book.

The XOR gate, like the NOT gate X, is a classical reversible gate. The XOR is the classical analog of the CNOT quantum gate discussed later. This is the reason, as will be seen in later discussions, all classical reversible gates can be finally expressed in terms of the XOR and NOT gates.

Table 3.4 XOR truth table

Input	Target	Input	Output
A	B	A	Z
0	0	0	0
0	1	0	1
1	0	1	1
1	1	1	0

The XOR gate, with input A and target B, has the following output Z

- If A or B is 1, but not both, then $Z = 1$.
- Otherwise, $Z = 0$.

The definition of the X_{OR} gate yields the following

$$X_{OR}|AB\rangle = |A, Z = A \oplus B\rangle$$

The XOR truth table is given in Table 3.4.

The matrix representation of X_{OR} gate can be directly verified by acting on the state vector representation of the 2-bits strings to be given by

$$X_{OR} = \begin{bmatrix} 1 & 0 & 0 & 0 \\ 0 & 1 & 0 & 0 \\ 0 & 0 & 0 & 1 \\ 0 & 0 & 1 & 0 \end{bmatrix} = \begin{bmatrix} \mathbb{I} & 0 \\ 0 & X \end{bmatrix} \tag{3.10}$$

The X_{OR} gate is reversible and hence has a direct generalization to the quantum gate denoted by CNOT. The X_{OR} gate and its quantum version are among the most useful gates since the X_{OR} gate carries out the operation of binary addition. More precisely, denoting binary addition by $\oplus$, the truth table of the XOR gate given in Table 3.4, is reproduced by binary addition since

$$A \oplus B = Z \quad \Rightarrow \quad 0 \oplus 0 = 0 \; ; \quad 0 \oplus 1 = 1 \; ; \quad 1 \oplus 0 = 1 \; ; \quad 1 \oplus 1 = 0 \; ;$$

What is missing in the XOR is the carry of 1 for the addition of $1 \oplus 1 = 0$ plus carry 1. And hence a full binary addition entails using two 1-bits and yields $1 \oplus 1 = 10$. Hence, XOR is used for the half-Adder that ignores the carry and one has to employ another NAND gate in conjunction with the XOR gate to obtain the full-Adder, discussed later in Sect. 3.8.

Similar to Eq. 3.9, the X_{NOR} is defined by applying a NOT gate to the output of X_{OR} and yields

$$X_{NOR} = (\mathbb{I} \otimes X)X_{OR} = X_2 X_{OR} = X_2 \begin{bmatrix} \mathbb{I} & 0 \\ 0 & X \end{bmatrix} = \begin{bmatrix} X & 0 \\ 0 & \mathbb{I} \end{bmatrix} \tag{3.11}$$

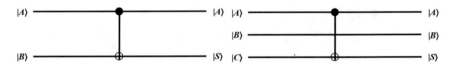

Fig. 3.7 X_{NOR} gate acting on 2- and 3-bits string. Published with permission of © Belal E. Baaquie and L. C. Kwek. All Rights Reserved

3.4.1 X_{OR}: 3-Bits String

The X_{OR} acts on a 2-bits string; as indicated in Fig. 3.7, the action is given by

$$X_{OR}|AB\rangle = |AS\rangle$$

In a more complex circuit, the X_{OR} acts on a **3-bits string**; as indicated in Fig. 3.7, the action is given by

$$X_{OR}|ABC\rangle = |ABS\rangle$$

Note that the X_{OR} does not act on the $|B\rangle$ bit; this fact is represented in Fig. 3.7 by the vertical line crossing the $|B\rangle$ bit without any empty or heavy dot.

A general feature of the notation for gates and bits, both classical and quantum, is the following:

- A vertical line represents the gate and acts on bits represented by horizontal lines.
- Bits that are affected by the gate are marked by a heavy dot or empty circle; in case of Fig. 3.7, affected bits are the bits A and C, marked by a heavy dot and empty circle respectively.
- For bits on which the gates don't act, the vertical line crosses these without any marking; in Fig. 3.7, bit B is left unchanged.

We need to extend the action of X_{OR}, which requires at least a 2-bits string, to a circuit with n-bits string—and in particular to a 3-bits string. To extend X_{OR} to a multibits string, we write X_{OR} in terms of the outer product of the classical bits. The truth table for X_{OR} given in Table 3.4, using the notation $|output\rangle\langle input|$ acting on only **two 1-bits** (drawing on the left in Fig. 3.7) yields

$$\begin{aligned}
X_{OR}(A, B) &= |00\rangle\langle00| + |01\rangle\langle01| + |11\rangle\langle10| + |10\rangle\langle11| \\
&= |0\rangle|\langle0| \otimes |0\rangle\langle0| + |0\rangle|\langle0| \otimes |1\rangle\langle1| + |11\rangle\langle10| + |10\rangle\langle11| \\
&= |0\rangle|\langle0| \otimes \left(|0\rangle\langle0| + \otimes|1\rangle\langle1|\right) + |1\rangle\langle1| \otimes |1\rangle\langle0| + |1\rangle\langle1| \otimes |0\rangle\langle1| \\
&= |0\rangle|\langle0| \otimes \mathbb{I} + |1\rangle\langle1| \otimes \left(|0\rangle\langle1| + |0\rangle\langle1|\right) \\
&= |0\rangle|\langle0| \otimes \mathbb{I} + |1\rangle\langle1| \otimes X
\end{aligned} \tag{3.12}$$

where $X = |0\rangle\langle1| + |0\rangle\langle1|$ is the NOT gate. Equation 3.12 shows that X_{OR} acts as a NOT gate on the second bit B when the first bit A is equal to $|1\rangle$.

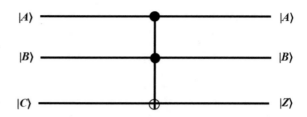

To evaluate $X_{OR}(A, B, C)$ as shown in Fig. 3.7, the gate X_{OR} has no action of the second bit $|B\rangle$ and acts only on the first and last bit, given by $|A\rangle$ and $|C\rangle$, respectively. The action of gate $X_{OR}(A, B, C)$ on the second bit B is the identity matrix $\mathbb{I}$. Hence, $X_{OR}(A, B, C)$ acting on a 3-bits string, from Eq. 3.12 is given by

$$X_{OR}(A, B, C) = |0\rangle|\langle 0| \otimes \mathbb{I} \otimes \mathbb{I} + |1\rangle\langle 1| \otimes \mathbb{I} \otimes X \qquad (3.13)$$

3.5   3-Bits String: Toffoli Gate

The Toffoli gate, denoted by $\mathcal{T}$, is a universal reversible logic gate, shown in Fig. 3.8, that acts on three incoming bits: two input bits A, B and one target bit C—and produces unchanged the two input bits A, B and a third bit Z which is the transformation of the target bit C. It can be shown that any reversible classical circuit can be constructed from Toffoli gates. It is shown in Eq. 3.20 of Sect. 3.6 that the Toffoli gate is equivalent to the unitary realization of the classical AND acting on a 3-bits string.

The Toffoli gate acts on three bits A, B, C and gives an output also of three bits A, B, Z. The Toffoli gate action on a 3-bits string and is shown in Fig. 3.8, with heavy dots indicating the conditioning of the output given by the Z bit.

The action of the Toffoli gate $\mathcal{T}$ on the incoming bits yields the following output

$$\mathcal{T}|ABC\rangle = |AB, Z\rangle \qquad (3.14)$$

Denoting binary multiplication by AB and addition by $\oplus$, the Toffoli gate is given by

$$\mathcal{T}|ABC\rangle = |AB, Z = C \oplus AB\rangle \qquad (3.15)$$

Using the notation for the 3-bits as enumerated in Eq. 2.46, from the definition of the Toffoli gate given in Eq. 3.15, the truth table is given by Table 3.5. The inputs, target and output bits of the Toffoli gate are given in Table 3.5.

There are a number of special cases of the Toffoli gate that can be read off from its definition. More precisely,

Table 3.5 Toffili gate truth table

Input	Input	Target	Input	Input	Output
A	B	C	A	B	Z
0	0	0	0	0	0
0	0	1	0	0	1
0	1	0	0	1	0
0	1	1	0	1	1
1	0	0	1	0	0
1	0	1	1	0	1
1	1	0	1	1	1
1	1	1	1	1	0

- Recall from Eq. 3.15

$$T|ABC\rangle = |AB, Z = C \oplus AB\rangle$$

- Equivalent to the AND gate for A, B

$$T|AB0\rangle = |AB, Z = AB\rangle$$

Hence, acting on a bit with $C = 0$ shows that T is equivalent to an AND gate for A, B.
- Equivalent to the NOT gate for C

$$T|11C\rangle = |AB, Z = C \oplus 1\rangle$$

For the case of $A = 1 = B$, T is equivalent to NOT gate for C.
- Equivalent to the XOR gate for A, C or B, C

$$T|A1C\rangle = |AB, Z = C \oplus A\rangle$$

The Toffoli gate for $B = 1$ is equivalent to the XOR gate acting on A, C. For $A = 1$, the Toffoli gate is equivalent to the XOR gate for B, C.

Note that for the first six row entries in Table 3.5, corresponding to the two input bits A, B not **both** having 1 as their value, the output Z is and same as the incoming target bit C and hence unchanged; only the last two output bits are changed by a NOT gate. Hence, for the eight-dimensional state vectors for the 3-bits strings, one has the 8×8 matrix representation of the Toffoli gate given by

$$
T = \begin{bmatrix}
1 & 0 & 0 & 0 & 0 & 0 & 0 & 0 \\
0 & 1 & 0 & 0 & 0 & 0 & 0 & 0 \\
0 & 0 & 1 & 0 & 0 & 0 & 0 & 0 \\
0 & 0 & 0 & 1 & 0 & 0 & 0 & 0 \\
0 & 0 & 0 & 0 & 1 & 0 & 0 & 0 \\
0 & 0 & 0 & 0 & 0 & 1 & 0 & 0 \\
0 & 0 & 0 & 0 & 0 & 0 & 0 & 1 \\
0 & 0 & 0 & 0 & 0 & 0 & 1 & 0
\end{bmatrix} \quad ; \quad \det T = -1 \tag{3.16}
$$

It can be seen that the Toffoli matrix given in Eq. 3.16 is invertible, and hence, the Toffoli gate represents a reversible transformation. It can be shown that all circuits of a classical computer can be built up from the Toffoli gate, and it is called the universal gate. The Fredkin gate, which is also a universal classical gate, is equivalent to the Toffoli gate [2].

The Toffoli gate can be written in the outer product formalism based on the truth table given in Table 3.5. The action of T gives the following mapping of the input to output for the classical bits:

$$
\begin{aligned}
\text{input} &\rightarrow \text{ouput} \\
|000\rangle &\rightarrow |000\rangle \\
|001\rangle &\rightarrow |001\rangle \\
|010\rangle &\rightarrow |010\rangle \\
|011\rangle &\rightarrow |011\rangle \\
|100\rangle &\rightarrow |100\rangle \\
|101\rangle &\rightarrow |101\rangle \\
|110\rangle &\rightarrow |111\rangle \\
|111\rangle &\rightarrow |110\rangle
\end{aligned}
$$

Hence, using the notation $T \simeq |\text{ouput}\rangle\langle\text{input}|$ yields

$$
\begin{aligned}
T = {}& |000\rangle\langle000| + |001\rangle\langle001| + |010\rangle\langle010| + |011\rangle\langle011| \\
&+ |100\rangle\langle100| + |101\rangle\langle101| + |111\rangle\langle110| + |110\rangle\langle111|
\end{aligned} \tag{3.17}
$$

Anticipating our derivation of the full adder, recall that the bits are written in the notation $|ABC\rangle = |A\rangle \otimes |B\rangle \otimes |C\rangle$, where A, B, C are the first, second and third bit, respectively. To separate out the third bit, note that the outer product of a ket and a bra vector yields the following

$$
\begin{aligned}
&|AB\rangle\langle DE| \equiv \Big(|A\rangle \otimes |B\rangle\Big)\Big(\langle D| \otimes \langle E|\Big) = |A\rangle\langle D| \otimes |B\rangle\langle E| \\
&\Rightarrow |ABC\rangle\langle DEF| = |AB\rangle\langle DE| \otimes |C\rangle\langle F| \\
&\Rightarrow |000\rangle\langle000| = |00\rangle\langle00| \otimes |0\rangle\langle0| \; ; \; |001\rangle\langle001| = |00\rangle\langle00| \otimes |1\rangle\langle1| \; ; \cdots
\end{aligned}
$$

Hence, $\mathcal{T}$ can be rewritten as follows

$$
\begin{aligned}
\mathcal{T} &= |00\rangle\langle 00| \otimes |0\rangle\langle 0| + |00\rangle\langle 00| \otimes |1\rangle\langle 1| \\
&\quad + |01\rangle\langle 01| \otimes |0\rangle\langle 0| + |01\rangle\langle 01| \otimes |1\rangle\langle 1| + |10\rangle\langle 10| \otimes |0\rangle\langle 0| \\
&\quad + |10\rangle\langle 10| \otimes |1\rangle\langle 1| + |11\rangle\langle 11| \otimes |1\rangle\langle 0| + |11\rangle\langle 11| \otimes |0\rangle\langle 1| \\
&= |00\rangle\langle 00| \otimes \Big(|0\rangle\langle 0| + |1\rangle\langle 1| \Big) + |01\rangle\langle 01| \otimes \Big(|0\rangle\langle 0| + |1\rangle\langle 1| \Big) \\
&\quad + |10\rangle\langle 10| \otimes \Big(|1\rangle\langle 0| + |0\rangle\langle 1| \Big) + |11\rangle\langle 11| \otimes \Big(|1\rangle\langle 0| + |0\rangle\langle 1| \Big)
\end{aligned}
$$

Using $\mathbb{I} = |0\rangle\langle 0| + |1\rangle\langle 1|$ and $X = |0\rangle\langle 1| + |1\rangle\langle 0|$, which is the NOT gate, the derivation above yields the final result

$$
\mathcal{T} = \Big(|00\rangle\langle 00| + |01\rangle\langle 01| + |10\rangle\langle 10| \Big) \otimes \mathbb{I} + |11\rangle\langle 11| \otimes X \qquad (3.18)
$$

Note Eq. 3.18 is the matrix representation of the fact that the Toffoli gate, acting on the 3-bits string $|ABC\rangle$ a) leaves the $|C\rangle$ bit unchanged when the first two bits $|AB\rangle\langle AB|$ have the values $|00\rangle\langle 00|$; $|01\rangle\langle 01|$; $|10\rangle\langle 10|$ and b) acts on $|C\rangle$ as a NOT gate for the bit $|AB\rangle\langle AB| = |11\rangle\langle 11|$.

The Toffoli being a universal gate implies that all other gates given in truth table Fig. 3.4 are special cases. The key feature of the Toffoli gate is that it provides a reversible realization of all the classical computer gates, including such gates as the AND and NAND gates that are irreversible. Hence, the Toffoli gate is the basis for the generalization of all classical gates to the case of quantum gates. The universal quantum gate is slightly more general than the Toffoli gate. The universal quantum gate, in addition to $\mathcal{T}$, requires one more qubit—and is discussed later.

3.6 Unitary AND, NAND and NOT Gates for 3-Bits

To represent the classical gates that are not unitary, like the AND, NAND, OR gates, one needs to define the gates using three bits and not just two bits. There seems to be a lot of redundancy in the assignment, but if one demands that the gate be unitary, there are some additional constraints. As in the case of a Toffoli gate, consider three incoming bits given by $|ABC\rangle$, with two input bits A, B and a new *auxiliary* target bit C; the outgoing bits are denoted by $|AB, Z\rangle$, with Z being the output bit. For any arbitrary Boolean function f, the action of the gate $\mathcal{G}$ that is one-to-one and unitary is given by

$$
\mathcal{G}|ABC\rangle = |AB, Z = C \oplus f(A, B)\rangle
$$

Recall for the Toffoli gate, from Eq. 3.15

$$
\mathcal{T}|ABC\rangle = |AB, Z = C \oplus AB\rangle
$$

Note that for $C = 0$, the Toffoli gate yields

$$T|AB, C = 0\rangle = |AB, Z = AB\rangle \tag{3.19}$$

which is the result given earlier for the AND gate in Eq. 3.1. Hence, Eq. 3.19 is a unitary extension of the AND to three bits, with initial value of the target bit being set to be equal to $C = 0$ and $|AB0\rangle$ are the incoming bits.

If one re-examines the truth table of the Toffoli gate given in Table 3.5 and looks at only the rows for which $C = 0$, one can see it is equal to the truth table for the AND gate given in Table 3.2. The last line of the truth table, Table 3.5, is non-trivial and fixed by the requirement that the 3-bit gate be unitary.

The AND gate also appears in the middle of the circuit, where it acts not only on the incoming target bit but other bits as well. One can extend the definition of the AND gate for $C = 0, 1$ to be equal to the Toffoli gate given in Eq. 3.15. Hence,

$$\mathcal{A}_3 = T \tag{3.20}$$

Note that the 3-bits AND gate $\mathcal{A}_3$ has the same truth table as the Toffoli gate given in Table 3.5.

The NAND gate also has an extension to a 8×8 matrix, acting on a 3-bit string. A direct way of obtaining the 3-bits expression for the NAND gate is to generalize Eq. 3.9. The NAND gate is obtained from the AND gate by acting with a NOT gate only on the target bit. Hence, we apply the NOT gate to the third output bit, similar to the case done earlier in Eq. 3.7. Define the NOT gate acting only on the third output bit, in 2×2 block matrix notation with identity $\mathbb{I}$, as follows

$$X_3 \equiv \mathbb{I} \otimes \mathbb{I} \otimes X = \begin{bmatrix} X & 0 & 0 & 0 \\ 0 & X & 0 & 0 \\ 0 & 0 & X & 0 \\ 0 & 0 & 0 & X \end{bmatrix} \; ; \; X = \begin{bmatrix} 0 & 1 \\ 1 & 0 \end{bmatrix}$$

Then one has the following definition for the NAND gate

$$\mathcal{N}_3 = (\mathbb{I} \otimes \mathbb{I} \otimes X)\mathcal{A}_3 = X_3 T \tag{3.21}$$

The Toffoli gate in block matrix notation is given by the following

$$T = \begin{bmatrix} \mathbb{I} & 0 & 0 & 0 \\ 0 & \mathbb{I} & 0 & 0 \\ 0 & 0 & \mathbb{I} & 0 \\ 0 & 0 & 0 & X \end{bmatrix} \tag{3.22}$$

Hence, using $X^2 = \mathbb{I}$, Eq. 3.21 yields

Table 3.6 NAND gate truth table for 3-bits strings

Input	Input	Target	Input	Input	Output	Mapping
A	B	C	A	B	Z	
0	0	0	0	0	1	$|000\rangle \mapsto |001\rangle$
0	0	1	0	0	0	$|001\rangle \mapsto |000\rangle$
0	1	0	0	1	1	$|010\rangle \mapsto |011\rangle$
0	1	1	0	1	0	$|011\rangle \mapsto |010\rangle$
1	0	0	1	0	1	$|100\rangle \mapsto |101\rangle$
1	0	1	1	0	0	$|101\rangle \mapsto |100\rangle$
1	1	0	1	1	0	$|110\rangle \mapsto |110\rangle$
1	1	1	1	1	1	$|111\rangle \mapsto |111\rangle$

$$\mathcal{N}_3 = \begin{bmatrix} X & 0 & 0 & 0 \\ 0 & X & 0 & 0 \\ 0 & 0 & X & 0 \\ 0 & 0 & 0 & X \end{bmatrix} \cdot \begin{bmatrix} \mathbb{I} & 0 & 0 & 0 \\ 0 & \mathbb{I} & 0 & 0 \\ 0 & 0 & \mathbb{I} & 0 \\ 0 & 0 & 0 & X \end{bmatrix} = \begin{bmatrix} X & 0 & 0 & 0 \\ 0 & X & 0 & 0 \\ 0 & 0 & X & 0 \\ 0 & 0 & 0 & \mathbb{I} \end{bmatrix} \tag{3.23}$$

Writing out the 8×8 matrix yields

$$\mathcal{N}_3 = \begin{bmatrix} 0 & 1 & 0 & 0 & 0 & 0 & 0 & 0 \\ 1 & 0 & 0 & 0 & 0 & 0 & 0 & 0 \\ 0 & 0 & 0 & 1 & 0 & 0 & 0 & 0 \\ 0 & 0 & 1 & 0 & 0 & 0 & 0 & 0 \\ 0 & 0 & 0 & 0 & 0 & 1 & 0 & 0 \\ 0 & 0 & 0 & 0 & 1 & 0 & 0 & 0 \\ 0 & 0 & 0 & 0 & 0 & 0 & 1 & 0 \\ 0 & 0 & 0 & 0 & 0 & 0 & 0 & 1 \end{bmatrix} \tag{3.24}$$

One can directly verify using the representation of $\mathcal{N}_3$ required for 3-bits is given in Eq. 3.24; one needs to only look at the result for $C = 0$ to obtain the truth table for NAND for the 2−bit case, given in Table 3.3. More precisely, the entire truth table is given below; only the target bit with $C = 0$ is relevant for the 2−bits NAND gate, and indicated by the red color (Table 3.6).

The incoming lines have been grouped in pairs to show how unitarity fixes some of the elements. Note that the first three values of $C = 0$ come in pairs to ensure unitarity. The last 3-bits string 100 is unchanged and hence does not have a pairing. The last entry 111 does not enter the definition of NAND and has no spare bit to be mapped into– and hence remains unchanged.

The examples of AND and NAND show that the embedding of the gates acting on 2-bits has a unique extension to unitary gates acting on 3-bits strings.

3.7 Unitary OR and NOR Gates

From the truth table for the OR gate, it can be verified that it has the following matrix representation

$$OR_2 = \begin{bmatrix} 1\,0\,0\,0 \\ 0\,1\,0\,0 \\ 0\,0\,0\,0 \\ 0\,0\,1\,1 \end{bmatrix} : \text{ OR gate} \tag{3.25}$$

Similar to the NAND gate, the NOR gate can be obtained from the OR gate by applying the NOT on the second bit, which is given by X_2, and yields

$$\mathcal{N}OR_2 = X_2 \cdot OR = (\mathbb{I} \otimes X) \cdot OR$$

$$= \begin{bmatrix} 0\,1\,0\,0 \\ 1\,0\,0\,0 \\ 0\,0\,0\,1 \\ 0\,0\,1\,0 \end{bmatrix} \begin{bmatrix} 1\,0\,0\,0 \\ 0\,1\,0\,0 \\ 0\,0\,0\,0 \\ 0\,0\,1\,1 \end{bmatrix}$$

$$= \begin{bmatrix} 0\,1\,0\,0 \\ 1\,0\,0\,0 \\ 0\,0\,1\,1 \\ 0\,0\,0\,0 \end{bmatrix} : \text{ NOR gate} \tag{3.26}$$

As expected, the OR_2 and $\mathcal{N}OR_2$ gates, given in Eqs. 3.25 and 3.26, respectively, are not reversible and since a number of 0's and 1's are not equal for the output bit. Embedding them in the 3-bits string states yields, similar to AND gate, in 2×2 block notation, yields the following

$$OR_3 = \begin{bmatrix} \mathbb{I}\,0\,0\,0 \\ 0\,X\,0\,0 \\ 0\,0\,X\,0 \\ 0\,0\,0\,X \end{bmatrix} \tag{3.27}$$

Similar to the NAND gate's derivation from the AND gate, we have the following

$$\mathcal{N}OR_3 = X_3 \cdot OR_3 = \begin{bmatrix} X\,0\,0\,0 \\ 0\,\mathbb{I}\,0\,0 \\ 0\,0\,\mathbb{I}\,0 \\ 0\,0\,0\,\mathbb{I} \end{bmatrix} \tag{3.28}$$

3.8 Classical Binary Addition

Binary numbers form the foundation of all computations carried out by both classical and quantum computers. Indeed, the bedrock of all arithmetic calculations is the addition of two binary numbers. Hence, binary addition is an important place to start the application of the gates and circuits that we have discussed so far.

The derivation in this section is for the half and full adder of the classical computer. The half adder is done with a logical circuit that is not reversible and hence has no direct generalization to the half adder for the quantum case. In contrast, the derivation of the full adder for the classical case is done with a reversible circuit that can in fact be generalized to the quantum case. The derivations are done to illustrate the differences in classical and quantum circuits, as well as to foreground the derivation for the quantum case done in Sect. 7.4.

Binary addition of two string bits is shown in Fig. 3.9. Just like in ordinary addition, the addition mod 2 is carried out column by column and there is a carry bit. In general, for a column in the middle of the string, there is an incoming carry bit, denoted by C_{in} and an outgoing carry bit, denoted by C_{out}.

For a non-reversible computation, the initial state $|ABC_{in}\rangle$ is transformed by the full adder $\mathcal{F}$ into $|ASC_{out}\rangle$, where S is the binary sum; hence

$$\mathcal{F}|ABC_{in}\rangle = |ASC_{out}\rangle$$

The classical circuit for the half and full adder is given in Fig. 3.10a and b. These circuits are not reversible since as one can see from the figure, information about the input bits A, B is lost in the output.

3.9 Half-Adder

Instead of using the notation for classical circuit for the half adder given in Fig. 3.10a, a different notation for the half adder is used for the same circuit and is given in Fig. 3.11. The notation has three incoming and three outgoing lines and introduces us to the one used for the reversible circuits, which in turn can be generalized to the quantum case. In the derivations for the full adder, a reversible classical circuit is later given in Fig. 3.13 and is used instead of the one given in Fig. 3.10b. Figure 3.13 can be directly generalized to a quantum circuit.

Fig. 3.9 Binary addition. Published with permission of © Belal E. Baaquie and L. C. Kwek. All Rights Reserved

```
            1  1  1  1      ← Carry-out

            1  1  0  1  1

         +  1  1  0  1
         _____

         1  0  1  0  0  0
```

(a)   (b)

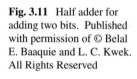

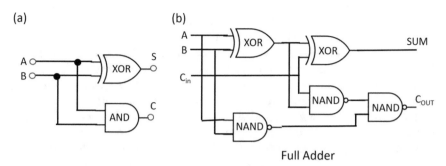

Full Adder

Fig. 3.10 Classical circuits for adding two bits using **a** half and **b** full-adder. Published with permission of © Belal E. Baaquie and L. C. Kwek. All Rights Reserved

Fig. 3.11 Half adder for adding two bits. Published with permission of © Belal E. Baaquie and L. C. Kwek. All Rights Reserved

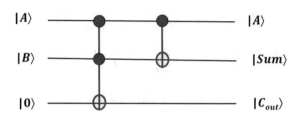

Table 3.7 Logic table for the half adder

A	B	Sum	Carry-out
0	0	0	0
0	1	1	0
1	0	1	0
1	1	0	1

To simplify our discussion and demonstrate the main features of binary addition, we start on the extreme right hand of the binary addition in Fig. 3.9—at the very beginning of the addition, with the incoming carry $C_{in} = 0$. This case is known as the half adder, since it takes into account only the outgoing carry but does not take into account an arbitrary incoming carry. Since addition is carried out column by column, it is sufficient to consider adding two 1-bits, called A and B; adding two multibit strings will be seen to be reducible to the addition of two 1-bits.

The logic table for the half adder is shown in Table 3.7. An irreversible circuit for the half adder has inputs A, B and $C_{in} = 0$ and outputs A, Sum S and carry C_{out}, as is given in Fig. 3.11.

The half-adder gate $\mathcal{H}$ is given by

$$\mathcal{H}|AB0\rangle = |ASC_{out}\rangle$$

where C_{out} denotes the carry-out bit. Since the value of the bit B has been lost in the output, the circuit in Fig. 3.11 is irreversible.

From our earlier discussion on gates, we know that the AND gate for a 3-bits string, denoted by $\mathcal{A}_3$, has the same truth tables as for a Toffoli gate given in Table 3.5— and hence is the gate required for the carry-out since it gives a nonzero result for output Z only if both bits are 1; that is, if $A = 1 = B$, then the output of the $\mathcal{A}_3$ gate is $|ABC_{out}\rangle$, with $C_{out} = Z = 1$. Furthermore, we know that X_{OR} gate does binary addition (which recall ignores the carry), and hence on the output of the $\mathcal{A}_3$ gate given by $|ABC_{out}\rangle$, we need to apply the X_{OR} gate only on the first two bits A and B, and the gate for the 3-bits string is $X_{OR} \otimes \mathbb{I}$, and yields the sum $A \oplus B$.

In summary, the logic of the half adder, shown in Fig. 3.11, is the following:

- The gate $\mathcal{A}_3(A, B, 0)$ gives an output of 1 only if both A and B are 1. Hence, the carry out $C_{out} = \mathcal{A}_3(A, B, 0) = \mathcal{T}(A, B, 0) = AB$
- The gate $X_{OR}(A, B) = A \oplus B$, which is the binary addition of bits A and B.

The half adder yields both the sum and carry for the bits A and B. Hence, for $S = A \oplus B$, the half-adder is given by

$$\mathcal{H}|AB0\rangle = (X_{OR} \otimes \mathbb{I})\mathcal{A}_3|AB0\rangle = (X_{OR} \otimes \mathbb{I})|ABC_{out}\rangle = |ASC_{out}\rangle \quad (3.29)$$

Recall that the explicit matrix representation of the Toffoli gate is given in both Eqs. 3.16 and 3.22. The explicit representation of $X_{OR} \otimes \mathbb{I}$ is required for verifying the result obtained for the half-adder.

Using the definition of the tensor product yields the following. In block 2×2 notation

$$X_{OR} \otimes \mathbb{I} = \begin{bmatrix} \mathbb{I} & 0 \\ 0 & X \end{bmatrix} \otimes \mathbb{I} \; ; \; \mathbb{I} = \begin{bmatrix} 1 & 0 \\ 0 & 1 \end{bmatrix} \; ; \; X = \begin{bmatrix} 0 & 1 \\ 1 & 0 \end{bmatrix}$$

The explicit derivation without using the block 2×2 notation is given by the following.

$$X_{OR} \otimes \mathbb{I} = \begin{bmatrix} 1 & 0 & 0 & 0 \\ 0 & 1 & 0 & 0 \\ 0 & 0 & 0 & 1 \\ 0 & 0 & 1 & 0 \end{bmatrix} \otimes \begin{bmatrix} 1 & 0 \\ 0 & 1 \end{bmatrix} = \begin{bmatrix} 1 & 0 & 0 & 0 & 0 & 0 & 0 & 0 \\ 0 & 1 & 0 & 0 & 0 & 0 & 0 & 0 \\ 0 & 0 & 1 & 0 & 0 & 0 & 0 & 0 \\ 0 & 0 & 0 & 1 & 0 & 0 & 0 & 0 \\ 0 & 0 & 0 & 0 & 0 & 0 & 1 & 0 \\ 0 & 0 & 0 & 0 & 0 & 0 & 0 & 1 \\ 0 & 0 & 0 & 0 & 1 & 0 & 0 & 0 \\ 0 & 0 & 0 & 0 & 0 & 1 & 0 & 0 \end{bmatrix} \quad (3.30)$$

In block 2×2 notation

$$X_{OR} \otimes \mathbb{I} = \begin{bmatrix} \mathbb{I} & 0 & 0 & 0 \\ 0 & \mathbb{I} & 0 & 0 \\ 0 & 0 & 0 & \mathbb{I} \\ 0 & 0 & \mathbb{I} & 0 \end{bmatrix} \tag{3.31}$$

We can verify that Eq. 3.29 is correct by checking the equation for the special case of $A = B = 1$ and $C_{in} = 0$; from binary addition given in Fig. 3.11

$$A \oplus B = 1 \oplus 1 = 0 = S \; ; \; C_{out} = 1$$

Hence, the half-adder should yield

$$\mathcal{H}|110\rangle = \mathcal{H}|ABC_{in}\rangle = |ASC_{out}\rangle = |101\rangle \tag{3.32}$$

The incoming and outgoing 3-bits string are

$$|110\rangle \equiv |1\rangle \otimes |1\rangle \otimes |0\rangle = \begin{bmatrix} 0 \\ 1 \end{bmatrix} \otimes \begin{bmatrix} 0 \\ 1 \end{bmatrix} \otimes \begin{bmatrix} 1 \\ 0 \end{bmatrix} = \begin{bmatrix} 0 \\ 1 \end{bmatrix} \otimes \begin{bmatrix} 0 \\ 0 \\ 1 \\ 0 \end{bmatrix}$$

$$\Rightarrow |110\rangle = \begin{bmatrix} 0 & 0 & 0 & 0 & 0 & 0 & 1 & 0 \end{bmatrix}^T$$

and similarly

$$|101\rangle = \begin{bmatrix} 0 & 0 & 0 & 0 & 0 & 1 & 0 & 0 \end{bmatrix}^T$$

The expression for $\mathcal{A}_3$ given in Eq. 3.16 yields

$$\mathcal{A}_3|110\rangle = \begin{bmatrix} 0 & 0 & 0 & 0 & 0 & 0 & 0 & 1 \end{bmatrix}^T$$

The half-adder, using Eq. 3.30, is given by the following

$$\mathcal{H}|110\rangle = \begin{bmatrix} \mathbb{I} & 0 & 0 & 0 \\ 0 & \mathbb{I} & 0 & 0 \\ 0 & 0 & 0 & \mathbb{I} \\ 0 & 0 & \mathbb{I} & 0 \end{bmatrix} \begin{bmatrix} 0 & 0 & 0 & 0 & 0 & 0 & 0 & 1 \end{bmatrix}^T$$

$$= \begin{bmatrix} 0 & 0 & 0 & 0 & 0 & 1 & 0 & 0 \end{bmatrix}^T \quad \Rightarrow \quad \mathcal{H}|110\rangle = |101\rangle$$

Hence, Eq. 3.32 has been explicitly verified.

Table 3.8 Logic table for a reversible half adder. The ancillary bit and the carry-in bit have been set to 0

A	B	Ancilla	Carry-in	A	B	Sum	Carry-out
0	0	0	0	0	0	0	0
0	1	0	0	0	1	1	0
1	0	0	0	1	0	1	0
1	1	0	0	1	1	0	1

One can construct a reversible half-adder by adding an ancillary bit, and hence in the reversible representation, the half-adder has *four* incoming lines and *four* outgoing lines and provides a reversible circuit. Table 3.8 is a logical truth table for a reversible half-adder. The reversible half-adder circuit is part of the circuit for a full-adder and is shown in Fig. 3.13.

3.10 Full-Adder

In classical computer, we add two (or three) bits using an exclusive OR gate with the carry-out bit being an AND gate in a half adder (see Table 3.10). A full adder with the carry-in as a bit, the picture is slightly more involved. One needs to consider the extra carry-in bit.

The need for the full adder arises if the incoming bit C_{in} can be either 0 or 1, instead of 0 as is the case for the half adder. The full-adder acts on two 1-bits and, in contrast to the half-adder, keeps track of the incoming and outgoing carry.

It is not difficult to verify that we want a gate that can perform the task given in the truth table, Table 3.9. The full adder uses the half-adder two times, once to add the numbers that in principle can generate a carry, and the second time using the half-adder again to add the incoming carry to the number generated by adding the two numbers. See Fig. 3.12.

Note that the truth table Table 3.9 yields an irreversible circuit, whereas a quantum circuit is reversible. Since a quantum gate is unitary, it is necessary to devise a gate that takes in the same number of inputs as the outputs. Note the important fact that a classical full adder is the special case of the quantum full adder–the only difference between the two does not lie in the gates, but in the fact that the classical bit is replaced by the quantum qubit. Hence, we will analyze the classical full adder using gates that are valid for both the classical and quantum full adder.

The logic of the full adder is similar to the half adder. We have the following:

- The sum bit: $S = (A \text{ XOR } B) \text{ XOR } C_{in} = (A \oplus B) \oplus C_{in}$
- The carry-out bit: $C_{out} = A \text{ AND } B \text{ OR } C_{in}(A \text{ XOR } B) = A \cdot B + C_{in}(A \oplus B)$

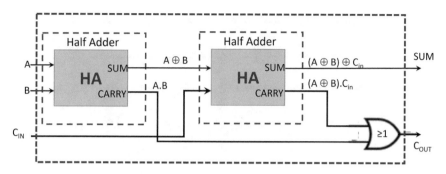

Fig. 3.12 Full adder equal using two Half adders (HA). Published with permission of © Belal E. Baaquie and L. C. Kwek. All Rights Reserved

Table 3.9 Truth table for full adder

Input 1	Input 2	Input 3	Output 1	Output 2
A	B	C_{in}	C_{out}	S
0	0	0	0	0
0	1	0	0	1
1	0	0	0	1
1	1	0	1	0
0	0	1	0	1
0	1	1	1	0
1	0	1	1	0
1	1	1	1	1

Table 3.10 Logic table for a reversible classical and quantum gate for the full adder

Input 1	Input 2	Input 3	Input 4	Output 1	Output 2	Output 3	Output 4
A	B	C_{in}	S	A	B	C_{out}	S
0	0	0	0	0	0	0	0
0	0	1	0	0	0	0	1
0	1	0	0	0	1	0	1
0	1	1	0	0	1	1	0
1	0	0	0	1	0	0	1
1	0	1	0	1	0	1	0
1	1	0	0	1	1	1	0
1	1	1	0	1	1	1	1

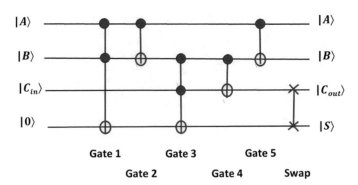

The logic of the full adder can be implemented by the logic table given in Table 3.10. We implement Table 3.10 with the quantum circuit in Fig. 3.13.

In Fig. 3.13, there are exactly five gates: Gates 1 to 5 and a Swap gate at the end for the third and fourth bits. Gate 1 is a AND gate on the first, second and fourth bits. On a quantum computer, we can do the same arithmetic with the circuit shown in Fig. 3.13. The circuit has four incoming bits and four outgoing bits given by $|ABC_{in}0\rangle$ and the adder yields the following

$$\mathcal{F}|ABC_{in}0\rangle = |ABC_{out}S\rangle \; ; \; A, B = 0 \textbf{ or } 1 \tag{3.33}$$

Note that the classical full adder, given above in Eq. 3.33 has $|A\rangle$, $|B\rangle$ as classical 1-bits; the input state is a definite combination of four bits, and the output is also a combination of four bits.

As mentioned earlier, the full adder gate $\mathcal{F}$ is the same for the classical and quantum case; the crucial difference between adding bits versus qubits lies in the 4-bits string $|ABC_{in}0\rangle$. For the quantum case, $|A\rangle$, $|B\rangle$ are both qubits and the result of the addition is not a single determinate state, as in Eq. 3.33, but a superposition of states. The quantum addition of two qubits is discussed in Section 7.4.

In Fig. 3.13, there are exactly five gates: Gate 1 to 5 say and a swap gate at the end for the third and fourth qubits. Gate 1 is a CCNot gate on the first, second and fourth qubits.

Using the result from Eqs. 3.13 and 3.18 and that $\mathcal{A}_3 = \mathcal{T}$, Gate 1 for the full-adder given in Fig. 3.13 is given by the gate acting as the identity matrix $\mathbb{I}$ on the third bit and hence yielding the following 16×16 matrix:

$$\text{Gate } 1 = \mathcal{A}_3(A, B, C, 0)$$
$$= \Big(|00\rangle\langle00| + |01\rangle\langle01| + |10\rangle\langle10|\Big) \otimes \mathbb{I} \otimes \mathbb{I} + |11\rangle\langle11| \otimes \mathbb{I} \otimes X$$

Note that the bit $|C\rangle$ for Gate1 and Gate2 is a placeholder for the third bit required for these gates. In Fig. 3.13 for Gate 1 this placeholder bit $|C\rangle = |0\rangle$ but it has a nontrivial value for Gate 3 onwards.

Using results from Eq. 3.13 yields the following 16×16 matrices for the other Gates for the full-adder given in Fig. 3.13:

$$Gate\ 2 = X_{OR}(A, B) \otimes \mathbb{I} \otimes \mathbb{I}$$
$$Gate\ 3 = \mathcal{A}_3(A, B, C_{in}, S) = \mathbb{I} \otimes T(B, C_{in}, S)$$
$$Gate\ 4 = \mathbb{I} \otimes X_{OR}(B, C_{in}) \otimes \mathbb{I}$$
$$Gate\ 5 = X_{OR}(A, B) \otimes \mathbb{I} \otimes \mathbb{I}$$
$$Swap = \mathbb{I} \otimes \mathbb{I} \otimes Swap12(C_{in}, S)$$

where

$$\mathbb{I} = \begin{bmatrix} 1 & 0 \\ 0 & 1 \end{bmatrix} \ ; \ X = \begin{bmatrix} 0 & 1 \\ 1 & 0 \end{bmatrix} \ ; \ X_{OR} = \begin{bmatrix} \mathbb{I} & 0 \\ 0 & X \end{bmatrix}$$

$$Swap12 = \begin{bmatrix} 1 & 0 & 0 & 0 \\ 0 & 0 & 1 & 0 \\ 0 & 1 & 0 & 0 \\ 0 & 0 & 0 & 1 \end{bmatrix}$$

The explicit 16×16 matrices for the Gates and the Swap are given in Sect. 7.4 on the quantum circuit for the full adder.

The full adder is given by

$$\mathcal{F} = Swap \cdot Gate5 \cdot Gate4 \cdot Gate3 \cdot Gate2 \cdot Gate1$$

with the full 16×16 matrix given in Sect. 3.11. Collecting all the results, the full-adder gate $\mathcal{F}$ is given by

$$\mathcal{F}|ABC_{in}0\rangle = |ABC_{out}S\rangle \tag{3.34}$$

Example

Consider the binary addition of two bits, with and without C_{in} and C_{out}. One can explicitly verify, as given below, Eq. 3.34 using the following examples of the addition of two binary bits.

- $A = 1; B = 1; C_{in} = 0 \Rightarrow 1 \oplus 1 \oplus C_{in} = 0 = S$ and $C_{out} = 1$
  $\Rightarrow \mathcal{F}|1100\rangle = |1110\rangle$
- $A = 1; B = 1; C_{in} = 1 \Rightarrow 1 \oplus 1 \oplus C_{in} = 1 = S$ and $C_{out} = 1$
  $\Rightarrow \mathcal{F}|1110\rangle = |1111\rangle$
- $A = 0; B = 0; C_{in} = 1 \Rightarrow 0 \oplus 0 \oplus C_{in} = 1 = S$ and $C_{out} = 0$
  $\Rightarrow \mathcal{F}|0010\rangle = |0001\rangle$

Hence, we have the following summary of our results

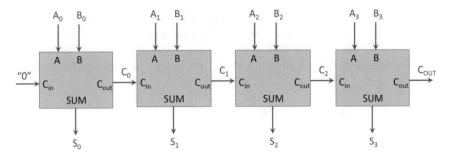

$$\mathcal{F}|1100\rangle = |1110\rangle \; ; \; \mathcal{F}|1110\rangle = |1111\rangle \; ; \; \mathcal{F}|0010\rangle = |0001\rangle \qquad (3.35)$$

One verifies, using the matrix representations given in Sect. 3.11, the results obtained in Eq. 3.35.

To add two binary numbers that are n-bits strings, one can cascade the result of adding two 1-bits by adding the n-bits strings column by column, passing the carry from one column to the next. This 'ripple' effect in adding two 4-bits strings is shown in Fig. 3.14.

If one is adding $n-$bits string to a $m-$bits string, with $n \leq m$, then the full adder gate $\mathcal{F}$ is a matrix of size 16×16 that has to be used successively $n-$times; this creates a 'ripple' effect of the output sum and carries out of the full adder for step j being the input and carry in for the next addition at step $j + 1$. Hence, for adding $n-$bits string

$$\mathcal{F}|ABC_{\text{in}}0\rangle = \mathcal{F}_n \cdots \mathcal{F}_{j+1}\mathcal{F}_j \cdots \mathcal{F}_2\mathcal{F}_1|ABC_{\text{in}}0\rangle = |ABC_{\text{out}}S\rangle$$

Let A, B be two 4-bits string

$$|AB\rangle = |A_0 B_0\rangle|A_1 B_1\rangle|A_2 B_2\rangle|A_3 B_3\rangle$$

with the initial $C_{\text{in}} = 0$. As shown in Fig. 3.14, each box is a full adder, adding two 1-bits A_i, B_i and an incoming carry C_{i-1} and an outgoing carry C_i. The result of sequentially applying the full adder starting from the least significant binary numbers A_0, B_0 and moving to the most significant the most significant binary A_3, B_3 yields the following result

$$\mathcal{F}|A, B, 0, 0\rangle = |AB\rangle|C_{\text{out}} S_3 S_2 S_1 S_0\rangle$$

3.11 Matrices of Full-Adder Gates

In matrix representation, Gate 1 is given by the 16 by 16 matrix:

$$\text{Gate } 1 = \Big(|00\rangle\langle00| + |01\rangle\langle01| + |10\rangle\langle10|\Big) \otimes \mathbb{I} \otimes \mathbb{I} + |11\rangle\langle11| \otimes \mathbb{I} \otimes X$$

$$= \begin{pmatrix}
1&0&0&0&0&0&0&0&0&0&0&0&0&0&0&0\\
0&1&0&0&0&0&0&0&0&0&0&0&0&0&0&0\\
0&0&1&0&0&0&0&0&0&0&0&0&0&0&0&0\\
0&0&0&1&0&0&0&0&0&0&0&0&0&0&0&0\\
0&0&0&0&1&0&0&0&0&0&0&0&0&0&0&0\\
0&0&0&0&0&1&0&0&0&0&0&0&0&0&0&0\\
0&0&0&0&0&0&1&0&0&0&0&0&0&0&0&0\\
0&0&0&0&0&0&0&1&0&0&0&0&0&0&0&0\\
0&0&0&0&0&0&0&0&1&0&0&0&0&0&0&0\\
0&0&0&0&0&0&0&0&0&1&0&0&0&0&0&0\\
0&0&0&0&0&0&0&0&0&0&1&0&0&0&0&0\\
0&0&0&0&0&0&0&0&0&0&0&1&0&0&0&0\\
0&0&0&0&0&0&0&0&0&0&0&0&0&1&0&0\\
0&0&0&0&0&0&0&0&0&0&0&0&1&0&0&0\\
0&0&0&0&0&0&0&0&0&0&0&0&0&0&0&1\\
0&0&0&0&0&0&0&0&0&0&0&0&0&0&1&0
\end{pmatrix}.$$

The other gates are given by

$$\text{Gate } 2 = X_{\text{OR}} \otimes \mathbb{I} \otimes \mathbb{I}$$

$$= \begin{pmatrix}
1&0&0&0&0&0&0&0&0&0&0&0&0&0&0&0\\
0&1&0&0&0&0&0&0&0&0&0&0&0&0&0&0\\
0&0&1&0&0&0&0&0&0&0&0&0&0&0&0&0\\
0&0&0&1&0&0&0&0&0&0&0&0&0&0&0&0\\
0&0&0&0&1&0&0&0&0&0&0&0&0&0&0&0\\
0&0&0&0&0&1&0&0&0&0&0&0&0&0&0&0\\
0&0&0&0&0&0&1&0&0&0&0&0&0&0&0&0\\
0&0&0&0&0&0&0&1&0&0&0&0&0&0&0&0\\
0&0&0&0&0&0&0&0&0&0&0&1&0&0&0&0\\
0&0&0&0&0&0&0&0&0&0&0&0&1&0&0&0\\
0&0&0&0&0&0&0&0&0&0&0&0&0&1&0&0\\
0&0&0&0&0&0&0&0&0&0&0&0&0&0&0&1\\
0&0&0&0&0&0&0&0&1&0&0&0&0&0&0&0\\
0&0&0&0&0&0&0&0&0&1&0&0&0&0&0&0\\
0&0&0&0&0&0&0&0&0&0&1&0&0&0&0&0\\
0&0&0&0&0&0&0&0&0&0&0&0&1&0&0&0
\end{pmatrix} ;$$

Gate $3 = \mathbb{I} \otimes \mathcal{T}(B, C_{in}, 0)$

$$
= \begin{pmatrix}
1 & 0 & 0 & 0 & 0 & 0 & 0 & 0 & 0 & 0 & 0 & 0 & 0 & 0 & 0 & 0 \\
0 & 1 & 0 & 0 & 0 & 0 & 0 & 0 & 0 & 0 & 0 & 0 & 0 & 0 & 0 & 0 \\
0 & 0 & 1 & 0 & 0 & 0 & 0 & 0 & 0 & 0 & 0 & 0 & 0 & 0 & 0 & 0 \\
0 & 0 & 0 & 1 & 0 & 0 & 0 & 0 & 0 & 0 & 0 & 0 & 0 & 0 & 0 & 0 \\
0 & 0 & 0 & 0 & 1 & 0 & 0 & 0 & 0 & 0 & 0 & 0 & 0 & 0 & 0 & 0 \\
0 & 0 & 0 & 0 & 0 & 1 & 0 & 0 & 0 & 0 & 0 & 0 & 0 & 0 & 0 & 0 \\
0 & 0 & 0 & 0 & 0 & 0 & 0 & 1 & 0 & 0 & 0 & 0 & 0 & 0 & 0 & 0 \\
0 & 0 & 0 & 0 & 0 & 0 & 1 & 0 & 0 & 0 & 0 & 0 & 0 & 0 & 0 & 0 \\
0 & 0 & 0 & 0 & 0 & 0 & 0 & 0 & 1 & 0 & 0 & 0 & 0 & 0 & 0 & 0 \\
0 & 0 & 0 & 0 & 0 & 0 & 0 & 0 & 0 & 1 & 0 & 0 & 0 & 0 & 0 & 0 \\
0 & 0 & 0 & 0 & 0 & 0 & 0 & 0 & 0 & 0 & 1 & 0 & 0 & 0 & 0 & 0 \\
0 & 0 & 0 & 0 & 0 & 0 & 0 & 0 & 0 & 0 & 0 & 1 & 0 & 0 & 0 & 0 \\
0 & 0 & 0 & 0 & 0 & 0 & 0 & 0 & 0 & 0 & 0 & 0 & 1 & 0 & 0 & 0 \\
0 & 0 & 0 & 0 & 0 & 0 & 0 & 0 & 0 & 0 & 0 & 0 & 0 & 1 & 0 & 0 \\
0 & 0 & 0 & 0 & 0 & 0 & 0 & 0 & 0 & 0 & 0 & 0 & 0 & 0 & 0 & 1 \\
0 & 0 & 0 & 0 & 0 & 0 & 0 & 0 & 0 & 0 & 0 & 0 & 0 & 0 & 1 & 0 \\
\end{pmatrix} ;
$$

Gate $4 = \mathbb{I} \otimes X_{\mathrm{OR}} \otimes \mathbb{I}$

$$
= \begin{pmatrix}
1 & 0 & 0 & 0 & 0 & 0 & 0 & 0 & 0 & 0 & 0 & 0 & 0 & 0 & 0 & 0 \\
0 & 1 & 0 & 0 & 0 & 0 & 0 & 0 & 0 & 0 & 0 & 0 & 0 & 0 & 0 & 0 \\
0 & 0 & 1 & 0 & 0 & 0 & 0 & 0 & 0 & 0 & 0 & 0 & 0 & 0 & 0 & 0 \\
0 & 0 & 0 & 1 & 0 & 0 & 0 & 0 & 0 & 0 & 0 & 0 & 0 & 0 & 0 & 0 \\
0 & 0 & 0 & 0 & 0 & 0 & 1 & 0 & 0 & 0 & 0 & 0 & 0 & 0 & 0 & 0 \\
0 & 0 & 0 & 0 & 0 & 0 & 0 & 1 & 0 & 0 & 0 & 0 & 0 & 0 & 0 & 0 \\
0 & 0 & 0 & 0 & 1 & 0 & 0 & 0 & 0 & 0 & 0 & 0 & 0 & 0 & 0 & 0 \\
0 & 0 & 0 & 0 & 0 & 1 & 0 & 0 & 0 & 0 & 0 & 0 & 0 & 0 & 0 & 0 \\
0 & 0 & 0 & 0 & 0 & 0 & 0 & 0 & 1 & 0 & 0 & 0 & 0 & 0 & 0 & 0 \\
0 & 0 & 0 & 0 & 0 & 0 & 0 & 0 & 0 & 1 & 0 & 0 & 0 & 0 & 0 & 0 \\
0 & 0 & 0 & 0 & 0 & 0 & 0 & 0 & 0 & 0 & 1 & 0 & 0 & 0 & 0 & 0 \\
0 & 0 & 0 & 0 & 0 & 0 & 0 & 0 & 0 & 0 & 0 & 1 & 0 & 0 & 0 & 0 \\
0 & 0 & 0 & 0 & 0 & 0 & 0 & 0 & 0 & 0 & 0 & 0 & 0 & 0 & 1 & 0 \\
0 & 0 & 0 & 0 & 0 & 0 & 0 & 0 & 0 & 0 & 0 & 0 & 0 & 0 & 0 & 1 \\
0 & 0 & 0 & 0 & 0 & 0 & 0 & 0 & 0 & 0 & 0 & 0 & 1 & 0 & 0 & 0 \\
0 & 0 & 0 & 0 & 0 & 0 & 0 & 0 & 0 & 0 & 0 & 0 & 0 & 1 & 0 & 0 \\
\end{pmatrix} ;
$$

Gate 5 $= X_{OR} \otimes \mathbb{I} \otimes \mathbb{I}$

$$
=
\begin{pmatrix}
1 & 0 & 0 & 0 & 0 & 0 & 0 & 0 & 0 & 0 & 0 & 0 & 0 & 0 & 0 & 0 \\
0 & 1 & 0 & 0 & 0 & 0 & 0 & 0 & 0 & 0 & 0 & 0 & 0 & 0 & 0 & 0 \\
0 & 0 & 1 & 0 & 0 & 0 & 0 & 0 & 0 & 0 & 0 & 0 & 0 & 0 & 0 & 0 \\
0 & 0 & 0 & 1 & 0 & 0 & 0 & 0 & 0 & 0 & 0 & 0 & 0 & 0 & 0 & 0 \\
0 & 0 & 0 & 0 & 1 & 0 & 0 & 0 & 0 & 0 & 0 & 0 & 0 & 0 & 0 & 0 \\
0 & 0 & 0 & 0 & 0 & 1 & 0 & 0 & 0 & 0 & 0 & 0 & 0 & 0 & 0 & 0 \\
0 & 0 & 0 & 0 & 0 & 0 & 1 & 0 & 0 & 0 & 0 & 0 & 0 & 0 & 0 & 0 \\
0 & 0 & 0 & 0 & 0 & 0 & 0 & 1 & 0 & 0 & 0 & 0 & 0 & 0 & 0 & 0 \\
0 & 0 & 0 & 0 & 0 & 0 & 0 & 0 & 0 & 0 & 0 & 0 & 1 & 0 & 0 & 0 \\
0 & 0 & 0 & 0 & 0 & 0 & 0 & 0 & 0 & 0 & 0 & 0 & 0 & 1 & 0 & 0 \\
0 & 0 & 0 & 0 & 0 & 0 & 0 & 0 & 0 & 0 & 0 & 0 & 0 & 0 & 1 & 0 \\
0 & 0 & 0 & 0 & 0 & 0 & 0 & 0 & 0 & 0 & 0 & 0 & 0 & 0 & 0 & 1 \\
0 & 0 & 0 & 0 & 0 & 0 & 0 & 0 & 1 & 0 & 0 & 0 & 0 & 0 & 0 & 0 \\
0 & 0 & 0 & 0 & 0 & 0 & 0 & 0 & 0 & 1 & 0 & 0 & 0 & 0 & 0 & 0 \\
0 & 0 & 0 & 0 & 0 & 0 & 0 & 0 & 0 & 0 & 1 & 0 & 0 & 0 & 0 & 0 \\
0 & 0 & 0 & 0 & 0 & 0 & 0 & 0 & 0 & 0 & 0 & 1 & 0 & 0 & 0 & 0
\end{pmatrix} ;
$$

The matrix products of the gates that describe the full-adder circuit are given by

$\mathcal{G} =$ Product of Gates $=$ Gate 5 $\cdot$ Gate 4 $\cdot$ Gate 3 $\cdot$ Gate 2 $\cdot$ Gate 1

$$
=
\begin{pmatrix}
1 & 0 & 0 & 0 & 0 & 0 & 0 & 0 & 0 & 0 & 0 & 0 & 0 & 0 & 0 & 0 \\
0 & 1 & 0 & 0 & 0 & 0 & 0 & 0 & 0 & 0 & 0 & 0 & 0 & 0 & 0 & 0 \\
0 & 0 & 1 & 0 & 0 & 0 & 0 & 0 & 0 & 0 & 0 & 0 & 0 & 0 & 0 & 0 \\
0 & 0 & 0 & 1 & 0 & 0 & 0 & 0 & 0 & 0 & 0 & 0 & 0 & 0 & 0 & 0 \\
0 & 0 & 0 & 0 & 0 & 0 & 0 & 1 & 0 & 0 & 0 & 0 & 0 & 0 & 0 & 0 \\
0 & 0 & 0 & 0 & 0 & 0 & 1 & 0 & 0 & 0 & 0 & 0 & 0 & 0 & 0 & 0 \\
0 & 0 & 0 & 0 & 1 & 0 & 0 & 0 & 0 & 0 & 0 & 0 & 0 & 0 & 0 & 0 \\
0 & 0 & 0 & 0 & 0 & 1 & 0 & 0 & 0 & 0 & 0 & 0 & 0 & 0 & 0 & 0 \\
0 & 0 & 0 & 0 & 0 & 0 & 0 & 0 & 0 & 0 & 0 & 0 & 1 & 0 & 0 & 0 \\
0 & 0 & 0 & 0 & 0 & 0 & 0 & 0 & 0 & 0 & 1 & 0 & 0 & 0 & 0 & 0 \\
0 & 0 & 0 & 0 & 0 & 0 & 0 & 0 & 1 & 0 & 0 & 0 & 0 & 0 & 0 & 0 \\
0 & 0 & 0 & 0 & 0 & 0 & 0 & 0 & 0 & 1 & 0 & 0 & 0 & 0 & 0 & 0 \\
0 & 0 & 0 & 0 & 0 & 0 & 0 & 0 & 0 & 0 & 0 & 0 & 0 & 1 & 0 & 0 \\
0 & 0 & 0 & 0 & 0 & 0 & 0 & 0 & 0 & 0 & 0 & 0 & 1 & 0 & 0 & 0 \\
0 & 0 & 0 & 0 & 0 & 0 & 0 & 0 & 0 & 0 & 0 & 0 & 0 & 0 & 0 & 1 \\
0 & 0 & 0 & 0 & 0 & 0 & 0 & 0 & 0 & 0 & 0 & 0 & 0 & 0 & 1 & 0
\end{pmatrix}
$$

Finally, we need to apply the SWAP gate on the last two qubits. Recall the SWAP gate in computational basis is given by the matrix

$$
\text{SWAP12} =
\begin{pmatrix}
1 & 0 & 0 & 0 \\
0 & 0 & 1 & 0 \\
0 & 1 & 0 & 0 \\
0 & 0 & 0 & 1
\end{pmatrix}
$$

But since this SWAP gate acts only on the last two qubits, it is described by

$$\mathcal{S} = \text{SWAP Gate} = \mathbb{I} \otimes \mathbb{I} \otimes \text{SWAP12}$$

$$= \begin{pmatrix}
1 & 0 & 0 & 0 & 0 & 0 & 0 & 0 & 0 & 0 & 0 & 0 & 0 & 0 & 0 & 0 \\
0 & 0 & 1 & 0 & 0 & 0 & 0 & 0 & 0 & 0 & 0 & 0 & 0 & 0 & 0 & 0 \\
0 & 1 & 0 & 0 & 0 & 0 & 0 & 0 & 0 & 0 & 0 & 0 & 0 & 0 & 0 & 0 \\
0 & 0 & 0 & 1 & 0 & 0 & 0 & 0 & 0 & 0 & 0 & 0 & 0 & 0 & 0 & 0 \\
0 & 0 & 0 & 0 & 1 & 0 & 0 & 0 & 0 & 0 & 0 & 0 & 0 & 0 & 0 & 0 \\
0 & 0 & 0 & 0 & 0 & 0 & 1 & 0 & 0 & 0 & 0 & 0 & 0 & 0 & 0 & 0 \\
0 & 0 & 0 & 0 & 0 & 1 & 0 & 0 & 0 & 0 & 0 & 0 & 0 & 0 & 0 & 0 \\
0 & 0 & 0 & 0 & 0 & 0 & 0 & 1 & 0 & 0 & 0 & 0 & 0 & 0 & 0 & 0 \\
0 & 0 & 0 & 0 & 0 & 0 & 0 & 0 & 1 & 0 & 0 & 0 & 0 & 0 & 0 & 0 \\
0 & 0 & 0 & 0 & 0 & 0 & 0 & 0 & 0 & 0 & 1 & 0 & 0 & 0 & 0 & 0 \\
0 & 0 & 0 & 0 & 0 & 0 & 0 & 0 & 0 & 1 & 0 & 0 & 0 & 0 & 0 & 0 \\
0 & 0 & 0 & 0 & 0 & 0 & 0 & 0 & 0 & 0 & 0 & 1 & 0 & 0 & 0 & 0 \\
0 & 0 & 0 & 0 & 0 & 0 & 0 & 0 & 0 & 0 & 0 & 0 & 1 & 0 & 0 & 0 \\
0 & 0 & 0 & 0 & 0 & 0 & 0 & 0 & 0 & 0 & 0 & 0 & 0 & 0 & 1 & 0 \\
0 & 0 & 0 & 0 & 0 & 0 & 0 & 0 & 0 & 0 & 0 & 0 & 0 & 1 & 0 & 0 \\
0 & 0 & 0 & 0 & 0 & 0 & 0 & 0 & 0 & 0 & 0 & 0 & 0 & 0 & 0 & 1
\end{pmatrix}$$

The full adder is given by the matrix

$$\mathcal{F} = \mathcal{S} \cdot \mathcal{G} \tag{3.36}$$

$$= \text{SWAP Gate} \cdot \text{Gate 5} \cdot \text{Gate 4} \cdot \text{Gate 3} \cdot \text{Gate 2} \cdot \text{Gate 1} \tag{3.37}$$

$$= \begin{pmatrix}
1 & 0 & 0 & 0 & 0 & 0 & 0 & 0 & 0 & 0 & 0 & 0 & 0 & 0 & 0 & 0 \\
0 & 0 & 1 & 0 & 0 & 0 & 0 & 0 & 0 & 0 & 0 & 0 & 0 & 0 & 0 & 0 \\
0 & 1 & 0 & 0 & 0 & 0 & 0 & 0 & 0 & 0 & 0 & 0 & 0 & 0 & 0 & 0 \\
0 & 0 & 0 & 1 & 0 & 0 & 0 & 0 & 0 & 0 & 0 & 0 & 0 & 0 & 0 & 0 \\
0 & 0 & 0 & 0 & 0 & 0 & 0 & 1 & 0 & 0 & 0 & 0 & 0 & 0 & 0 & 0 \\
0 & 0 & 0 & 0 & 1 & 0 & 0 & 0 & 0 & 0 & 0 & 0 & 0 & 0 & 0 & 0 \\
0 & 0 & 0 & 0 & 0 & 0 & 1 & 0 & 0 & 0 & 0 & 0 & 0 & 0 & 0 & 0 \\
0 & 0 & 0 & 0 & 0 & 1 & 0 & 0 & 0 & 0 & 0 & 0 & 0 & 0 & 0 & 0 \\
0 & 0 & 0 & 0 & 0 & 0 & 0 & 0 & 0 & 0 & 0 & 1 & 0 & 0 & 0 & 0 \\
0 & 0 & 0 & 0 & 0 & 0 & 0 & 0 & 1 & 0 & 0 & 0 & 0 & 0 & 0 & 0 \\
0 & 0 & 0 & 0 & 0 & 0 & 0 & 0 & 0 & 0 & 1 & 0 & 0 & 0 & 0 & 0 \\
0 & 0 & 0 & 0 & 0 & 0 & 0 & 0 & 0 & 1 & 0 & 0 & 0 & 0 & 0 & 0 \\
0 & 0 & 0 & 0 & 0 & 0 & 0 & 0 & 0 & 0 & 0 & 0 & 0 & 1 & 0 & 0 \\
0 & 0 & 0 & 0 & 0 & 0 & 0 & 0 & 0 & 0 & 0 & 0 & 0 & 0 & 0 & 1 \\
0 & 0 & 0 & 0 & 0 & 0 & 0 & 0 & 0 & 0 & 0 & 0 & 1 & 0 & 0 & 0 \\
0 & 0 & 0 & 0 & 0 & 0 & 0 & 0 & 0 & 0 & 0 & 0 & 0 & 0 & 1 & 0
\end{pmatrix} \tag{3.38}$$

To get the adder, one applies the full-adder matrix $\mathcal{F}$ to the inputs. The output does not make sense unless one notes that the fourth qubit is always set to the zero computational bit. We therefore only need to look at those inputs whose fourth qubit is $|0\rangle$. We remind the readers that the encoding is

$$|(\text{input A})(\text{input B})(\text{carry in or out})(\text{Sum})\rangle$$

This process effectively provides us with the mapping of classical bits as:

$$|0000\rangle \rightarrow |0000\rangle$$
$$|0010\rangle \rightarrow |0001\rangle$$
$$|0100\rangle \rightarrow |0101\rangle$$
$$|0110\rangle \rightarrow |0110\rangle$$
$$|1000\rangle \rightarrow |1001\rangle$$
$$|1010\rangle \rightarrow |1010\rangle$$
$$|1100\rangle \rightarrow |1110\rangle$$
$$|1110\rangle \rightarrow |1111\rangle$$

as described by the full-adder matrix $\mathcal{F}$ for the circuit given in Eq. 3.36.

References

1. Feynman RP (2018) Lectures on computation. CRC Press, USA
2. Mermin ND (2007) Quantum computer science. Cambridge University Press, UK

Chapter 4
Principles of Quantum Mechanics

The principles and formalism of quantum mechanics are reviewed as these provide the basis for quantum computers and quantum algorithms. This chapter is based on the Copenhagen interpretation of quantum mechanics pioneered by Werner Heisenberg, Niels Bohr and Max Born, with the derivations following the approach of Baaquie [1]. There are many other interpretations of quantum mechanics, but by and large, the Copenhagen interpretation is the one that is followed by most practitioners of quantum computing [2].

This chapter starts with a general discussion on the degree of freedom and its Hilbert space and Hermitian operators (matrices): these three ingredients play a central role in the description of quantum systems. The Schroödinger equation is introduced, but not discussed in any detail since it plays only an indirect role in quantum algorithms. The measurement procedure and process of a quantum system are discussed in some detail as the key difference between a classical and quantum system lies in the significance of quantum measurements. In particular, the cardinal difference between a classical and quantum algorithm lies in the essential role played by measurements in quantum mechanics. The mathematical formalism of quantum mechanics is based on linear vector spaces, and Chap. 2 should be reviewed for the background material required for this and later chapters.

The description and dynamics of a quantum entity, based on the Copenhagen interpretation of quantum mechanics [1, 2], can be summarized as follows. Quantum mechanics is built on **five main conceptual pillars** that are shown in Fig. 4.1 and given below [1].

- The quantum degree of freedom space $\mathcal{F}$.
- The quantum state space $\mathcal{V}$, which is a Hilbert space.
- State vector $\psi(\mathcal{F})$, which is an element of Hilbert space. Time evolution of $\psi(\mathcal{F})$ is given by the Schrödinger equation.
- Hermitian operators $\mathcal{O}_i(\mathcal{F}); i = 1, \ldots, I$ that encode all the observable properties of the state vector $\psi(\mathcal{F})$.

© The Author(s), under exclusive license to Springer Nature Singapore Pte Ltd. 2023   71
B. E. Baaquie and L.-C. Kwek, *Quantum Computers*,
https://doi.org/10.1007/978-981-19-7517-2_4

Degree of freedom	State space	Dynamics	Operators	Observation
$\mathcal{F}$	$\mathcal{V}(\mathcal{F})$	$\dfrac{\partial \psi(t,\mathcal{F})}{\partial t}$	$\hat{O}(\mathcal{F})$	$E_v[\hat{O}(\mathcal{F})]$

Fig. 4.1 Theoretical schema of quantum mechanics, based on five fundamental ingredients. Published with permission of © Belal E. Baaquie. All Rights Reserved

- The process of measurement, with repeated observations yielding the expectation value of the operators, namely $E_V[\mathcal{O}_i(\mathcal{F})]$.

4.1 Degrees of Freedom: Indeterminate

The 'ground' on which quantum mechanics stands is the **degree of freedom**. For quantum computers, the single 1-bit of the classical computer $x = \{0, 1\}$ is elevated in quantum mechanics to the quantum binary degree of freedom. The two values of a bit $x = \{0, 1\}$ taken together, as a two element set, yield the binary degree of freedom $\mathcal{F} = \{0, 1\}$. For the case of *two* 1-bits, all possible combinations yield the degree of $\mathcal{F}^{\otimes 2} = \{0, 1\} \otimes \{0, 1\} = (x_1, x_2)$ consisting of $2 \times 2 = 2^2$ number of points; written out in full detail, the degrees of freedom $\mathcal{F}^{\otimes 2}$ is the following four element set

$$\mathcal{F}^{\otimes 2} = \{0, 1\} \otimes \{0, 1\} = (x_1, x_2) = \{00, 01, 10, 11\}; \quad x_1, x_2 = 0, 1$$

The case for $n = 3$, all possible combinations yield the discrete set of eight points for the space of the degrees of freedom $\mathcal{F}^{\otimes 3}$

$$\mathcal{F}^{\otimes 3} = \{0, 1\}^{\otimes 3} = \{0, 1\} \otimes \{0, 1\} \otimes \{0, 1\} \tag{4.1}$$
$$= \{000, 001, 010, 011, 100, 101, 110, 111\}$$

The $2^3 = 8$ points are organized as the corners of a three-dimensional cube and shown in Fig. 4.2.

If the system has n-bits, then the quantum degrees of freedom $\mathcal{F}^{\otimes n}$ is

$$\mathcal{F}^{\otimes n} = \{0, 1\} \otimes \{0, 1\} \cdots \{0, 1\} = \{0, 1\}^{\otimes n}$$

The specific values n-degrees of freedom $\mathcal{F}^{\otimes n}$ are denoted by

$$(x_1, x_2, \ldots, x_n) : x_i = 0, 1; \quad i = 1, 2, \ldots, n$$

The n-binary degrees of freedom $\mathcal{F}^{\otimes n}$ consist of a discrete space of 2^n points that form the corners of a 2^n hypercube and is denoted by

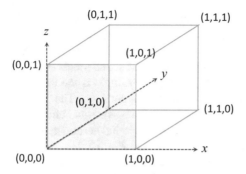

Fig. 4.2 Discrete space $\mathcal{F}^{\otimes 3} = \{0, 1\}^{\otimes 3}$ for 3-binary degrees of freedom, consisting of $2^3 = 8$ points at the corners of the cube. The coordinates of the corners are the values taken by the binary degrees of freedom. Published with permission of © Belal E. Baaquie and L. C. Kwek. All Rights Reserved

$$\mathcal{F}^{\otimes m} = \{0, 1\}^{\otimes n}$$

The degrees of freedom are **indeterminate**: what this means is that the degrees of freedom exist as the entire space (set of points) $\mathcal{F}^{\otimes m} = \{0, 1\}^{\otimes n}$, and not as a specific point of that space [1]. In other words, the degrees of freedom never have a specific value; for example, in the case of the single binary degree of freedom $x = \{0, 1\}$, the degree of freedom always *simultaneously* has both values of $x = 0$ **and** 1. The specific and determinate values of the degrees of freedom are the points that constitute the space $\mathcal{F}^{\otimes m}$.

The degree of freedom $\mathcal{F}$ can never, in principle, be observed in its totality: one can only observe the specific and determinate values of the degrees of freedom, and that too, one by one: no experiment ever has been able to simultaneously detect two or more determinate values of the degree of freedom.[1] All objects in spacetime have a determinate and specific value: the degree of freedom $\mathcal{F}$ cannot be directly observed in spacetime, as such, since it is indeterminate. What one observes in experiments are the specific and determinate values of the degrees of freedom. In particular, the quantum degrees of freedom given by $\{0, 1\}^{\otimes n}$ always remain indeterminate [1]. Any device attempting to measure the degrees of freedom encounters the state vector of the degrees of freedom. This aspect of the degree of freedom is discussed later in Sect. 4.2 and shown in see Fig. 4.3.

If the degree of freedom can never be directly observed by any experiment, then how does one 'know' about it? This is best answered by an example. The position of a particle is the basis of classical physics: at every instant t, the particle has a position

---

[1] One of us (BEB) had an interesting discussion with Nobel Laureate Anthony Leggett. It was mentioned to him, over dinner, that the quantum mechanical degree of freedom can never, in principle, be observed: one could only observe the specific values of the degrees of freedom. His view on this was sought. He stopped having his dinner, put down his knife and fork, and responded after some thought that: 'Indeed, this is rather strange, but true'. Later discussions with Leggett confirmed this aspect of quantum mechanics.

$x(t)$. From the quantum mechanical point of view, all possible 'positions' of the particle are interpreted as specific and determinate values of the degree of freedom. All the determinate values of the position of the particle, taken together, constitute the continuous position degree of freedom and yields the degree of freedom space $\mathcal{F} = \{x : -\infty \leq x \leq +\infty\}$. Another example is the case of the binary degree of freedom, a concept central to quantum algorithms. The 1-bit of a classical computer is given by 0 and 1; as mentioned earlier, the quantum binary degree of freedom combines these two values into a single indeterminate entity that simultaneously consists of 0,1 and constitutes the space $\mathcal{F} = \{0, 1\}$.

What observes on measuring the position degree of freedom are the specific and determinate values of the degree of freedom. Experiments can observe **only one** of the possible determinate values of the degree of freedom. The outcome of measuring the degree of freedom is uncertain, discussed in Sect. 4.2, and what one can finally obtain, after many observations, is the expected (average) value of the position degree of freedom.

In general, independent classical variables can be generalized to quantum degrees of freedom by collecting all possible values of the classical variable into a single space, which is the space $\mathcal{F}$ of the degree of freedom that corresponds to the classical variable. The degree of freedom is a theoretical construct that 'exists', for human beings, as a mathematical form in the human mind; its ontological significance is discussed in [1].

4.2 Hilbert Space; State Vectors

The degrees of freedom space $\mathcal{F}$ is the 'ground', the **foundation** on which the Hilbert space is built. The degrees of freedom space $\mathcal{F}$ exists outside Hilbert space. The term state space is used interchangeably with Hilbert space throughout this book, and the terms **state vector and wave function** are also used interchangeably.

Hilbert space $\mathcal{V}$ is a linear vector space of all possible mappings of space (degree of freedom) $\mathcal{F}$ to the complex numbers $\mathbb{C}$ that have a norm of unity.

$$\mathcal{V} : \mathcal{F} \to \mathbb{C}$$

The norm of a state vector is defined in Eq. 2.9.

Hilbert space consists of all possible functions of the degrees of freedom space $\mathcal{F}$ that have unit norm. Elements of Hilbert space are the *quantum state vectors*. The state vector, denoted by ψ, is a function of time t and the degrees of freedom and is given by $\psi(t, \mathcal{F}) \in \mathcal{V}$. Hence, as given in Eq. 2.14

$$\psi(t, \mathcal{F}) : \mathcal{F} \to \mathbb{C}; \quad |\psi|^2 = 1$$

The norm of a state vector $|\psi|^2$ is defined more precisely later on.

Fig. 4.3 Degree of freedom $\mathcal{F}$ enclosed by the state vector $\Psi(\mathcal{F})$. Published with permission of © Belal E. Baaquie 2012. All Rights Reserved

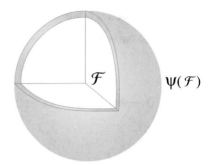

All information about the degrees of freedom is encoded in the state vector $\psi(t, \mathcal{F})$: the information has to be extracted by quantum measurements. The only information that one can extract from the degrees of freedom is what is the **likelihood** that the degrees of freedom can be observed taking its various particular values—while always remaining indeterminate. The degree of freedom, on being observed, leaves a 'click' in the detector: the collection of 'clicks' yields the probability of the indeterminate degrees of freedom taking different values.

The process of quantum measurement, discussed in some detail in Sect. 4.5, is required for obtaining the results of experiments. The probability of the detector having a reading corresponding to a determinate value x_0 of the degree of freedom is given by $|\psi(x_0)|^2$. Note that before the measurement is made, unlike a classical random variable, the quantum degree of freedom is indeterminate and, unlike a classical random variable, does not have a definite value; this is the result of the famous Bell theorem [1].

The degree of freedom is completely 'veiled' by the state vector $\psi(x)$: the degree of freedom is, metaphorically speaking, enclosed and encompassed by its Hilbert space, as shown in Fig. 4.3: the Hilbert space constitutes an impenetrable barrier and any experiment that attempts to directly observe the degree of freedom encounters the state vector enclosing the degree of freedom. This is the reason that the degree of freedom, as such, cannot be directly observed. The state vector $\psi(x)$ mediates the process of measurement–interposed between the experiment and the degree of freedom.

The Hilbert space, in general, is an infinite dimensional function space, with the additional requirement that the norm of all its state vectors is 1. For discrete degrees of freedom, as is the case for the binary degree of freedom, a Hilbert space is a finite dimensional linear vector space consisting of all possible unit norm state vectors.[2]

Let the degrees of freedom be given by x_j, $j = 1, 2, \ldots, n$ and let the basis state vectors be given by

$$\langle x|\psi_i\rangle = \psi_i(x) = \psi_i(x_1, x_2, \ldots, x_n), \quad i = 1, 2, \ldots, m$$

---

[2] To be mathematically rigorous, a Hilbert space is a vector space endowed with an inner product. It is also a complete metric space with respect to the distance function induced by the inner product.

The Hilbert space $\mathcal{V}$ is a linear vector space that has the following properties.

- The basis states are orthonormal

$$\langle \psi_i | \psi_j \rangle = \delta_{i-j}$$

- For α_i complex numbers, an arbitrary state vector has the following expansion

$$|\psi\rangle = \sum_{i=1}^{m} \alpha_i |\psi_i\rangle : \text{Superposition}$$

- All the state vectors belonging to Hilbert space $\mathcal{V}$ have norm of unity

$$|\psi|^2 = \langle \psi | \psi \rangle = 1 \Rightarrow \langle \psi | \psi \rangle = 1 = \sum_{i=1}^{m} |\alpha_i|^2$$

- One can form the tensor product of the Hilbert space to yield the product Hilbert space

$$\mathcal{V}_P = \mathcal{V}^{\otimes m} = \underbrace{\mathcal{V} \otimes \mathcal{V} \otimes \cdots \mathcal{V} \otimes \mathcal{V}}_{m\text{-times}}$$

- The tensor product Hilbert space is a mapping of $\mathcal{F}^{\otimes m}$ to the complex numbers $\mathbb{C}$; hence

$$\mathcal{V}^{\otimes m} : \mathcal{F}^{\otimes m} \to \mathbb{C}$$

- The tensor product of state vectors exists in the product space $\mathcal{V}_P$

$$\mathcal{F}^{\otimes k} = x^{\otimes k}; \ |\Psi\rangle_P = \sum_{ij...k=1}^{m} \alpha_{ij...k} |\psi_i\rangle \otimes |\psi_j\rangle \cdots \otimes |\psi_k\rangle : \text{Entanglement}$$

- The coordinate representation of a general product state vector is given by

$$\langle x^{\otimes m} | \Psi \rangle_P = \Psi_P(x) = \sum_{ij...k=1}^{m} \alpha_{ij...k} \psi_i(x_i) \psi_j(x_j) \ldots \psi_k(x_k)$$

Superposition and entanglement of state vectors are properties of Hilbert space quite independent of how the states evolve in time, which is determined by the Schrödinger equation. A key feature of the Schrödinger equation, discussed later, is that it is a linear differential equation, and hence, its solutions can be superposed to yield new solutions. Hence, the fact that Hilbert space allows for the superposition of its elements turns out to play a key role in quantum mechanics. This feature also plays a key role in quantum algorithms.

Entangled states are not necessarily solutions of the Schrödinger equation. However, one can prepare initial states that are entangled and these states can be evolved in time so that they continue to be entangled. These entangled states have a very special role to play both in quantum algorithms and, quantum cryptography and quantum communication [3, 4]. Entanglement, similar to superposition of state vectors, is a property of Hilbert space and results in phenomena that are forbidden both in classical physics and classical algorithms.

4.2.1 Continuous Degrees of Freedom

For ease of notation, consider $z_1, z_2, \ldots, z_n$ to be real variables taking values in the n-dimensional Euclidean space given by $\mathcal{R}^n \equiv \mathcal{R} \otimes \cdots \otimes \mathcal{R}$; the z_i's are continuous *degrees of freedom*. The quantum state vector $|\psi\rangle$ of $z_1, z_2, \ldots, z_n$ degrees of freedom is given by

$$|\psi\rangle = \int_{\mathcal{R}^n} dz \psi(z_1, z_2, \ldots, z_n)|z_1, z_2, \ldots, z_n\rangle : dz = \prod_{i=1}^{n} dz_i \qquad (4.2)$$

where the probability amplitude is given by the complex-valued wave function $\psi(z_1, z_2, \ldots, z_n)$. All information about a quantum system is contained in $\psi(z_1, z_2, \ldots, z_n)$; measurement needs to be carried out repeatedly by devices that respond to particular values of the degrees of freedom (see Sect. 4.5 on measurements). On measuring the degrees of freedom, the device in the laboratory records an observation corresponding to a particular and determinate value of the degree of freedom, denoted by $z_1^{(0)}, z_2^{(0)}, \ldots, z_n^{(0)}$. The measurement process is the following

$$|\psi\rangle \ \rightarrow \ |z_1^{(0)}, z_2^{(0)}, \ldots, z_n^{(0)}\rangle$$

The probability of a measurement resulting in the detector recording the said outcome—as postulated by Max Born [2]—is given by

$$|\psi(z_1^{(0)}, z_2^{(0)}, \ldots, z_n^{(0)})|^2$$

Note that both the following state vector

$$\psi(z_1^{(0)}, z_2^{(0)}, \ldots, z_n^{(0)}) \text{ and } e^{i\phi}\psi(z_1^{(0)}, z_2^{(0)}, \ldots, z_n^{(0)})$$

lead to the same probabilities and hence are equivalent. For this reason, in many of the derivations, the overall phase factor is dropped or ignored.

For the probabilistic interpretation, the total probability of observing some determinate value of the degree of freedom must be unity; hence, the functions in Hilbert space $\psi(z_1, z_2, \ldots, z_n)$ are normalized functions such that

$$\int_{\mathcal{R}} dz |\psi(z_1, z_2, \ldots, z_n)|^2 = 1$$

4.2.2 Discrete Degrees of Freedom: Qubits

Recall the *discrete binary degree of freedom* is the set of two possible values of a classical bit, namely 0, 1, and hence is represented by the binary set of two points $x = \{0, 1\}$. The single qubit is a state vector of this binary degree of freedom, denoted by $|q\rangle = |q(x)\rangle$. The Hilbert space of the qubit is $\mathcal{H}_q$ is a function of the binary degree of freedom $\{0, 1\}$, with qubits given by the vectors in $\mathcal{H}_q$.

We will see that the representation of the most general single qubit (given explicitly in Eq. 6.3) is the superposition of the binary degree of freedom's basis states $|0\rangle$, $|1\rangle$, which is called the computational basis states, and can be taken to be two-dimensional column vectors. The qubit is given by

$$|\psi\rangle = |q(\theta, \phi)\rangle = \cos\left(\frac{\theta}{2}\right)|0\rangle + e^{i\phi} \sin\left(\frac{\theta}{2}\right)|1\rangle; \quad \langle q(\theta, \phi)|q(\theta, \phi)\rangle = 1 \quad (4.3)$$

Recall that functions in Hilbert space that differ by a phase (i.e., along a 'ray') are equivalent in quantum mechanics since they do not contribute to the probability. Up to this ray equivalence, the Hilbert space of a single qubit consists of all possible values of θ, ϕ—and which forms the surface of a two-dimensional sphere and hence $\mathcal{H}_q \equiv S^2$, as shown in Fig. 6.1.

The n-qubit of a quantum computer are the state vectors of n-binary discrete degrees of freedom $\mathcal{F}^{\otimes n}$. The degree of freedom space is the n-fold tensor product $\mathcal{F}^{\otimes n} = \{0, 1\} \otimes \cdots \otimes \{0, 1\} = \{0, 1\}^{\otimes n}$ and is a discrete set consisting of 2^n points that are the corners of a 2^n-dimensional hypercube. It is the analog of the degree of freedom space $\mathcal{R}^n$ for n-real variables. The state vector of n input and output qubits are elements of *Hilbert space*. For n-binary discrete degrees of freedom $\mathcal{F}^{\otimes n}$, the Hilbert space is a finite dimensional linear vector space and consists of all functions defined on the set $\mathcal{F}^{\otimes n}$ with unit norm. The quantum state vector of n binary degrees of freedom $\mathcal{F} = \{0, 1\}^{\otimes n}$ is given by

$$|\psi(t, \mathcal{F})\rangle = |q\rangle = |q(x_1, x_2, \ldots, x_n)\rangle$$

For the n-binary discrete degrees of freedom—which is the foundation of quantum computing—a widely used and convenient basis is the computational basis states, given by $|x\rangle$, $x = 0, 1, 2, \ldots, 2^n - 1$. The state vector $|\psi\rangle$ of the n-binary degrees of freedom can be represented by the following expansion,

$$|\psi\rangle = \sum_{x=0}^{2^n-1} \alpha(x)|x\rangle; \quad \sum_{x=0}^{2^n-1} |\alpha(x)|^2 = 1 \tag{4.4}$$

The complex numbers $\alpha(x)$ are the coordinates of the n-qubit state vector in the 2^n-dimensional Hilbert space. $\alpha(x)$ is the *amplitude* of the n-qubit state to be in quantum state, $|x\rangle$ and $|\alpha(x)|^2$ is the probability of the state $|x\rangle$ being observed by a measuring device when $|\psi\rangle$ is subjected to repeated measurements.

Another representation of state vector for n-binary discrete degrees of freedom is given by

$$|\psi\rangle = \left[\prod_{i=0}^{2^n-1} \sum_{x_i=0}^{1} \right] \psi(x_0, x_1, \ldots, x_{n-1})|x_0, x_1, \ldots, x_{n-1}\rangle \tag{4.5}$$

$$\left[\prod_{i=0}^{2^n-1} \sum_{x_i=0}^{1} \right] |\psi(x_0, x_1, \ldots, x_{n-1})|^2 = 1 \tag{4.6}$$

Both these representations are used throughout this book.

The 2^n dimensional Hilbert space $\mathcal{H}_n$ is given by

$$\mathcal{H}_n \equiv \underbrace{S^2 \otimes \cdots \otimes S^2}_{n\text{-fold tensor product}} \tag{4.7}$$

As will become clear in the discussion on superposition and entanglement, the intermediate steps in the quantum algorithm all take place in Hilbert space and cannot, in principle, be observed: all that one can observe is the initial state and the final output state. Hence, to understand the mathematics of quantum algorithms, it is necessary to study the properties of Hilbert space.

4.3 Hermitian and Unitary Operators

Every degree of freedom $\mathcal{F}$ defines a state space V and operators $\mathcal{O}$ that act on that state space. All operators $\mathcal{O}$ are mathematically defined to be *linear mappings* of the state space V into itself, shown in Fig. 4.4, and yield

$$\mathcal{O}: |\psi\rangle \rightarrow \mathcal{O}|\psi\rangle \Rightarrow \mathcal{O}: V \rightarrow V$$
$$\mathcal{O}\Big(a|\psi_1\rangle + b|\psi_2\rangle\Big) = a\mathcal{O}|\psi_1\rangle + b\mathcal{O}|\psi_2\rangle : \text{linear operator}$$

where a, b are constants.

Fig. 4.4 An operator $\mathcal{O}$
acting on element $|\psi\rangle$ of the
state space $\mathcal{V}$ and mapping it
to $\mathcal{O}|\psi\rangle$. Published with
permission of © Belal E.
Baaquie 2012. All Rights
Reserved

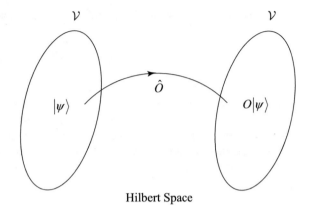

An operator is an element of the space formed by the outer product of $\mathcal{V}$ with its
dual $\mathcal{V}_D$, that is

$$\widehat{O} \in V \otimes V_D$$

All physical quantities of a degree of freedom such as position, momentum,
energy, angular momentum, spin, charge and so on are represented by Hermitian
operators. Hermitian operators $\mathcal{O}$ map the space $\mathcal{V}$ onto itself.

For a single binary degree of freedom, the state space is given by the Bloch sphere,
discussed in Sect. 6.3, and operators are 2×2 complex-valued Hermitian matrices.
A Hermitian matrix is only defined for complex *square* matrices $N \times N$ with M_{ij}
with $i, j = 1, 2, \ldots, N$ and M_{ij} satisfies

$$M^\dagger = (M^T)^* = M \;\Rightarrow\; M_{ji}^* = M_{ij}; \quad i, j = 1, 2, \ldots, N : \text{Hermitian} \quad (4.8)$$

Note the crucial point that unless i, j have the same range, the equality in Eq. 4.8
cannot hold for all i, j. Hermitian operators on linear vector state space are infinite dimensional generalizations of $N \times N$ matrices, with $N \to \infty$ and have new
properties that are absent in finite matrices.

The range over which the finite index i takes values for a finite dimensional matrix
has a generalization for state vectors in an infinite dimensional Hilbert space $\mathcal{V}$. The
domain of the operator $\mathcal{O}$ denoted by $D(\mathcal{O}) \subset \mathcal{V}$ is defined by all elements $|\psi\rangle$ in
$\mathcal{V}$ such that $\mathcal{O}|\psi\rangle \in \mathcal{V}$. Similarly, the vector $|\chi\rangle$ is in $D(\mathcal{O}^\dagger)$, the domain of $\mathcal{O}^\dagger$, if
$\mathcal{O}^\dagger|\chi\rangle \in \mathcal{V}$.

The analog of Hermitian conjugation, being defined only for a square matrix, is
that for operators on Hilbert space, the adjoint (Hermitian conjugate) operator can
be defined for *only* those operators $\mathcal{F}$ for which the domain of the operator and its
adjoint are isomorphic, or in other words, $D(\mathcal{O}) = D(\mathcal{O}^\dagger)$.

Hermitian conjugation of operators on state space is defined by

$$\langle\psi|\mathcal{O}^\dagger|\chi\rangle \equiv \langle\chi|\mathcal{O}|\psi\rangle^* : \text{Hermitian conjugation}$$

for any two vectors $|\psi\rangle, |\chi\rangle \in \mathcal{V}$. Once the domains of the operator and its conjugate are isomorphic, the form of the operator has to be invariant under conjugation, that is, $\mathcal{O} = \mathcal{O}^\dagger$, for the operator to be self-adjoint.[3] More precisely, an operator is Hermitian if its Hermitian conjugate operator is equal to the operator itself, that is, if

$$\mathcal{O}^\dagger = \mathcal{O} \Rightarrow \langle\psi|\mathcal{O}|\chi\rangle \equiv \langle\chi|\mathcal{O}|\psi\rangle^* : \text{Hermitian operator} \qquad (4.9)$$

Note that all the *diagonal elements* of a Hermitian operator $\mathcal{O}$ are real since for any arbitrary state vector $|\psi\rangle$, the diagonal element is as shown below

$$\langle\psi|\mathcal{O}|\psi\rangle = \langle\psi|\mathcal{O}^\dagger|\psi\rangle = \langle\psi|\mathcal{O}|\psi\rangle^* : \text{Real} \qquad (4.10)$$

Furthermore, similar to matrices, Hermitian conjugation is a linear operation and yields, for a sum and products of operators, the following

$$(c_1\mathcal{O}_1 + c_2\mathcal{O}_2 + \cdots)^\dagger = c_1^*\mathcal{O}_1^\dagger + c_2^*\mathcal{O}_2^\dagger \ldots; \quad (\mathcal{O}_1\mathcal{O}_2\ldots)^\dagger = \ldots \mathcal{O}_2^\dagger\mathcal{O}_1^\dagger$$

The trace operation for an operator $\mathcal{O}$ similar to matrices is defined as a sum of all its 'diagonal elements'. To make this statement more precise, one needs a resolution of the identity operator on state space $\mathcal{V}$. Consider for concreteness, the continuous degree of freedom with the completeness equation given by

$$\mathbf{I} = \int_{-\infty}^{\infty} dx |x\rangle\langle x|$$

Trace is a *linear* operation on $\mathcal{O}$ and is defined by

$$\text{Tr}(\mathcal{O}) = \text{Tr}(\mathcal{O}\mathbf{I}) = \int_{-\infty}^{\infty} dx\, \text{Tr}\Big(\mathcal{O}|x\rangle\langle x|\Big) = \int_{-\infty}^{\infty} dx \langle x|\mathcal{O}|x\rangle \qquad (4.11)$$

The properties of the trace operation are summarized below.

---

[3] One of the reasons for studying Hermitian (self-adjoint) operators is because one need not ascertain the space that an operator acts on, namely whether it acts on $\mathcal{V}$ or on its dual $\mathcal{V}_D$. For non-Hermitian operators, and these are the ones that occur in describing classical random systems such as those that occur in finance [5], the difference is important.

$$\text{Tr}[\sum_i c_i \mathcal{O}_i] = \sum_i c_i \text{Tr}[\mathcal{O}_i]$$

$$\text{Tr}\left[\sum_i c_i \mathcal{O}_i\right]^\dagger = \sum_i c_i^* \text{Tr}^*[\mathcal{O}_i]$$

$$\text{Tr}(\mathcal{O}_1 \mathcal{O}_2 \mathcal{O}_3) = \text{Tr}(\mathcal{O}_3 \mathcal{O}_1 \mathcal{O}_2) : \text{cyclic}$$

Cyclicity of the trace makes it invariant under unitary transformations U, namely

$$\text{Tr}[U \mathcal{O} U^\dagger] = \text{Tr}[\mathcal{O} U^\dagger U] = \text{Tr}[\mathcal{O}]$$

A unitary operator, the generalization of the exponential function $\exp i\phi$, is given in terms of a Hermitian operator $\mathcal{O}$ by the following

$$U = e^{i\phi\mathcal{O}} \Rightarrow UU^\dagger = \mathbb{I}$$
$$V = \frac{1 - ia\mathcal{O}}{1 + ia\mathcal{O}} \Rightarrow VV^\dagger = \mathbb{I}$$

Every Hermitian operator defines a unitary operator. The most important Hermitian operator in quantum mechanics is the Hamiltonian. In quantum algorithms, Hermitian operators rarely occur since all quantum gates are represented by unitary operators.

4.4   The Schrödinger Equation

The Schrödinger equation determines the time evolution of the state function $|\psi(t)\rangle$, where the label t denotes the time parameter. To write down the Schrödinger equation one first needs to specify the degrees of freedom of the system in question, which in turn specifies its state space $\mathcal{V}$; one also needs to specify the Hamiltonian $\mathcal{H}$ of the system that describes the range and form of the possible energies the quantum system can have.

The celebrated Schrödinger equation is given by

$$-\frac{\hbar}{i}\frac{\partial|\psi(t)\rangle}{\partial t} = \mathcal{H}|\psi(t)\rangle \tag{4.12}$$

Consider a quantum particle with mass m moving in one dimension in a potential $V(x)$; the 'position' of a classical particle x is elevated to a continuous degree of freedom $x \in \mathbb{R}$. The Schrödinger equation is given as follows,

$$-\frac{\hbar}{i}\left\langle x \Big| \frac{\partial}{\partial t}|\psi(t)\rangle \right\rangle = \langle x|\mathcal{H}|\psi(t)\rangle \Rightarrow -\frac{\hbar}{i}\frac{\partial\psi(t,x)}{\partial t}$$

$$= \mathcal{H}\left(x, \frac{\partial}{\partial x}\right)\psi(t,x) \tag{4.13}$$

The Hamiltonian operator acts on the dual basis states. In the position basis, the state vector is

$$\langle x | \psi(t) \rangle = \psi(t, x)$$

The Hamiltonian for the important case of a quantum particle moving in one dimension is given by

$$\mathcal{H} = -\frac{\hbar}{2m} \frac{\partial^2}{\partial x^2} + V(x) \tag{4.14}$$

A variety of techniques have been developed for solving the Schrödinger equation for a wide class of potentials as well as for multiparticle quantum systems [6].

Let $|\psi\rangle$ be the initial value of the state vector at $t = 0$ with $\langle \psi | \psi \rangle = 1$. Equation 4.12 can be integrated to yield the following formal solution

$$|\psi(t)\rangle = e^{-\frac{i}{\hbar} t \mathcal{H}} |\psi\rangle = U(t) |\psi\rangle \tag{4.15}$$

The Hamiltonian $\mathcal{H}$ is an operator that translates the initial state vector in time, as in Eq. 4.15. The evolution operator $U(t)$ is defined by

$$U(t) = e^{-\frac{i}{\hbar} t \mathcal{H}}; \quad U^\dagger(t) = e^{\frac{i}{\hbar} t \mathcal{H}}$$

$U(t)$ is unitary since H is Hermitian; more precisely

$$U(t) U^\dagger(t) = \mathbb{I}$$

The unitarity of $U(t)$, and by implication the Hermiticity of $\mathcal{H}$, is crucial for the conservation of probability. The total probability of the quantum system is conserved over time since unitarity of $U(t)$ ensures that the normalization of the state function is time-independent; more precisely,

$$\langle \psi(t) | \psi(t) \rangle = \langle \psi | U^\dagger(t) U(t) | \psi \rangle = \langle \psi | \psi \rangle = 1$$

The operator $U(t)$ is the exponential of a Hermitian operator that in many cases, as given in Eq. 4.14, is a differential operator. The Feynman path integral is a mathematical tool for analyzing $U(t)$ and is discussed in [7].

As mentioned earlier, in quantum algorithms the Hamiltonian $\mathcal{H}$ or the other Hermitian operators seldom appear. The reason is that for quantum algorithms, all the quantum gates required for transforming the state vector from one state to the next involve an evolution over a *finite* time Δt and hence are all the quantum gates $\mathcal{G}$ are unitary operators and are generically of the form

$$\mathcal{G} = U(\Delta t) = \exp\left\{ -\frac{i}{\hbar} \Delta t \mathcal{H} \right\}$$

The role of the Hamiltonian $\mathcal{H}$ is central for constructing a physical device for executing a quantum algorithm and has been discussed in some detail in [8, 9].

As discussed in Chap. 3, a circuit for a classical computer is a physical circuit built from conducting materials through which electrical currents flow. All the computational steps of a classical algorithm take place in the electrical circuits of a classical computer. Any intermediate step in the classical algorithm can be observed directly.

A quantum circuit is also a physical object with physical process that evolves the quantum state from its initial state vector to the final output state vector. However, the computational process of a quantum circuit 'exists' in Hilbert space and *all the steps in a quantum computation are transformations of the state vector of n-qubits—given by* $|\psi(t)\rangle$*—that take place in Hilbert space.* The input qubits state vector $|\psi_I\rangle$, given at initial time t_I, is evolved by $|\psi(t)\rangle = U(t)|\psi_I\rangle$, to the state vector of the output qubits $|\psi_F\rangle$ at final time t_F and given by

$$|\psi_F\rangle = \exp\left\{-\frac{i}{\hbar}(t_F - t_I)\mathcal{H}\right\}|\psi_I\rangle \qquad (4.16)$$

This is further discussed in Sect. 4.9 and shown in Figs. 4.6 and 4.7.

4.4.1 Key Features of the Schrödinger Equation

- The state vector $|\psi(t)\rangle$ is a complex-valued vector and an element of Hilbert space. This is the reason that one can entangle state vectors depending on different degrees of freedom. Note, entanglement is independent of the Hamiltonian as well as of the Schrödinger equation: it is solely a result of state vectors existing in Hilbert space. Entanglement of state vectors is an important resource in the efficiency of quantum algorithms.

- The Schrödinger equation is a first-order differential equation in time, in contrast to Newton's equation of motion that is a second-order differential equation in time. At $t = 0$, the Schrödinger equation requires that the initial state function for *all values* of the degree of freedom be specified, namely $|\psi(\mathfrak{R})\rangle$, whereas in Newton's law, only the position and velocity *at the starting point* of the particle is required.

- At each instant, Schrödinger's equation specifies the state function for all values of the *indeterminate* degree of freedom. This is the reason that quantum algorithms are executed in Hilbert space.

- The linearity of the Schrödinger equation is the reason that all the state vectors $|\psi(t)\rangle$ are elements of a linear vector space $\mathcal{V}$. Since Eq. 4.12 is a linear equation; two solutions $|\psi(t)\rangle$ and $|\chi(t)\rangle$ of the Schrödinger equation can be added to yield yet another solution given by $a|\psi(t)\rangle + b|\chi(t)\rangle$.

- The linearity of the Schrödinger equation is the reason that state vectors obey the rule of superposition – a crucial resource in quantum algorithms.

- The Schrödinger equation is the first equation in natural science in which complex numbers are essential and not just a convenient mathematical tool for representing real quantities.

4.5 Quantum Measurement: Born Rule

The quantum theory of measurement that is discussed in this section, and the one used in this book is based on the Copenhagen interpretation of quantum mechanics. In general, in quantum mechanics physical quantities like energy, angular momentum and so on are represented by Hermitian operators. Consider a state vector $|\psi\rangle$ belonging to state space $\mathcal{V}$ that describes the degrees of freedom $\mathcal{F}$. The measurement of a physical quantity is mathematically represented by the measuring operator $\mathcal{O}$; we use the term measuring device and the measuring operator representing the measuring device interchangeably. Repeated measurements yield the expectation (average) value of the Hermitian operator $\mathcal{O}$, given by

$$E_\mathcal{V}\big[\mathcal{O}(\mathcal{F})\big] = \langle\psi|\mathcal{O}|\psi\rangle$$

To mathematically represent a measurement, one needs to form the density matrix ρ. The density matrix is discussed in Sect. 5.4. Let the state vector be denoted by $|\psi\rangle$; the density matrix, from Eq. 5.6, is defined by

$$\rho = |\psi\rangle\langle\psi| \qquad (4.17)$$

Measurements on the state vector $|\psi\rangle$ using a device represented by the Hermitian operator $\mathcal{O}$ yield the result

$$E_\psi\big[\mathcal{O}\big] = \text{Tr}(\mathcal{O}\rho)$$

The Born rule states that the process of measurement leads to the quantum state vector $|\psi\rangle$ **undergoing a discontinuous change**: this discontinuous change leads to the **collapse of the state vector** to an eigenstate of the measuring operator. The measurement is realized by applying **measurement gates**, which are projection operators, on the state vector $|\psi\rangle$. The measurement gates are the only *irreversible gates* in a quantum circuit. Representing the process of measurement by irreversible gates is based on the Copenhagen interpretation of quantum mechanics and is explained below [1, 2].

Unitary transformation can only represent continuous changes in the state vector. Because the result of a measurement is discontinuous, the transformation of the initial pre-measurement state to the final post-measurement state cannot be represented by a unitary transformation. This is the reason that the measurement gates are non-unitary and irreversible. The process of measurement cannot be represented by the Schrödinger equation since it requires processes existing outside the smooth evolution of the state vectors. The mechanism of the collapse of the state vector is not known even after more than a century of quantum mechanics and remains one of the outstanding unsolved mysteries of quantum mechanics.

On a more granular level, the process of measurement **entangles** the eigenstates of measuring device with the quantum state $|\psi\rangle$ (entanglement is discussed in Sect. 5.7). The measurement results in the measuring device being put into a determinate eigenstate and which implies with certainty, due to entanglement, that state $|\psi\rangle$ is put into the corresponding determinate eigenstate. The process of measurement, including entanglement, is discussed rigorously for the spin 1/2 case using the Stern–Gerlach experiment [1].

The measurement process can use a variety of basis states. Consider a measurement operator $\mathcal{O}$ with the following expansion,

$$\mathcal{O} = \sum_{n=1}^{N} \lambda_n |\psi_n\rangle\langle\psi_n|; \quad \sum_{n=1}^{N} \psi_n\rangle\langle\psi_n| = \mathbb{I}$$

The eigenvalues λ_n of $\mathcal{O}$ are not needed for our discussion as they refer to the properties of measuring device [1]. They are given by the operators that project the state vector to one of the eigenstates $|D_n\rangle$—which correspond to the eigenstates $|\psi_n\rangle$ of the measuring operator $\mathcal{O}$.

$$|D_n\rangle \iff |\psi_n\rangle$$

Hence, the measurement gates are given by

$$\mathcal{M}_n = |D_n\rangle\langle D_n|; \quad \sum_{n=1}^{N} |D_n\rangle\langle D_n| = \mathbb{I}$$

where $|D_n\rangle$ are eigenstates of the measuring device. The key requirement of the gates $|D_n\rangle$ is that they provide a complete set of basis states. The expansion of state vector $|\psi\rangle$ in terms of the eigenfunctions $|\psi_n\rangle$ of the measurement operator $\mathcal{O}$ yields, from Eq. 4.17, the following density matrix

$$|\psi\rangle = \sum_{n=1}^{N} c_n |\psi_n\rangle \Rightarrow \rho = \sum_{m,n=1}^{N} c_n c_m^* |\psi_n\rangle\langle\psi_m|$$

The preparation of the state for the measurement results in entangling the state $|\psi\rangle$ with the device yields

$$|\psi\rangle \rightarrow \text{preparation} \rightarrow \sum_{n=1}^{N} c_n |\psi_n\rangle |D_n\rangle$$

and results in the following transformation of the density matrix

$$\rho \rightarrow \text{preparation} \rightarrow \rho_{\text{out}} = \sum_{m,n=1}^{N} c_n c_m^* |\psi_n\rangle\langle\psi_m| \otimes |D_n\rangle\langle D_m|$$

The collapse of the wave function due to a measurement results is an unpredictable outcome and the Born rule states that the state function collapses to a specific eigenstate $|D_n\rangle$ of the measuring device. The final result of the measurement is the following [1]

$$\rho_{\text{out}} \rightarrow \text{Measurement} \rightarrow \rho_{\text{final}} = \sum_{n=1}^{N} |c_n^2| D_n\rangle\langle D_n|$$

Hence, the process of measurement yields the following

$$\rho \rightarrow \text{preparation} \rightarrow \rho_{\text{out}} \rightarrow \text{Measurement} \rightarrow \rho_{\text{final}} = \sum_{n=1}^{N} |c_n^2| D_n\rangle\langle D_n|$$

The irreversibility of the measurement is due to decoherence of ρ_{final}, since all the off-diagonal $m \neq n$ terms of ρ_{out} have been eliminated by the process of measurement. Furthermore, due to decoherence, the density matrix ρ_{final} has no quantum indeterminacy—and represents a classical random system with random outcomes for the measuring gates.

The density matrix ρ_{final} means that observing $|\psi\rangle$ results in the state vector collapsing to a gate $|D_n\rangle$, with the probability of the collapse given by

$$\text{Tr}\Big(|D_n\rangle\rho_{\text{final}}\Big) = \langle D_n|\psi\rangle = |c_n|^2; \quad n = 0, 1, \dots, N \tag{4.18}$$

Summarizing our discussion, applying all the measurement gates on the state vector $|\psi\rangle$ causes the state vector to undergo the following collapse [1, 2].

$$|\psi\rangle \rightarrow \text{Measurement} \rightarrow |D_n\rangle$$
$$\text{Probability of outcome} = |c_n|^2 \tag{4.19}$$

We focus henceforth on the role of quantum measurements for a quantum computer. All quantum algorithms provide the sought for solution of a problem in the form of an output state, and with the quantum computation being terminated by irreversible *measurement gates* $\mathcal{M}$.

The n-binary degrees of freedom **computational basis states** are represented by

$$|z_0, z_1, \dots, z_{n-1}\rangle$$

where the bits $z_i = 0, 1 : i = 1, \dots, n$ have 2^n possible determinate values. The n-qubits string state vector $|\psi\rangle$ is given by Eq. 4.4

$$|\psi\rangle = \sum_{x=0}^{2^n-1} \alpha(x)|x\rangle; \quad \sum_{x=0}^{2^n-1} |\alpha(x)|^2 = 1 \qquad (4.20)$$

The density matrix is the following

$$\rho_\psi = \sum_{x,x'=0}^{2^n-1} \alpha(x)\alpha^*(x')|x\rangle\langle x'| = \sum_{x,x'=0}^{2^n-1} \rho(x,x')|x\rangle\langle x'|$$

where

$$\rho(x,x') = \alpha(x)\alpha^*(x'); \quad x,x' = 0,1,2,\ldots,2^n-1$$

Consider *measurements* performed on the n binary degrees of freedom quantum state $|\psi\rangle$. For the measurements to be performed using the computational basis states $|z_0, z_1, \ldots, z_{n-1}\rangle$, we need a physical device that can measure all values of $z_i = 0, 1; i = 1, 2, \ldots, n-1$ that are in the range of the degrees of freedom. The physical device consists of a collection of measurement gates. The mathematical representation of the measurement gates is Hermitian projection operators:

$$\mathcal{M}; \quad \mathcal{M} = \mathcal{M}^2$$

The Hermitian operator representing the measurement process in Hilbert space is given by

$$\mathcal{O} = \sum_z \lambda(z)\mathcal{M}_z; \quad \sum_z \mathcal{M}_z = \mathbb{I}$$

The eigenvalues $\lambda(z)$ are not required for the measurements. In the computational basis states, there are 2^n projection operators $\mathcal{M}_z$ that are given by

$$\mathcal{M}_z = |z_0, z_1, \ldots, z_{n-1}\rangle\langle z_0, z_1, \ldots, z_{n-1}| : z_i = 0, 1; \quad i = 1, 2, \ldots, n-1$$

The post-measurement state is completely arbitrary and, in principle, cannot be predicted. What quantum mechanics tells us is that if the measurement process is **identically repeated** many times (in principle, infinitely many times), then the **probability** of the outcome is given by the state vector and the measurement operator.

In summary, the result of a measurement is that, with some likelihood, the state vector $|\psi\rangle$ collapses to a particular eigenstate of the operator $\mathcal{O}$ representing the measuring device and is given by

$$|z_0^{(0)}, z_1^{(0)}, \ldots, z_{n-1}^{(0)}\rangle \qquad (4.21)$$

The degree of freedom's state vector in Hilbert space is reduced to the state vector given in Eq. 4.21 and the measuring also has the same state vector due to entanglement. In other words, in the computational basis the result of the measurement is that the device's counters have **determinate** value, given by $z_i^{(0)}; i = 0, 1; i =$ $1, 2, \ldots, n - 1$, in which particular eigenstate (device's counter-readings) the state vector collapses completely random and unpredictable. One can only know the likelihood of (a) the collapse in Hilbert space to one the eigenstates of the operator $\mathcal{O}$ and (b) which corresponds, for the device, the collapse to determinate counter-readings.

After repeated measurements on identically prepared quantum states, the number of times a particular computational basis state has been observed in the measuring device is collected for all the gates: the normalized frequency table yields the probability of a measurement gate registering a reading. In principle, the experiment needs to be carried out infinitely many times, but in most cases a large sample is sufficient. As mentioned earlier, the measurement process collapses the quantum state vector to one of the possible computational basis states. Recall from Eq. 4.5

$$|\psi\rangle = \left[\prod_{i=0}^{2^n-1} \sum_{x_i=0}^{1} \right] \psi(x_0, x_1, \ldots, x_{n-1}) |x_0, x_1, \ldots, x_{n-1}\rangle \qquad (4.22)$$

The probability of the quantum state $|\psi\rangle$ collapsing to a specific computational basis state is given by

$$\text{Tr}(\mathcal{M}_z \rho_{\text{final}}) = |\langle z_0^{(0)}, z_1^{(0)}, \ldots, z_{n-1}^{(0)} | \psi \rangle|^2 = |\psi(z_0^{(0)}, z_1^{(0)}, \ldots, z_{n-1}^{(0)})|^2$$

The quantum theory of measurement is summed up as follows: the quantum state $|\psi\rangle$, when subjected to a measurement, leads to the following non-unitary and discontinuous collapse

$$|\psi\rangle \;\rightarrow\; \text{Measurement} \;\rightarrow\; |z_0^{(0)}, z_1^{(0)}, \ldots, z_{n-1}^{(0)}\rangle \qquad (4.23)$$
$$\text{Probability of outcome} = |\psi(z_0^{(0)}, z_1^{(0)}, \ldots, z_{n-1}^{(0)})|^2$$

If another measurement is performed immediately after the collapse of the state vector to $|\psi_n\rangle$, then the system will still be in the state $|z_0^{(0)}, z_1^{(0)}, \ldots, z_{n-1}^{(0)}\rangle$—and remains in that state as long as there is no interaction of the system with the outside environment.

The quantum state of n-qubits subjected to the measurement results in the measuring device being in a state such that all the n 1-bits in the measuring device have definite values. For example, one possible outcome for the measuring device is given by

$$|z_0^{(0)}, z_1^{(0)}, \ldots, z_{n-1}^{(0)}\rangle = \underbrace{|10001110110\ldots01101\rangle}_{n\text{-bits computational basis state}}$$

4.6   Quantum Measurements and Degrees of Freedom

We review, for greater clarity, Born's rule on the role of measurements and of the degrees of freedom in the quantum theory of measurement.

The measurement process results in an unpredictable final state for the degree of freedom's state vector as well correspondingly for measuring device—that is recorded by one of the measurement gates. The indeterminate nature of the quantum degrees of freedom results in a **different** final state vector every time a measurement is performed.

In the quantum mechanical formalism, as discussed earlier, the *degrees of freedom are always indeterminate and unobservable*, regardless of whether a quantum measurement is carried out on them or not. The quantum measurement performed on $|\psi\rangle$, as expressed in Eq. 4.23, results in the post-measurement state $|z_0^{(0)}, z_1^{(0)}, \ldots, z_{n-1}^{(0)}\rangle$ that is a specific quantum state: an eigenfunction of the projection operators. The cardinal point to note is that before and after the measurement, the degrees of freedom remain indeterminate.

All that changes as a result of the measurement is that the input state vector $|\psi\rangle$—which determines all the information that can be extracted about the degree of freedom—is altered to another specific state $|z_0^{(0)}, z_1^{(0)}, \ldots, z_{n-1}^{(0)}\rangle$ as a result of the measurement.

Although the degrees of freedom cannot be directly observed, what one can observe in the device are the specific and determinate possible values of the degrees of freedom; for example, the following n-binary string

$$z_0^{(0)}, z_1^{(0)}, \ldots, z_{n-1}^{(0)}$$

is one possible determinate value of the degrees of freedom. In other words, no experiment can observe, as such, the indeterminate form of the degrees of freedom: instead, experiments—repeatedly carried out—can enumerate, one by one, all the possible determinate values of the degrees of freedom. The design of the experiments required for carrying out the observations on, as well as the enumeration of, the possible determinate values are largely guided by the mathematical models made for the degrees of freedom. In effect, quantum measurements sample the possible determinate values of the indeterminate degrees of freedom.

It is important to note that the enumeration of the possible determinate values of the indeterminate degrees of freedom is not unique: rather, it depends on the choice of the eigenstates of the measuring gates. For the case of a quantum computer, another useful basis for the measuring gates are the Bell states, discussed in Sect. 6.4. Using the Bell states to measure the state vector will result in a different enumeration of the possible determinate values of the degrees of freedom. All the distinct possible enumerations of the possible determinate values of the indeterminate degrees of freedom are related by unitary transformations.

Due to the irreversibility of the measurement process, the post-measurement state $|z_0^{(0)}, z_1^{(0)}, \ldots, z_{n-1}^{(0)}\rangle$ has less information about the degrees of freedom than the pre-measurement state vector $|\psi\rangle$: in particular, $|\psi\rangle$ contains the superposition of many projection operators, as shown in Eq. 4.22. Repeated measurements (of identically prepared states) result in the counters giving readings of a specific post-measurement state, $|z_0^{(0)}, z_1^{(0)}, \ldots, z_{n-1}^{(0)}\rangle$, with the likelihood of occurrence given by $|\psi(z_0^{(0)}, z_1^{(0)}, \ldots, z_{n-1}^{(0)})|^2$.

Experimental devices exist in spacetime and Hermitian operators representing the devices act on the degree of freedom's Hilbert space, which is a mathematical construct that 'exists' outside spacetime.

Quantum measurements are realized by Hermitian operators acting on quantum states—and collapsing these states in Hilbert space; due to the entanglement of the devices' state vectors with the state being observed, the readings of the measuring device record the corresponding collapsed state of Hilbert space. The likelihood of the measurement observing a particular state vector of the degrees of freedom is determined by the Born rule that depends on the operators and state vectors of the device being used for making the measurement.

Once a measurement is completed, the particular observed state vector persists; if a second measurement is performed, it will find the same previously observed state vector. The observed state vector persists until it is affected by an interaction; in the real world, the state vector persists for a very short time, and this lack of persistence is called decoherence.

4.7 No-Cloning Theorem

The result of any measurement of the degrees of freedom results in an uncertain outcome. To find the probabilities of the various outcomes, one needs to repeat the experiment many times. If it was possible to easily make many copies of the quantum state, then one could simultaneously carry out experiments on all the copies and obtain the result in, effectively, a single measurement. Unfortunately, making exact copies of the state vector is forbidden in quantum mechanics and is called the no-cloning theorem [2].

This theorem essentially states that it is impossible to create an independent and identical copy of an arbitrary unknown quantum state. The no-cloning theorem is a direct result of the structure of Hilbert space, and all operations on the elements of Hilbert space need to be carried out using unitary operators.

The history of how the no-cloning theorem was discovered makes a good read [10]. Let us first presume that we indeed have a copying machine that faithfully copies the state $|0\rangle$ and $|1\rangle$; i.e., there exists a unitary transformation that realizes

$$
\begin{aligned}
U|0\rangle|\xi\rangle &\rightarrow |0\rangle|0\rangle, \\
U|1\rangle|\xi\rangle &\rightarrow |1\rangle|1\rangle,
\end{aligned}
\tag{4.24}
$$

for some ancillary state $|\xi\rangle$. In standard quantum mechanics, all unitary transformations are linear. Suppose we have an unknown state

$$|\psi\rangle = \alpha|0\rangle + \beta|1\rangle; \quad |\alpha|^2 + |\beta|^2 = 1$$

The question is whether we can use the unitary transformation U to change it to two copies of $|\psi\rangle$, i.e., $|\psi\rangle \otimes |\psi\rangle$. Cloning requires

$$U\Big(|\psi\rangle|\xi\rangle\Big) \rightarrow |\psi\rangle|\psi\rangle$$

$$= \alpha^2|0\rangle|0\rangle + \alpha\beta|0\rangle|1\rangle + \alpha\beta|1\rangle|0\rangle + \beta^2|1\rangle|1\rangle \qquad (4.25)$$

However

$$U\Big(|\psi\rangle|\xi\rangle\Big) = U\Big[\,(\alpha|0\rangle + \beta|1\rangle)\,\Big)|\xi\rangle\Big]$$

$$= \alpha U\Big(|0\rangle|\xi\rangle\Big) + \beta U\Big(|1\rangle|\xi\rangle\Big) \text{ by linearity}$$

$$= \alpha|0\rangle|0\rangle + \beta|1\rangle|1\rangle \qquad (4.26)$$

Comparing Eqs. 4.25 and 4.26, one sees that consistency requires that α or β be zero. This is sometimes known as the no-cloning theorem.

4.8 Copenhagen Interpretation: Open Questions

There a number of interpretations of quantum mechanics besides the Copenhagen one and have been summarized in [1]. All these interpretations arise from the differing views on the ontological status of the state vector, as well as the underlying assumptions that connect the results of measurements with the state vector. The central mystery of quantum mechanics, and which has led to many of its various interpretations, is the so-called collapse of the state vector. Some of the questions that arise in quantum mechanics are discussed, with the responses based on the Copenhagen interpretation.

- What constitutes a measurement? Since all physical entities should obey the laws of quantum mechanics, how does the measuring device yield determinate values for the degrees of freedom being observed? The response of Niels Bohr was that the measuring device needs to be treated like a classical determinate object—which is not a very satisfactory answer since there are no classical objects in a quantum mechanical Universe.
- A more precise answer has been given by John Bell, whose view is that any object with more than 10^7 atoms can function effectively as a measuring device. The reason being the state vector of an object with 10^7 atoms has destructive quantum interference and results in decoherence that leads, in effect, to the collapse of the

state vector: the object's collapsed state vector functions as a measuring device, with the degree of freedom, on being observed, having a determinate value [1].

- Unlike the statements made by some people, there is no need for a conscious observer in quantum mechanics: once the result of a measurement is recorded in a device, the process of measurement has been completed. It is of no consequence whether a human mind reads or does not read the result of a measurement.

- The state vector is in general a non-local entity spread over space. In the collapse of the state vector, its value becomes zero everywhere except for a particular value, where the degree of freedom is observed. The question is: does the collapse happen instantaneously everywhere, or does it take place over a finite time?

  Quantum mechanics requires an instantaneous collapse everywhere and an experiment to test this shows that the lower limit for the speed of collapse is 1550 times the speed of light.[4] Hence, the collapse being instantaneous seems to be consistent with experiment.

- Since the collapse of the state vector seems to be instantaneous, the question arises: does the infinite speed required for collapsing the state vector simultaneously everywhere contradict the special theory of relativity? The answer is: No.

- When the state vector is subjected to a measurement, what is the mechanism that causes the state vector to collapse—so that the degree of freedom is observed to have a determinate value? The answer to this question is not known. This is probably what Richard Feynman had in mind when he stated: *It is safe to say that nobody understands quantum mechanics.*[5]

- Note that the state vector $\psi(x)$: x-indeterminate can never be observed, and what is observed in an experiment is $|\psi(x_0)|^2$: x_0-determinate. So the question arises: do the state vector $\psi(x)$ and Hilbert space 'exist', or are they only mathematical entities that carry information about the degree of freedom x.

  One can take many different consistent positions on this question. For Bohr and Heisenberg the only thing that is 'real' are the results obtained by measurements. They had an agnostic view regarding Hilbert space and the state vector: since one cannot measure $\psi(x)$, there was no need to discuss the existence or otherwise of the state vector.

  There are other views such as the Many-World interpretation where the state vector is taken to be real, but the state vector does not collapse when subjected to measurements. Another view, adopted by Baaquie, is that the state vector does, in fact, undergo a collapse when it is observed—but it also has an ontological existence; for this view one needs to define precisely what one means by the concept: 'to exist' [1].

The philosophical question—of whether the degree of freedom and Hilbert space actually 'exist' or are simply mathematical constructs for computing experimental results—does not affect any of the derivations or conclusions regarding quantum computers and quantum algorithms. The procedure and processes for measurements and computations are the same for all the interpretations. There have been proposals

---

[4] https://www.nature.com/articles/s41598-019-48387-8.pdf.

[5] https://physicscourses.colorado.edu/phys3220/phys3220_fa08/quotes.html.

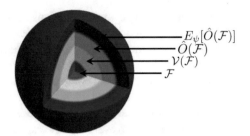

Fig. 4.5 Degree of freedom $\mathcal{F}$ enclosed by the Hilbert space $\mathcal{V}$. Repeated application of Operator $\widehat{O}(\mathcal{V})$ on state vector $\psi(\mathcal{V}) \in \mathcal{V}$ yields the observed expectation value of the operator given by $E_\psi[\widehat{O}(\mathcal{V})]$. Published with permission of © Belal E. Baaquie 2012. All Rights Reserved

to set up experiments for differentiating between the various interpretations,[6] but so far there have been no unequivocal results.[7] For those who are interested, the various interpretations of quantum mechanics are discussed in [1].

4.9 Summary of Quantum Mechanics

In this section, a summary is given of results that are based on the Copenhagen interpretation of quantum mechanics. Piecing together the various ingredients of quantum mechanics discussed so far yields two different representations of the schema of quantum mechanics, which reflect the following two different points of view [1]:

1. The relation of the degree of freedom $\mathcal{F}$ to the layered superstructure it carries, shown in Fig. 4.5.
2. The relation of the degree of freedom $\mathcal{F}$ with the quantum theory of measurement and observable, given in Fig. 4.6.

- **Superstructure of the Degrees of Freedom**: Fig. 4.5.

   1. The superstructure of the degree of freedom shown in Fig. 4.3 is completed in Fig. 4.5.
   2. The core of quantum mechanics is the degree of freedom that is intrinsically indeterminate and in principle unobservable—and constitutes a discrete set or a continuous space.
   3. The first shell encompassing the degree of freedom $\mathcal{F}$ is Hilbert space its $\mathcal{V}(\mathcal{F})$, first shown in Fig. 4.3. All possible information about the degree of freedom is carried by the state vectors $|\psi\rangle$ of Hilbert space $\mathcal{V}(\mathcal{F})$.
   4. The second shell consists of operators $\widehat{\mathcal{O}}(\mathcal{F})$ that are mappings of Hilbert space onto itself.

---

[6] https://www.sciencedirect.com/science/article/pii/S135521981530023X.

[7] https://www.scientificamerican.com/article/the-many-interpretations-of-quantum-mechanics/.

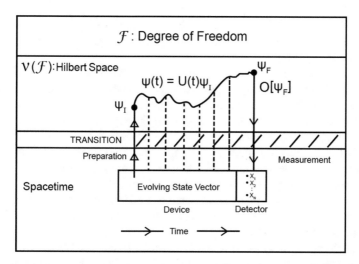

Fig. 4.6 Graphical representation of the schema for the theory of measurements in quantum mechanics. Published with permission of © Belal E. Baaquie 2012. All Rights Reserved

5. The last outer shell is devices that are designed to represent the operators $\widehat{\mathcal{O}}(\mathcal{F})$; applying the device on the state vector $|\psi\rangle$ of the degree of freedom causes the detector to collapse to a determinate reading in the device, reflecting the collapse, in Hilbert space, of the state vector $|\psi\rangle$ to the observed state vector.
6. Repeatedly applying the measuring device on identically and independently prepared state vector $|\psi\rangle$ results in the observed expectation $E_\psi[\widehat{\mathcal{O}}(\mathcal{F})] = \langle\psi|\widehat{\mathcal{O}}(\mathcal{F})|\psi\rangle$.

- **Copenhagen Theory of Quantum Measurement**: Fig. 4.6.

  1. From the point of quantum measurements, there are four distinct domains that result from the structure and superstructure of quantum mechanics as discussed in the context of Fig. 4.5.
  2. The foundation of quantum phenomenon is (the space of) the degree of freedom $\mathcal{F}$.
  3. The quantum states of Hilbert space $\mathcal{V}(\mathcal{F})$ provide all the information that can be obtained about the degree of freedom $\mathcal{F}$.
  4. The boundary between the degree of freedom $\mathcal{F}$ and Hilbert space $\mathcal{V}(\mathcal{F})$ is **impenetrable** and not accessible to any form of measurement, as shown in Fig. 4.6.
  5. The degree of freedom $\mathcal{F}$ and its Hilbert space $\mathcal{F}$ both exist *outside of spacetime*, as shown in Fig. 4.6.
  6. The Hilbert space $\mathcal{F}$ is connected to devices in spacetime by the process of measurement, as shown in Fig. 4.6.

7. The Hilbert space $\mathcal{F}$ and the device in spacetime have a permeable transitional domain that can be either sharp or fuzzy, depending on the measurement, as shown in Fig. 4.6 [1].

8. The experimental device prepares the initial state vector of the degree of freedom Ψ_I. Once the state vector is prepared and is not subject to any measurements, it evolves in Hilbert space, as shown in Fig. 4.6.

9. The state vector Ψ_I evolves according to the equation $\Psi(t) = U(t)\Psi_I$, until a measurement is performed at time $t_* > t$, when $\Psi(t_*) = \Psi_F$. The observed state Ψ_F can be, for example, the completion of a quantum algorithm.

10. The measuring devices are designed to represent the measure the states of the degrees of freedom using operators $\widehat{\mathcal{O}}(\mathcal{F})$. In Fig. 4.6, the measuring device has been designed to measure a degree of freedom that has N discrete values; distinct counter-readings $x_1, x_2, \ldots, x_N$ correspond to the possible values of the degrees of freedom.

11. To perform the measurement, the state vector is subjected to the experimental device by applying the operator $\widehat{\mathcal{O}}(\mathcal{F})$ on Ψ_F—and results in the states of the device becoming entangled with the state vector Ψ_F [1].

12. The operator $\widehat{\mathcal{O}}(\mathcal{F})$ being applied on the final state vector Ψ_F causes it to collapse to a definite state of the measuring device and shows a definite counter-reading, as shown in Fig. 4.6. Entanglement allows us to conclude that the state Ψ_F in Hilbert space is the same as the state observed in the device.

13. After repeating the experiment many times on samples prepared identically and independently, one obtains from the counter-readings of the measuring device $E_\psi[\widehat{\mathcal{O}}(\mathcal{F})] = \langle \psi_F | \widehat{\mathcal{O}}(\mathcal{F}) | \psi_F \rangle$, which is the expectation value the operator $\widehat{\mathcal{O}}(\mathcal{F})$.

The two figures given in Figs. 4.5 and 4.6 and their explanations, in short, are the theoretical foundations of quantum mechanics and the connection of quantum mechanics to empirically observed results [1].

4.10 Generalized Born Rule

The Born rule states that a measurement performed on all the degrees of freedom results in a post-measurement state vector that is an eigenvector of the measurement operators and yields a determinate reading of the measuring device [1, 2]. This rule has to be generalized to the case when the system consists of two or more degrees of freedom: if only a few—not all—of the degrees of freedom are subjected to a measurement, it leaves the remaining degrees of freedom in a quantum indeterminate state. To complete the Born rule, the post-measurement state vector of the degrees of freedom that are **not subjected** to a measurement needs to be specified. The generalized Born rule is not discussed in most of the standard books on quantum mechanics; we follow the derivation given by Mermin [2].

The generalized Born rule has important applications in quantum algorithms. For instance, an auxiliary degree of freedom is often introduced in a quantum algorithm, with a measurement being performed only on the input degrees of freedom—leaving the remaining auxiliary degrees of freedom untouched. Or sometimes, only one set of the degrees of freedom are measured, leaving the rest untouched by the measurement. The result of doing these measurements requires the generalized Born rule for its interpretation.

Consider a state vector of n-degrees of freedom given by

$$|\psi\rangle \;\Rightarrow\; \langle x_1, x_1, \ldots, x_n|\psi\rangle = \psi(x_1, x_2, \ldots, x_n)$$

There are two cases for state vectors depending on two or more degrees of freedom that are non-entangled or entangled (entanglement is discussed in Sect. 5.7).

- Non-entangled. The state vector can be completely factorized into a product of state vectors depending on one subset of degrees of freedom.

$$\psi(x_1, x_2, \ldots, x_n) = \psi_{12}(x_1, x_2) \ldots \psi_n(x_n)$$

- Entangled. The state vector is not factorizable into state vectors depending on separate degrees of freedom.

$$\psi(x_1, x_2, \ldots, x_n) \neq \psi(x_1)\psi(x_2) \ldots \psi(x_n)$$

Consider a **non-entangled** state vector depending on two degrees of freedom and given by

$$|\psi\rangle = |\psi_1\rangle|\psi_2\rangle \;\Rightarrow\; \langle x, x_2|\psi\rangle = \psi(x_1, x_2) = \psi_1(x_1)\psi_2(x_2)$$

Consider a device that measures only the x_1 degree of freedom. The process of measurement yields

$$|\psi\rangle \;\rightarrow\; \text{Measurement} \;\rightarrow\; |x_1\rangle|\psi_2\rangle$$

The value of the observed $|x_1\rangle$ is completely arbitrary; the probability of a specific value of $|x_1\rangle$ detected by the observation given by

$$\text{Tr}(M(x_1)|\psi\rangle\langle\psi|) = |\psi_1(x_1)|^2$$

where

$$\mathcal{M}(x_1) = |x_1\rangle\langle x_1| \otimes \mathbb{I}(x_2)$$

Since the degrees of freedom x_1, x_2 are non-entangled, each degree of freedom makes no reference to the other.

Consider now the case of an **entangled** state vector from Eq. 4.4, for a system with $n + m$ binary degrees of freedom; its state vector can be represented as follows

$$|\psi\rangle_{n+m} = \sum_{x=0}^{2^n-1} \sum_{y=0}^{2^m-1} \alpha_{xy} |x\rangle_n |y\rangle_m; \quad \sum_{xy} |\alpha_{xy}|^2 = 1 \tag{4.27}$$

Suppose one wants to measure only the first n degrees of freedom, leaving the other degrees of freedom untouched. The projection operators required to carry out the measurement process act on the first n degrees of freedom, leaving the other m degrees of freedom unchanged. Hence, the required projection operators are given by

$$\mathcal{M}_z = |z\rangle_{n\ n}\langle z| \otimes \mathbb{I}_m$$

Recall according to the Born rule, the result of the measurement results in the state vector for n-degrees of freedom being in an eigenstate of one of the projection operators. The generalized Born rule states that the remaining m-degrees of freedom are left undisturbed and in an indeterminate state and described by a state vector that depends only on the m-degrees of freedom [2]. Hence, the act of measurement yields the following

$$|\psi\rangle_{n+m} \ \rightarrow \ \text{Measurement} \ \rightarrow \ |z\rangle_n |\Phi(z)\rangle_m \tag{4.28}$$

The probability of observing the state $|z\rangle_n |\Phi(z)\rangle_m$ is given by

$$p(z) = \text{Tr}\Big(\mathcal{M}_z |\psi\rangle_{n+m\ n+m}\langle\psi|\Big) \tag{4.29}$$

$$= \sum_{xy,x'y'} \alpha_{xy} \alpha^*_{x'y'} \text{Tr}\Big[\Big(|z\rangle_{n\ n}\langle z| \otimes \mathbf{I}_m\Big)\Big(|x'\rangle_{n\ n}\langle x| \otimes |y'\rangle_{m\ m}\langle y|\Big)\Big]$$

$$\Rightarrow p(z) = \sum_y \alpha_{zy} \alpha^*_{zy} = \sum_y |\alpha_{zy}|^2 \tag{4.30}$$

The total probability of observing some particular state must be 1 and, using Eq. 4.27, we have the expected result

$$\sum_z p(z) = \sum_z \sum_y \alpha_{zy} \alpha^*_{zy} = \sum_z \sum_y |\alpha_{zy}|^2 = 1 \tag{4.31}$$

In summary, in general the measurement of x degrees of freedom results in a post-measurement quantum state yielding the *conditional probability amplitude* $|\Phi(z)\rangle_m$: an amplitude **given** that the state vector depending only on the n degrees of freedom are in an eigenstate $|z\rangle_n$. The post-measurement state is given by the generalization of Eq. 4.19 and from Eq. 4.28 yields

$$|\psi\rangle_{n+m} \;\rightarrow\; \text{Measurement} \;\rightarrow\; |x\rangle_n |\Phi(x)\rangle_m; \quad {}_m\langle\Phi(x)|\Phi(x)\rangle_m = 1 \quad (4.32)$$

The *key point* to note is that a partial measurement collapses the post-measurement n-degrees of freedom's state vector into an eigenstate $|x\rangle_n$ *and* the remaining m degrees of freedom continue to be described by the state vector that yields the (conditional) probability given by the *probability amplitude* $|\Phi(x)\rangle_m$. The normalization of $|\Phi(x)\rangle_m$ being equal to one reflects the fact that the state vector $|\Phi_m(x)\rangle$ yields the (conditional) probability for the outcome of measuring any of the remaining m degrees of freedom.

The conditional probability amplitude, due to the unit normalization given in Eq. 4.32, is the following

$$|\Phi_m(x)\rangle = \frac{1}{\sqrt{p(x)}} \sum_y \alpha_{xy} |y\rangle_m \;\Rightarrow\; {}_m\langle\Phi(x)|\Phi(x)\rangle_m = 1 \qquad (4.33)$$

4.10.1 Example

Consider the case of $n + 1$ *binary* degrees of freedom, and only one of the degrees of freedom is measured, with the rest of the n binary degrees of freedom left untouched. The general state vector is given by

$$|\psi\rangle_{n+1} = |0\rangle \sum_x a_x |x\rangle_n + |1\rangle \sum_x b_x |x\rangle_n = |\Psi_0\rangle_{n+1} + |\Psi_1\rangle_{n+1}$$

with

$$_{n+1}\langle\Psi_0|\Psi_1\rangle_{n+1} = 0$$

and

$$_{n+1}\langle\psi_x|\psi_x\rangle_{n+1} = 1 = \sum_x \left(|a_x|^2 + |b_x|^2 \right) \qquad (4.34)$$

The probability of observing the state $|i\rangle$, $i = 0, 1$ is given by

$$p(0) = \sum_x |a_x|^2; \quad p(1) = \sum_x |b_x|^2$$

From Eq. 4.34, and as required by Eq. 4.18, we have

$$p(0) + p(1) = \sum_x \left(|a_x|^2 + |b_x|^2 \right) = 1$$

In anticipation of the post-measurement state, we rewrite the initial state vector as follows

$$|\psi\rangle_{n+1} = \sqrt{p(0)}|0\rangle|\Phi_0\rangle_n + \sqrt{p(1)}|1\rangle|\Phi_1\rangle_n$$

where

$$|\Phi_0\rangle_n = \frac{1}{\sqrt{p(0)}} \sum_x a_x |x\rangle_n; \quad |\Phi_1\rangle_n = \frac{1}{\sqrt{p(1)}} \sum_x b_x |x\rangle_n$$
$$_n\langle\Phi_i|\Phi_i\rangle_n = 1 : i = 0, 1; \quad _n\langle\Phi_0|\Phi_1\rangle_n \neq 0$$

Note $|\Phi_0\rangle$, $|\Phi_1\rangle_n$ are not orthogonal since

$$_n\langle\Phi_0|\Phi_1\rangle_n = \frac{1}{\sqrt{p(0)p(1)}} \sum_x a_x^* b_x \neq 0$$

A measurement performed on the first binary degree of freedom has two possible outcomes and hence yields

$$|\psi\rangle_{n+1} \rightarrow \text{ Measurement } \rightarrow |z\rangle|\Phi(z)\rangle_n; \quad z = 0, 1 \qquad (4.35)$$

More precisely

- The measurement results in the eigenstate $|0\rangle$ with probability $p(0)$. The state vector undergoes the collapse yielding the following the post-measurement state vector

$$|\psi\rangle_{n+1} \rightarrow |0\rangle|\Phi_0\rangle_n; \quad _n\langle\Phi_0|\Phi_0\rangle_n = 1$$

- The measurement results in the eigenstate $|1\rangle$ with probability $p(1)$. The state vector undergoes the collapse yielding the following the post-measurement state vector

$$|\psi\rangle_{n+1} \rightarrow |1\rangle|\Phi_1\rangle_n; \quad _n\langle\Phi_1|\Phi_1\rangle_n = 1$$

Note that, as expected from the generalized Born rule, $|\Phi_0\rangle_n$, $|\Phi_1\rangle_n$ are the probability amplitudes for the remaining n degrees of freedom *after* the measurement of the first degree of freedom is made.

4.11 Consistency of Generalized Born Rule

The underlying reason for the validity of the generalized Born rule discussed in Sect. 4.10 is because the degrees of freedom are independent of each other: the order of measuring the degrees of freedom does not matter because of their independence. Hence, the results obtained for measuring a set of degree of freedom must not depend on the order in which the measurements are carried out on the degrees of freedom.

To check the consistency of the generalized Born rule, the following two procedures should yield the same result: take a collection of $n + m + s$ degrees of freedom and proceed in the following manner,

- First measure $n + m$ degrees of freedom directly.
- Alternatively, measure n degrees of freedom, obtain the conditional amplitude for $m + s$ degrees of freedom and then make a second measurement of the m degrees of freedom.

We start with the following $n + m + s$ degrees of freedom state vector

$$|\psi\rangle_{n+m+s} = \sum_{x=0}^{2^n-1} \sum_{y=0}^{2^m-1} \sum_{z=0}^{2^s-1} \alpha_{xyz} |x\rangle_n |y\rangle_m |z\rangle_s; \quad \sum_{xyz} |\alpha_{xyz}|^2 = 1 \quad (4.36)$$

Measuring $n + m$ degrees of freedom, the probability of observing $|x\rangle_n |y\rangle_m$ is given by

$$p(xy) = \sum_z |\alpha_{xyz}|^2$$

We obtain the post-measurement state vector, using the result given in Eq. 4.33

$$|x\rangle_n |y\rangle_m |\Phi(xy)\rangle_s$$

where

$$|\Phi(xy)\rangle_s = \frac{1}{\sqrt{p(xy)}} \sum_{z=0}^{2^s-1} \alpha_{xyz} |z\rangle_s \quad (4.37)$$

We now start with the state vector $|\psi_{n+m+s}\rangle$ given in Eq. 4.36 and measure only the $|x\rangle_n$ degrees of freedom; this yields the probability

$$p(x) = \sum_{yz} |\alpha_{xyz}|^2$$

The measurement leads to the collapse of the initial state vector to post-measurement conditional state vector given by

$$|\psi\rangle_{n+m+s} \rightarrow |x\rangle_n |\Phi(x)\rangle_{m+s}$$

where

$$|\Phi(x)\rangle_{m+s} = \frac{1}{\sqrt{p(x)}} \sum_{yz} \alpha_{xyz} |y\rangle_m |z\rangle_s \quad (4.38)$$

The second measurement is the result of measuring m degrees of freedom that collapses their state vector to $|y\rangle_m$—*given* that one has already measured the n degrees of freedom. The second measurement leads to post-measurement state vector that is given by

$$|x\rangle_n |\Phi(x)\rangle_{m+s} \;\to\; |x\rangle_n |y\rangle_m |\widetilde{\Phi}(y|x)\rangle_s$$

Hence, we obtain from the conditional amplitude given in Eq. 4.38 the conditional probability of observing state vector $|y\rangle_m$ in the second measurement given by

$$p(y|x) = \frac{1}{p(x)} \sum_z |\alpha_{xyz}|^2 = \frac{p(xy)}{p(x)} \tag{4.39}$$

and hence

$$p(xy) = p(y|x)p(x) \tag{4.40}$$

From Eqs. 4.38 and 4.40 we have

$$|\widetilde{\Phi}(y|x)\rangle_s = \frac{1}{\sqrt{p(y|x)}} \left[\frac{1}{\sqrt{p(x)}} \sum_{z=0}^{2^s-1} \alpha_{xyz} |z\rangle_s \right] = |\Phi(xy)\rangle_s \tag{4.41}$$

where the last equality above follows from Eq. 4.37. Hence, Eq. 4.41 shows the consistency of the generalized Born rule [2].

Note that Eq. 4.39 yields

$$p(y|x) = \frac{p(xy)}{p(x)}$$

Equation 4.39 is precisely the equation that one obtains from the theory of probability: it expresses the conditional probability $p(y|x)$ in terms of the joint probability function $p(xy)$ and the probability of the conditioning given by $p(x)$ [11].

4.12   Quantum Mechanics and Quantum Computers

The first quantum computer was proposed by Manin in 1980 [12] and Feynman in 1982 [13]. The first quantum code that showed, in principle, that a quantum algorithm is superior to any classical algorithm is attributed to a 1985 paper by Deutsch [14].

A fundamental question is the relation of computers, both classical and quantum, to a physical device. Classical computers are based on devices that operate according to the laws of classical physics, which is an approximation to the laws of quantum mechanics. Of course, all devices ultimately obey the laws of quantum mechanics, including the classical computer, so this is not what is meant by a quantum computer. A quantum computer is a physical device that carries out algorithms that are directly based on the laws of quantum mechanics. Quantum mechanics provides a new computational paradigm that, surprisingly, had not been contemplated prior to the 1980s.

The quantum computer demonstrates that the theory of computation can be generalized to indeterminate qubits based on the physics of quantum mechanics and provides a ground-breaking paradigm for computer science. It can be shown that all classical algorithms are special cases of quantum algorithms, and hence classical computers and information science are a sub-branch of quantum physics. Quantum information science is yet another arena where the mysterious and 'counterintuitive behavior' of quantum phenomena comes into play.

In physics, the state vector of a physical entity is determined by natural laws, and theoretical physics attempts to deduce the state vector by studying Nature. In contrast, quantum computers and algorithms have the following distinctive features in their use and application of the principles of quantum mechanics.

- For quantum computers, the state vector ψ is **not obtained** by studying Nature, but instead, the state vector is **designed** to carry the information required by the algorithms of information science. The quantum computer is a physical device that provides the binary degrees of freedom as well as the various unitary gates required for quantum algorithms. The physical quantum computer is subjected to diverse forms of physical interactions so that the state vector evolves in a manner required for carrying out the various steps of the quantum algorithm.
- The input of the quantum algorithm is a determinate state consisting of a n-binary bits string; however, although the output of a quantum algorithm is also a n-binary bits string, its value is **random** and **uncertain**: the output takes many possible values with different probabilities. The Schrödinger equation determines the probabilities for the different outputs.
- Unlike a classical computer, for which executing the classical algorithm only once is sufficient to obtain the required output, a quantum algorithm has to be run many times (in principle infinitely many times), with identical preparation, to ascertain the probabilities for the various outputs.
- All the intermediate steps of a classical computer can, in principle, be directly observed. Each updating of a classical algorithm processing n-binary bits strings produces a determinate state, which is a member of 2^n binary strings $|x\rangle$; hence, each of the binary strings has to be individually updated. This requires, in general, a memory storage device for all possible configurations of the n-binary bits strings, which is 2^n binary strings. Hence executing a classical algorithm entails having to store 2^n binary strings.
- In contrast, during the process of executing a quantum algorithm, the intermediate steps of a quantum algorithm cannot, in principle, be directly observed, as discussed in Sect. 5.2, and is the result of the principle of quantum superposition. A quantum computer, hence, needs a storage for only the input and output states, which requires at most two n-binary strings.
- In summary, the quantum algorithm is unlike a classical algorithm, since the quantum algorithm only requires the storage of two n-binary strings. This is unlike a classical computer that entails having to store 2^n-binary strings.
- In updating a classical computer, each step in the algorithm produces a determinate binary string $|x\rangle$; hence, each binary string has to be individually updated. In

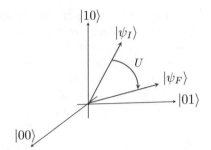

contrast, the state vector evolves as a single entity. The quantum device evolves the state vector $|\psi_I\rangle$ from initial time t_I to final time t_F with state vector $|\psi_F\rangle$ that is given, from Eq. 4.16, by

$$|\psi_F\rangle = \exp\left\{-\frac{i}{\hbar}(t_F - t_I)\mathcal{H}\right\}|\psi_I\rangle = U|\psi_I\rangle \qquad (4.42)$$

Equation 4.42 expresses the fact that the entire state vector is evolved in time. The time evolution of a state vector depending on three computational basis states is shown in Fig. 4.7.

Mathematically speaking, a vector is independent of its representation in terms of the basis states. For quantum algorithms, one is interested in specific information that is contained in the components of the state vector in the computational basis states; the components $a_I(x)$ of $|\psi_I\rangle$ in the computational basis states yield the following

$$|\psi_I\rangle = \sum_{x=0}^{2^n-1} a_I(x)|x\rangle$$

The initial value of each coefficient $a_I(x)$ is automatically evolved to its final value $a_F(x)$ since it is contained in $|\psi_F\rangle$, where

$$|\psi_F\rangle = \sum_{x=0}^{2^n-1} a_F(x)|x\rangle$$

The solution for the quantum algorithm is usually contained in the coefficients $a_F(x)$. In the evolution of the state vector, all the coefficients $a_I(x)$ are simultaneously updated to $a_F(x)$. This is another example of quantum parallelism.

- In summary, a quantum computer (hardware) processing a quantum algorithm based on n-binary bits strings requires a storage device for only two n-binary strings and not 2^n-binary strings that are required by a classical algorithm. Furthermore, all the coefficients that express the state vector as a quantum superposition of the computational basis states are simultaneously updated when the state vector is updated. These are two reasons that the quantum computer is far more efficient and sometimes exponentially faster than a classical computer.

- One can easily imagine a quantum computer with 1000 binary degrees of freedom being made within a year, and for which no possible physical device can store the information that the quantum computer is processing—since 2^{1000} is far greater than all the atoms in the Universe. This aspect of a quantum algorithm is further discussed in Sect. 19.3.

For computer scientists, the most striking feature of a quantum algorithm is that, for certain computational tasks of considerable practical interest, a quantum computer can be vastly more efficient than anything that is possible for a classical algorithm. As the size of the input is increased, the time it takes the quantum computer to accomplish such tasks scales up much more slowly compared to any classical computer. In fact, in some special cases, the quantum algorithm is **exponentially faster** than any possible classical algorithm.

References

1. Baaquie BE (2013) The theoretical foundations of quantum mechanics. Springer, UK
2. Mermin ND (2007) Quantum computer science. Cambridge University Press, UK
3. Nicolas G, Grégoire R, Wolfgang T, Hugo Z (2002) Quantum cryptography. Rev Mod Phys 74(1):145
4. Kwek L-C, Cao L, Luo W, Wang Y, Sun S, Wang X, Liu AO (2021) Chip-based quantum key distribution. AAPPS Bull 31(1):1–8
5. Baaquie BE (2004) Quantum finance: path integrals and Hamiltonians for options and interest rates. Cambridge University Press
6. Gottfried K, Yan T-M (2003) Quantum mechanics, 2nd edn. Springer, Germany
7. Baaquie BE (2014) Path integrals and Hamiltonians: principles and methods. Cambridge University Press
8. Nielsen MA, Chang IL (2000) Quantum computation and quantum information. Cambridge University Press, UK
9. Berman GP, Doolen GD, Mainieri R, Tsifrinovich VI (1998) Introduction to quantum computers. World Scientific, Singapore
10. Juan O (2018) Twelve years before the quantum no-cloning theorem. Am J Phys 86(3):201–205
11. Baaquie BE (2020) Mathematical methods and quantum mathematics for economics and finance. Springer, Singapore
12. Manin YI (1980) Computable and noncomputable. Soviet Radio 2(4):13–15
13. Feynman RP (1982) Simulating physics with computers. Int J Theor Phys 21:467–488
14. Deutsch D (1985) Quantum theory, the Church-Turing principle and the universal quantum computer. Roy Soc Lond Proc Ser A 400:97–117

Chapter 5
Quantum Superposition and Entanglement

Two properties of Hilbert space that are pivotal in making quantum algorithms faster than classical algorithms are *superposition* and *entanglement*, discussed in Sects. 5.1 and 5.7. A few special cases, discussed below, concretely illustrate the general principles of *superposition* and *entanglement* for quantum algorithms.

For a number of algorithms, a quantum computer is much faster than a classical computer—and in some cases is exponentially faster due to the features of superposition and entanglement being incorporated into the algorithm. The fact that all the possible states of many qubits can be updated simultaneously in the process of a quantum computation is due to the quantum principle of superposition. The property of quantum entanglement appears in many intermediate steps in the process of a quantum computation, a property not allowed for classical computers. The principles of quantum mechanics are required for understanding these counterintuitive and nontrivial aspects of superposition and entanglement.

To understand both the theory of measurement and entanglement requires the study of what is called the density matrix. The main properties of density matrices are reviewed, and the concept of pure, mixed and reduced density matrix are defined. Separable and entangled states are discussed. The binary degrees of freedom illustrate many key features of entanglement, and it is shown later, in Sect. 6.4, that the Bell states for two binary degrees of freedom are examples of maximally entangled quantum states.

The discussions in this chapter are based on the results given in Baaquie [1].

© The Author(s), under exclusive license to Springer Nature Singapore Pte Ltd. 2023   107
B. E. Baaquie and L.-C. Kwek, *Quantum Computers*,
https://doi.org/10.1007/978-981-19-7517-2_5

5.1   Quantum Superposition

The concept of the *indeterminacy* of *quantum paths* has been discussed in detail in Belal [1], and this concept is now analyzed using the two-slit experiment. Similar to the indeterminate degrees of freedom, quantum superposition—unlike classical superposition—is a result of indeterminate quantum paths that can never, in principle, be directly observed.

The two-slit experiment goes back to Young (1799), who showed that light going through two slits results in interference, and was crucial in demonstrating that light is a wave: the cardinal property of a wave is that it is spread over space. The two-slit experiment is one of the deepest and most important experiments in quantum mechanics and is discussed with the aim of demonstrating the mode of existence of indeterminate paths, as well as the role of measurement in causing a transition from the indeterminate form of the quantum entity to its empirical manifestation.

Quantum superposition is one of the bedrocks of quantum mechanics. Quantum superposition is also a major resource for quantum algorithms, and together with quantum entanglement, is the reason that, for some cases, quantum algorithms can perform exponentially faster than the corresponding classical algorithms.

The two-slit experiment provides one of the simplest illustrations of indeterminate paths and quantum superposition. As explained by Richard Feynman, one of the leading quantum theorists, the two-slit experiment '*has been designed to contain all of the mystery of quantum mechanics, to put you up against the paradoxes and mysteries and peculiarities of nature one hundred percent*' [2]. Feynman further explains that all the other paradoxical situations in quantum mechanics can always be explained by this experiment—which reveals '*nature in her most elegant and difficult form*' [2].

The simplest case of indeterminate paths is for the quantum particle to simultaneously exist in two distinct paths, as shown in Fig. 5.1 and which can be generalized to the case of the N-slit, shown in Fig. 5.2. In this section, the two-slit experiment is analyzed using electrons; it is shown that when a measurement determines the path of the electron, the path is empirical and determinate and the electron behaves like a classical particle; however, when the trajectory is *not observed*, the electron exists in an indeterminate state and exhibits the phenomenon of quantum interference.

The two-slit experiment is employed for analyzing the following topics.

- The concepts of the empirical and indeterminacy are applied to the time evolution of a quantum entity. In classical mechanics, the classical entity takes a determinate path, going through either the slit at position x_1 or through the slit at position x_2. The result of the two-slit experiment can be explained by postulating that when the quantum entity is *not observed*, the *path* taken by the quantum entity is *indeterminate*, with the quantum entity simultaneously taking *both* the paths.
- The superposition of the quantum state vector, which is due to the *linearity* of the Schrödinger equation, is also valid for indeterminate paths. For a quantum particle 'taking' indeterminate paths, the probability amplitude of going from an initial to a final position is shown to result from the quantum superposition of the determinate empirical paths.

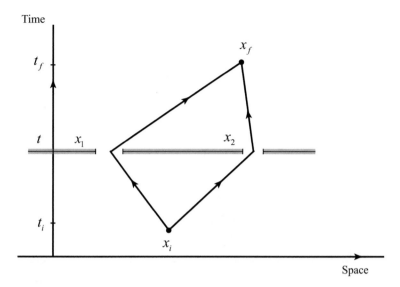

Fig. 5.1 A quantum particle is observed *only* at initial position x_i at time t_i and at final position x_f at time t_f. The paths taken from initial to final position are indeterminate. Published with permission of © Belal E. Baaquie 2012. All Rights Reserved

5.1.1 The Experiment

A quantum particle going through two slits, as given in Fig. 5.1, is realized experimentally by the arrangement shown in Fig. 5.3. Unlike Fig. 5.1, the time dependence of the paths is not shown in Fig. 5.3, where the emphasis is on the measurements being performed. The information about the paths, in particular whether they are empirical or indeterminate, is reconstructed from the experimental measurements.

The experiment consists of an electron gun (*source*) that sends identically prepared electrons, through a barrier with *two slits*, to a *screen* where a screen-detector keeps track of the point at which the electron hits the screen. Note that the electrons are sent toward the slits *one by one*, so that at any given time there is only *one* electron traveling from the electron gun to the screen.

The electron leaves the source, shown in Fig. 5.3a, b, with the initial position of the electron denoted by s; it is then observed at the screen at position denoted by x. There are two possible paths from source to screen, labeled path 1 going through slit 1 and path 2 going through slit 2, and shown in Fig. 5.3b.

The experiment is performed *with* detectors 1 and 2, as shown in Fig. 5.3a and *without* these two detectors, as shown in Fig. 5.3b. In effect, with detectors 1 and 2 switched on, the path taken by the electron is known, whereas in the case without the detectors the path information is not known.

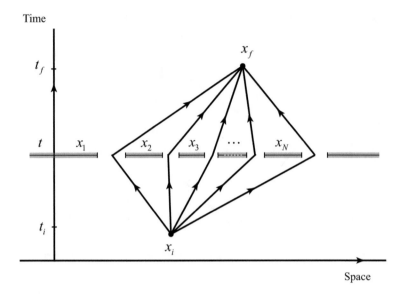

Fig. 5.2 A quantum particle is observed at first at initial position x_i at time t_i and a second time at final position x_f at time t_f. The quantum particle's path being indeterminate means that the single particle simultaneously *exists in all the allowed paths*. Published with permission of © Belal E. Baaquie 2012. All Rights Reserved

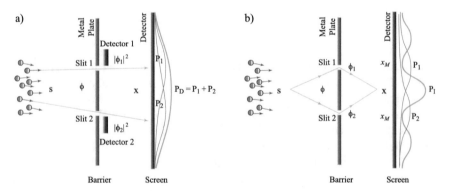

Fig. 5.3   a Two-slits experiment *with* determination of the path taken to reach the screen. **b** Two-slits experiment *without* determination of the path taken to reach the screen. Published with permission of © Belal E. Baaquie 2012. All Rights Reserved

5.1.2 Experiment with Detectors 1 and 2: No Interference

The experiment is shown in Fig. 5.3a; *both* slits 1 and 2 are open and detectors 1 and 2, at the back of the slits, record which slit the electron passes through. Since it is known which slit the electron goes through one can plot the following three distribution curves.

- Distribution curve P_1 for electrons that pass through slit 1
- Distribution curve P_2 for electrons that pass through slit 2
- Distribution curve P_D for electrons that pass through *either* slit 1 *or* 2.

The experimental result is the following.

$$P_D = P_1 + P_2 \; ; \quad \int dx \, P_D(x) = 1 \tag{5.1}$$

The behavior of the electron *with detection* of which path is taken is exactly the same as one would obtain for bullets being shot through a metal screen with two openings. When the trajectory of the electron from source to screen is *experimentally observed*, the electron received at the screen-detector came *either* through slit 1 *or* through slit 2. The two possible ways of getting to the screen are mutually exclusive, and these do not interact in any manner and do not generate any *interference* pattern.

The two-slit experiment has two possible electron paths, namely path 1 and path 2. The quantum mechanical description of the two-slit experiment is given by assigning a *probability amplitude* (a complex number) to each of the two paths. The probability amplitude is the analog of the state vector for paths and is derived from the quantum state of the electron's quantum degree of freedom and is discussed in more detail in Belal [1].

The following is a notation for the probability amplitude for the different possibilities

ϕ_1 : the probability amplitude for determinate path 1

ϕ_2 : the probability for amplitude determinate path 2

ϕ : the general probability amplitude

The probability of finding the electron at the screen *with detection* of path taken is labeled by P_D and *without detection* of path taken is labeled by P_I.[1]

The experimental result for P_D is shown in Fig. 5.3a, and the result for P_I is shown in Fig. 5.3b. Note the experimental results obtained for P_D and P_I are *qualitatively different* and need an explanation.

---

[1] Recall all measurements in quantum mechanics are of operators representing a physical quantity. 'Measuring the position' of the electron is a shorthand for the more precise statement that measurement is mathematically represented by applying the device that contains all the position projection operators (of the electron) on the electron's state vector.

The explanation of quantum mechanics for the experiment with detectors 1 and 2 switched on is the following. The electron leaves the source s with probability amplitude ϕ. When the probability amplitude encounters the detectors at the barrier, a measurement is performed by either detector 1 or 2, depending on which slit the electron is detected. The probability amplitude ϕ, similar to the state vector on being measured, *collapses* to its *empirical manifestation – to either* $|\phi_1|^2$ *or to* $|\phi_2|^2$, depending on whether it is detected at slit 1 or slit 2, respectively.

The electron arrives at the screen in an empirical state and the probability of being observed at a point on the screen is given by either $|\phi_1|^2$ or $|\phi_2|^2$. The probability of finding the electron at the screen-detector is the result of the two mutually exclusive possibilities and hence is given by their sum, namely

$$P_D = |\phi_1|^2 + |\phi_2|^2 = P_1 + P_2 \quad : \quad \text{collapse at barrier} \tag{5.2}$$

and one obtains the result given in Eq. 5.1.

The result given in Eqs. 5.1 and 5.2 is equivalent to the *classical result* obtained by shooting classical particles through the two slits. The reason is that observations made by detectors 1 and 2 at the two slits *determines* which slit the electron went through by collapsing the probability amplitude, and in doing so causes the electron to take a determinate path, namely going through *either* slit 1 *or* slit 2.

Having the information of which path is taken is equivalent to the classical description of a particle, since the classical particle always takes a unique path from its initial to final position.

5.1.3 Experiment Without Detectors 1 and 2: Indeterminate

Consider now the same experiment as before, but with detectors 1 and 2 at the barrier removed. The experiment is shown in Fig. 5.3b. The electrons are sent in one by one, and *no measurement* is made to determine which slit the electron goes through, and hence, the path taken by the electron is no longer known. As shown in Fig. 5.3b, the electron can take two possible paths to reach the point $|x\rangle$ at the screen.

The electron leaves the source with probability amplitude ϕ. On crossing the slits (barrier), since the electron is *not* observed, the path taken by the electron is not known and hence the path of the electron is indeterminate.

One needs to decide as to what is the probability amplitude past the barrier. Since the propagation of the probability amplitude ϕ is determined by the linear Schrödinger equation the probability amplitude obeys the *superposition principle*. The probability amplitude for the two possible paths both obeys the Schrödinger equation, and hence, their linear sum also obeys the Schrödinger equation; the probability amplitude for arriving at the screen is given by **quantum superposition**, namely *summing* the probability amplitude for the two determinate paths and yields

$$\phi = \phi_1 + \phi_2 \quad : \quad \text{trans-empirical probability amplitude at the screen}$$

Note that to obtain ϕ one is not superposing material displacements of a medium as is the case for classical waves, but instead one is superposing probability amplitudes ϕ_1 and ϕ_2 for determinate paths to obtain the probability amplitude $\phi = \phi_1 + \phi_2$ for the electron taking an indeterminate path; ϕ is said to be wave-like since it is spread over space.

Quantum superposition is qualitatively different from superposing, for example, water waves in which one is adding the physical displacement of the underlying water. Quantum superposition is the superposition of information carried by the probability amplitude, and it is this aspect of quantum superposition that comes to the forefront in quantum algorithms.

On reaching the screen, the measurement of the electron at the screen-detector collapses the probability amplitude ϕ to its empirical manifestation $|\phi|^2$ and yields the empirical probability P_I. Hence,

$$P_I = |\phi|^2 = |\phi_1 + \phi_2|^2 \quad : \quad \text{collapse at the screen}$$

$$\Rightarrow P_I = |\phi_1|^2 + |\phi_2|^2 + \phi_1\phi_2^* + \phi_1^*\phi_2 \; ; \quad \int dx\, P_I(x) = 1 \qquad (5.3)$$

The probability amplitude for the electron with indeterminate path has nonlocal information about the likelihood of occurrence everywhere in space. In particular, for the two-slit experiment, there are nodal points (minimas) of interference pattern, points for which the probability amplitude is zero, yielding zero likelihood of that the electron will be detected at those points. Since the electrons are sent in one by one, each electron, regardless of where on the screen it is detected, has the information about the entire screen since no electron ever hits the nodal points on the screen.

The interference pattern P_I shown in Fig. 5.3b has been verified by many experiments and shows that on repeatedly sending in the electrons—*sent in one by one*—and detecting the position of the electrons arriving at the screen, results in building up, step by step, an interference pattern given by P_I. The probability amplitude at the screen is $\phi = \phi_1 + \phi_2$ and shows that, when the path taken by the electron is *not detected*, the electron's path is indeterminate and trans-empirical, showing interference.[2]

The quantum superposition is lost in Fig. 5.3a because one *detects* the passage of the electron, by say shining light on the electron as it passes through the slit. The shining of light is a measurement process that causes a transition by collapsing the trans-empirical probability amplitude before the barrier, namely $\phi = \phi_1 + \phi_2$ to the empirical *probability* given by either $|\phi_1|^2$ or $|\phi_2|^2$ after the barrier—depending on which slit the electron is observed. The empirical expression $|\phi_1|^2$ or $|\phi_2|^2$ is said to be particle-like since it implies that the electron is following a definite trajectory. In other words, when the electron's path is measured, the nonlocal probability amplitude collapses to a (localized) empirically observed determinate state that is particle-like.

---

[2] It is important to note, as discussed in the next section, that the electron is interfering with *itself*—a completely non-classical and enigmatic phenomenon.

A cardinal point to note is the crucial role of *measurement* in producing two *qualitatively different* results—namely of P_D *with* the detection of the electron's path and of P_I *without* detection of its path. Measurement causes a transition of the electron from its indeterminate state to its empirical manifestation, and the two-slit experiment shows this difference in a stark and clear manner.

In conclusion, the two-slit experiment shows that both quantum states and quantum paths display empirical and indeterminate behavior—depending on whether a measurement is carried out or not. Furthermore, measurement has a central role in determining whether the electron behaves as an empirical entity or as an indeterminate state. When it is not observed, *the electron evolves on indeterminate paths and displays wave-like behavior* that reflects its indeterminate form; and when the electron is observed, it is an empirical condition and behaves like a classical particle.

5.2   Quantum Superposition and Quantum Algorithms

One of the main takeaways of quantum superposition for quantum algorithms, using the two-slit experiment as an exemplar, is the following.

- A quantum algorithm starts from an initial state, which is the analog of the electron starting at the initial point s. The two-slit experiment is the analog of a quantum algorithm having a single intermediate step before completion, the analog of the electron reaching the screen.
- The quantum algorithm is subjected to irreversible quantum gates at the screen and the result that is obtained is analogous to the interference pattern obtained as the screen at position x and given by P_I in Fig. 5.3b.
- One is forbidden to observe the quantum algorithm during any of the intermediate steps it is taking. A measurement of the quantum algorithm before its completion (analog of hitting the screen) is analogous to measuring which slit the electron is going through, as in Fig. 5.3a. This would result in the final result at the screen being the analog of P_D, which is clearly wrong since the actual result is the analog of P_I.
- The analog of the case of the N-slit experiment, shown in Fig. 5.2, is a quantum algorithm having N intermediates steps before completion. If any of the slits is observed to determine whether the electron took that slit or not will spoil the interference. Similarly, if the quantum algorithm is observed at any intermediate step, the algorithm will become invalid.
- In summary, the rules of quantum superposition, in particular its connection to measurement, are the reason that unlike a classical algorithm, the intermediate steps leading to the final answer for a quantum algorithm cannot, in principle, be observed. Such an observation would destroy the quantum algorithm.

5.3 Partial Trace for Tensor Products

Consider the outer product of two state vectors given by

$$\mathcal{O} = |\psi\rangle\langle\chi| \; ; \;\; \langle x'|\mathcal{O}|x\rangle = \psi(x')\chi^*(x)$$

In the position basis, for a continuous degree of freedom x, trace is defined in Eq. 4.11 as follows

$$\text{Tr}[\mathcal{O}] = \int dx \langle x|\mathcal{O}|x\rangle \tag{5.4}$$

In particular, the trace operation for the outer product of two states is, from Eq. 5.4, the following

$$\mathcal{O} = |\psi\rangle\langle\chi| \;\; \Rightarrow \;\; \text{Tr}[\mathcal{O}] = \int dx \, \psi(x)\chi^*(x)$$

If the state vectors are elements of an N-dimensional Euclidean space, then $\mathcal{O}$ is simply and $N \times N$ matrix and the trace of $\mathcal{O}$ is the sum of it's diagonal elements, since for matrix M_{ij}, trace is defined by $\sum_i M_{ii}$; hence, it follows that

$$\text{Tr}[\mathcal{O}] = \sum_{i=1}^{N} \langle i|\mathcal{O}|i\rangle = \sum_{i=1}^{N} \psi(i)\chi^*(i)$$

Consider a system with two degrees of freedom with state vectors $|\psi_1\rangle|\psi_2\rangle$; the outer product is given by

$$\mathcal{O} = |\psi_1\rangle\langle\psi_1| \otimes |\psi_2\rangle\langle\psi_2|$$

One can now perform a *partial trace* on $\mathcal{O}$, say over system 2, and yields

$$\text{Tr}_2(\mathcal{O}) = |\psi_1\rangle\langle\psi_1|\big[\langle\psi_2|\psi_2\rangle\big] = c|\psi_1\rangle\langle\psi_1| \;\; \text{with} \;\; \langle\psi_2|\psi_2\rangle = c$$

One can further generalize the concept of a partial trace; consider the following linear sum of the outer product of states

$$\mathcal{O} = \sum_{i=1}^{N} p_i |\psi_1^i\rangle|\langle\psi_1^i| \otimes |\psi_2^i\rangle\langle\psi_2^i|$$

where p_i are numbers. The partial trace over system 2, using the linearity of trace as given in Eq. 4.11, is defined as follows

$$\mathrm{Tr}_2(\mathcal{O}) = \sum_{i=1}^{N} p_i |\psi_1^i\rangle\langle\psi_1^i| \langle\psi_2^i|\psi_2^i\rangle$$

$$= \sum_{i=1}^{N} p_i c_i |\psi_1^i\rangle\langle\psi_1^i| \quad \text{with} \quad \langle\psi_2^i|\psi_2^i\rangle = c_i \qquad (5.5)$$

5.4 Density Matrix ρ

The density matrix was introduced in Sect. 4.5 as it plays a defining role in the theory of quantum measurements. The density matrix is the key to Heisenberg's operator formulation [1], with the Schrödinger state vector being replaced by the density matrix providing a fundamental description of the quantum system. The density matrix is a special Hermitian operator that has many applications and is the principal mathematical construction required for describing measurements in quantum mechanics, as discussed in Sect. 4.5. The density matrix also provides a quantum mechanical generalization of the concept of conditional probabilities for quantum mechanical degrees of freedom, as discussed in Sect. 4.10. As mentioned earlier, the density matrix is required for providing a mathematical criteria for the entanglement of quantum degrees of freedom.[3]

5.4.1 Pure Density Matrix

The pure density matrix is a Hermitian operator that is equivalent to the state vector and provides an operator description of the quantum entity.

For a state vector $|\chi\rangle$, the pure density matrix is defined by

$$\rho_P = |\chi\rangle\langle\chi| \qquad (5.6)$$

A pure density matrix ρ_P is a projection operator and has the following properties

$$\rho_P = |\chi\rangle\langle\chi| \; ; \; \rho_P^2 = \rho_P$$
$$\mathrm{Tr}(\rho_P^2) = \mathrm{Tr}(\rho_P) = 1 \; : \; \text{Pure state} \qquad (5.7)$$

Expressing the state vector in terms of a complete basis state given by $|\chi_i\rangle$ yields the following

---

[3] The density matrix should be termed the density operator since, in general, it is not a finite or infinite matrix; however, the term density matrix is so widely used that its proper definition is implicitly understood.

$$|\chi\rangle = \sum_i c_i|\chi_i\rangle \; ; \;\; \rho_P \equiv |\chi\rangle\langle\chi|$$

$$\Rightarrow \rho_P = \sum_{ij} c_i c_j^*|\chi_i\rangle\langle\chi_j| = \sum_i |c_i|^2|\chi_i\rangle\langle\chi_j| + \sum_{ij;\, i\neq j} c_i c_j^*|\chi_i\rangle\langle\chi_j| \quad (5.8)$$

The off-diagonal terms $i \neq j$ given in Eq. 5.8 are completely quantum mechanical in origin and are due to correlations between two different eigenstates $|\chi_i\rangle$ and $|\chi_j\rangle$.

The expectation value of any operator $\mathcal{O}$ in a state $|\psi\rangle$ can be obtained from the pure state density matrix and is discussed in Sect. 5.12. The density matrix for a pure state, namely ρ_P, is *equivalent* to the state vector $|\psi\rangle$ and encodes the result of all observations that can be made on the quantum system.

5.4.2 Mixed Density Matrix

A *mixed density matrix* is defined for a collection of orthonormal projection operators $|\psi_i\rangle\langle\psi_i|$ and is given by

$$\rho_M = \sum_{i=1}^{N} p_i|\psi_i\rangle\langle\psi_i| \quad (5.9)$$

$$0 < p_i < 1 \; ; \;\; \sum_{i=1}^{N} p_i = 1 \;\; \Rightarrow \;\; \text{Tr}(\rho_M) = \sum_{i=1}^{N} p_i = 1$$

Note the cardinal point that for the mixed density matrix, there are no off-diagonal terms such as the terms $|\chi_i\rangle\langle\chi_j|$, $i \neq j$ given in Eq. 5.8.

The mixed density matrix has the following defining property

$$\rho_M^2 = \sum_i p_i^2|\psi_i\rangle\langle\psi_i| \;\; \Rightarrow \;\; \text{Tr}(\rho_M^2) = \sum_i p_i^2 < 1 \quad (5.10)$$

Only for a pure state, where only one of the p_i is 1, is $\text{Tr}(\rho^2) = 1$. Hence, a definition of a mixed state is

$$\text{Tr}(\rho_M^2) < 1 \;\; : \;\; \text{Mixed state} \quad (5.11)$$

In Eq. 5.5, it was shown that if one starts with a pure state density matrix and a partial trace is performed over one of the degrees of freedom, then one obtains a mixed state density matrix. Performing a partial trace erases information about the degree of freedom and hence, the density matrix of a mixed state contains less information than a pure state.

The density matrix for a mixed state is required for mathematically representing the result of quantum measurements, discussed in Sect. 4.5, and is a precise measure of how much information is lost in performing an observation on a quantum system.

Another important application of the mixed density matrix is in the description of quantum mechanical states that, in addition to quantum indeterminacy, also have classical randomness—as is the case for the thermodynamics of a quantum system—and is discussed in Sect. 5.12.

5.4.3 Density Matrix for a Two-State System

The general expression for a ket vector $|\psi\rangle$ of a two-state system, parametrized by the Bloch sphere and discussed later in Sect. 6.3, is given by

$$|\psi\rangle = \cos\left(\frac{\theta}{2}\right)\begin{bmatrix} 1 \\ 0 \end{bmatrix} + e^{i\phi}\sin\left(\frac{\theta}{2}\right)\begin{bmatrix} 0 \\ 1 \end{bmatrix}$$

The density matrix for the pure state is

$$\rho_P = |\psi\rangle\langle\psi| = \frac{1}{2}[\mathbb{I} + i\sum_{i=1}^{3}\hat{n}_i\sigma_i] \; ; \; tr\rho_P^2 = \hat{n}^2 = 1 \qquad (5.12)$$

where the σ_i are the Pauli spin matrices given by

$$\sigma_1 = \begin{bmatrix} 0 & 1 \\ 1 & 0 \end{bmatrix}; \; \sigma_2 = \begin{bmatrix} 0 & -i \\ i & 0 \end{bmatrix}; \; \sigma_3 = \begin{bmatrix} 1 & 0 \\ 0 & -1 \end{bmatrix}; \; \text{Tr}(\sigma_i\sigma_j) = 2\delta_{i-j} \qquad (5.13)$$

The unit vector $\hat{n}$ is an arbitrary three-dimensional vector that lies *on* the Bloch sphere, shown in Fig. 6.1, and is given by

$$\hat{n} = (\sin\theta\cos\phi, \cos\theta\cos\phi, \sin\phi) \; ; \; 0 \le \theta \le \pi \; ; \; 0 \le \phi \le 2\pi$$

A vector lying *inside* the Bloch sphere is given by

$$a\hat{n} \; ; \; a \in [0, 1]$$

It can be shown that the most general two-state mixed density matrix is given by

$$\rho_M = \frac{1}{2}[\mathbb{I} + ia\sum_{i=1}^{3}\hat{n}_i\sigma_i] \; ; \; \text{Tr}\rho_M^2 = a^2\hat{n}^2 = a^2 < 1 \qquad (5.14)$$

For a mixed state, the density matrix is ρ_M with $0 \le a < 1$ and, hence, all the density matrices for *mixed states* lie *inside* the Bloch sphere. For a pure state $\text{Tr}\rho_p^2 = 1$ and which yields $a = 1$. Hence, all the density matrices for *pure states* are on the *surface* of the Bloch sphere.

The two-state density matrix has a major application that is the study of quantum information, in particular on studying the effect of measurements on qubits.

5.5 Reduced Density Matrix

The concept of reduced density matrix can be defined for a system having two or more degrees of freedom. Consider an experiment in which the projection operators for only one of the degrees of freedom are measured, with the projection operators for the other degrees of freedom being completely ignored. Clearly, there is a loss of information regarding the state of the other degrees of freedom. The reduced density matrix provides a precise measure on how much information is lost in such a 'partial' experiment.

Consider a quantum entity with only two different degrees of freedom; the state vectors $|\psi_i^I\rangle$ and $|\psi_i^{II}\rangle$ are state vectors for the two distinct degrees of freedom I and II, respectively. The general state vector, using the Schmidt decomposition discussed in Belal [1], for the system of two distinct degrees of freedom is given as follows

$$|\Psi\rangle = \sum_{i=1}^{N} c_i |\psi_i^I\rangle |\psi_i^{II}\rangle \;\; ; \;\; \sum_i |c_i|^2 = 1$$

$$\langle \psi_i^I | \psi_j^I \rangle = \delta_{i-j} = \langle \psi_i^{II} | \psi_j^{II} \rangle$$

The pure density matrix for the system is given by

$$\rho = |\Psi\rangle\langle\Psi| = \sum_{ij=1}^{N} c_i c_j^* |\psi_i^I\rangle\langle\psi_i^I| \otimes |\psi_i^{II}\rangle\langle\psi_i^{II}| \tag{5.15}$$

If measurements are made on only the degree of freedom I with state vectors ψ_i^I, then the loss of information encoded in state vectors ψ_i^{II} is mathematically realized by performing a partial trace over the II degrees of freedom, as discussed in Sect. 5.3. Performing the partial trace in Eq. 5.15 yields the reduced density matrix ρ_R, namely

$$\rho_R = \mathrm{Tr}_{II}(\rho) = \mathrm{Tr}_{II}\big(|\Psi\rangle\langle\Psi|\big)$$

$$= \sum_{ij=1}^{N} c_i^* c_j |\psi_i^I\rangle\langle\psi_j^I|\Big[\langle\psi_j^{II}|\psi_i^{II}\rangle\Big]$$

$$\Rightarrow \rho_R = \sum_{i=1}^{N} |c_i|^2 |\psi_i^I\rangle\langle\psi_i^I| \tag{5.16}$$

Hence, Eq. 5.16 shows that the loss of information for a pure density matrix, given in Eq. 5.15, yields a reduced density matrix that is a mixed density matrix, defined in Eq. 5.9.

The analysis for the reduced density matrix carried out for discrete degrees of freedom and given in Eq. 5.16 can also be done for continuous degrees of freedom. Consider, for concreteness, a quantum system with two degrees of freedom, for example, two particles with degrees of freedom x_1, x_2 (coordinates in one dimension) respectively, and with state vectors $\psi(x_1, x_2)$. Consider a non-factorizable state vector and its density matrix given by

$$\langle x_1, x_2 | \psi \rangle = \psi(x_1, x_2) \neq \psi_1(x_1)\psi_2(x_2)$$
$$\rho = |\psi\rangle\langle\psi| \;\; ; \;\; \langle x_1, x_2 | \rho | x_1', x_2' \rangle = \psi(x_1, x_2)\psi^*(x_1', x_2') \tag{5.17}$$

One can sum over one of the degrees of the freedom—in general, by performing a partial trace of ρ over a degree of freedom as was done in Eq. 5.5—say over the coordinate x_2 and obtain the *reduced density matrix* ρ_R that provides a complete description for all measurement carried out on only the degree of freedom x_1; in symbols (dropping the subscript 1 on x_1)

$$\rho_R = \text{Tr}_2\rho = \text{Tr}_2(|\psi\rangle\langle\psi|) = \int dx \rho_R(x, x')|x\rangle\langle x'|$$
$$\rho_R(x, x') = \langle x | \rho_R | x' \rangle = \int dx_2 \psi(x, x_2)\psi^*(x', x_2) \tag{5.18}$$

Equation 5.18 shows that the reduced density matrix provides a quantum mechanical generalization of the concept of marginal distribution of the classical theory of probability [3].

5.6   Separable Quantum Systems

A *separable* quantum system is defined to a system in which the degrees of freedom can be *exactly factorized* in the sense that the state vector for the degrees of freedom is a tensor product, as given below

$$|\psi\rangle = |\psi_I\rangle|\psi_{II}\rangle \;\; ; \;\; \langle x_1, x_2 | \psi \rangle = \psi_I(x_1)\psi_{II}(x_2) \tag{5.19}$$

and yields the pure density matrix given by

$$\rho_P = |\psi\rangle\langle\psi| = |\psi_I\rangle\langle\psi_I| \otimes |\psi_{II}\rangle\langle\psi_{II}|$$

The reduced density matrix, as discussed in Eq. 5.16, is obtained by performing a partial trace over the x_2 degree of freedom, and for the separable quantum system is given by[4]

$$\rho_{P,R} = \text{Tr}_2\big(|\psi\rangle\langle\psi|\big) = |\psi_I\rangle\langle\psi_I|\big(\langle\psi_{II}|\psi_{II}\rangle\big)$$
$$= |\psi_I\rangle\langle\psi_I| \; ; \; \text{Tr}(\rho_{P,R}) = 1 = \text{Tr}(\rho_{P,R}^2)$$

In other words, the reduced density matrix of a separable system is *also* a pure density matrix.

Consider two different systems with their own degrees of freedom with density matrices ρ_i^A and ρ_i^B such that

$$\text{Tr}(\rho_i^A) = 1 = \text{Tr}(\rho_i^B)$$

One can think of the density matrices as projection operators for the two different systems. A general representation of a composite system consisting of two *separable subsystems* is given by the following bipartite (mixed) density matrix

$$\rho_{AB} = \sum_{i=1}^{N} p_i \rho_i^A \otimes \rho_i^B \; \Rightarrow \; \text{Tr}(\rho_{AB}) = \sum_{i=1}^{N} p_i = 1 \; ; \; p_i \in [0, 1] \quad (5.20)$$

It is the condition of $\sum_{i=1}^{N} p_i = 1$ that implies that the system is separable, with a complete description of system A and B being contained solely in ρ_i^A and ρ_i^B, respectively. The bipartite density matrix represents a *separable quantum system* for which the degrees of freedom for A and B can be considered in isolation from each other. In other words, one can unambiguously separately measure the degrees of freedom for A and B and still obtain the correct result for the expectation value of all observables pertaining to only one of the systems.

The reduced density matrix for the separable system is given by

$$\rho_{A,R} = \text{Tr}_B(\rho_{AB}) = \sum_{i=1}^{N} p_i \rho_i^A \; ; \; \rho_{B,R} = \text{Tr}_A(\rho_{AB}) = \sum_{i=1}^{N} p_i \rho_i^B \quad (5.21)$$

5.7 Entangled Quantum States

In classical mechanics, the point particles obeying Newton's laws are *always* distinct entities. In contrast, the distinct 'identity' of a particular quantum mechanical degree of freedom is only meaningful for special cases.

---

[4] A similar result holds for taking a partial trace over the x_1 degree of freedom.

More precisely, if the state vector for two degrees of freedoms can be completely *factorized*, namely if the joint state vector is a tensor product of the individual state vectors of each degree of freedom, then one of the degrees of freedom can be observed independently of the other. However, if the joint state vectors cannot be factorized, which are called *entangled states*, the two degrees of freedom become *inseparable*, and one cannot consider either of the degrees of freedom independently of the other.

For example, the degrees of freedom of the state vector in Eq. 5.17 do not factorize, and hence, the two degrees of freedom cannot be studied in isolation: the expectation values for degree of freedom x_1 depend on the behavior of degree of freedom x_2. This is an example of an entangled state and indicates that the (two) degrees of freedom are inseparable, and the state given in Eq. 5.17 is an entangled state.

One needs a quantum system with two or more degrees of freedom to obtain an entangled state.

An entangled state vector does not have any dynamics, and the property of entanglement is purely kinematic; namely it pertains entirely to the structure of the state vector and not to how it evolves in time (dynamics). The quantum entity represented by an entangled state does not exist in classical physics and shows the rich structure of quantum mechanics.

Note that the basis states of state space are only defined up to a unitary transformation [3]. Hence, a state vector that is apparently not separable could, in fact, be separable if the basis states are transformed to a new basis. To provide a precise *basis independent* formulation of entangled states, one needs to express the quantum system in the language of the density matrix. Just such a general criterion is provided by the reduced density matrix and is derived below.

In the Schmidt decomposition, an entangled state vector of two degrees of freedom is given by [1]

$$|\Psi_E\rangle = \sum_{i=n} c_i |\psi_i^I\rangle |\psi_i^{II}\rangle \tag{5.22}$$

where, in general, c_i can depend on the state vector $|\psi_i^I\rangle$. The *pure density matrix* for the state vector $|\Psi_E\rangle$ given in Eq. 5.22 is the following

$$\rho_E = |\Psi_E\rangle\langle\Psi_E| = \sum_{ij=1}^N c_i c_j^* |\psi_i^I\rangle\langle\psi_j^I| \otimes |\psi_i^{II}\rangle\langle\psi_j^{II}| \;\; ; \;\; \mathrm{Tr}(\rho_P^2) = 1$$

As was the case for Eq. 5.16, performing a partial trace over the degree of freedom II yields, from Eqs. 5.5 and 5.22, the *reduced density matrix* for the entangled state as follows

$$\rho_{E,R} = \text{Tr}_{II}\Big(|\Psi_E\rangle\langle\Psi_E|\Big) = \sum_{i=1}^{N} |c_i|^2 |\psi_i^I\rangle\langle\psi_i^I| \tag{5.23}$$

$$\text{Tr}(\rho_{E,R}^2) = \sum_{i=1}^{N} |c_i|^4 < 1 \tag{5.24}$$

$\text{Tr}(\rho_{E,R}^2) < 1$ is a basis independent result, since a unitary change of basis leaves $\text{Tr}(\rho_{E,R}^2)$ invariant. $\text{Tr}(\rho_{E,R}^2) < 1$ leads to the conclusion that the state $|\Psi_E\rangle$ itself cannot be written, in *any* set of basis states, as a product state $|\psi_I\rangle|\psi_{II}\rangle$. This is because a partial trace of the product state would lead to a reduced matrix ρ_{ER} that would be a pure density matrix – and thus contradict the result that $\text{Tr}(\rho_{ER}^2) < 1$, obtained in Eq. 5.24.

In conclusion, for $c_i \neq 0$ and $N > 1$, we have

$$|\Psi_E\rangle = \sum_{i=1}^{N} c_i |\psi_i^I\rangle|\psi_i^{II}\rangle \neq |\chi_I\rangle|\chi_{II}\rangle : \text{ Entangled}$$

$|\Psi_E\rangle$ is an entangled state; in general, the two or more degrees of freedom for an entangled state need to be treated as one non-decomposable and inseparable system, with the identities of the individual degrees of freedom, taken in isolation, being meaningless. In contrast, for a separable system, each degree of freedom can be considered to be a distinct entity and separate from the other degrees of freedom.

5.8 Entanglement for Composite Systems

The criterion of entanglement for a pure density matrix $\rho_p = |\psi\rangle\langle\psi|$ is given by examining its reduced density matrix ρ_R; if $\text{Tr}(\rho_R^2) < 1$, then the state $|\psi\rangle$ is entangled. This criterion does not hold for density matrix of systems that are the composite of two or more different systems. In particular for bipartite states

$$\rho_{AB} = \sum_{i=1}^{N} p_i \rho_i^A \otimes \rho_i^B \Rightarrow \text{Tr}\big((\rho_{AB})^2\big) = \sum_{i=1}^{N} p_i^2 < 1$$

Although one has $\text{Tr}((\rho_{AB})^2) < 1$, this *does not* necessarily imply that either system A or B is entangled. Separable systems have been proven to satisfy, using definitions given in Eq. 5.21, the following two inequalities

$$\mathbb{I}_A \otimes \rho_{B,R} - \rho_{AB} \geq 0 \; ; \; \rho_{A,R} \otimes \mathbb{I}_B - \rho_{AB} \geq 0$$

The operator inequality means that all the eigenvalues of the operator are non-negative. If any one of these two conditions are violated, then ρ_{AB} represents a composite system that is entangled. This is called the reduction criterion [4].

5.9 Entangled State: Two Binary Degrees of Freedom

Consider a pair of binary degrees of freedom (two state systems), which could be equal to two values of classical bits, with basis states $|u_1\rangle, |d_1\rangle$ and $|u_2\rangle, |d_2\rangle$ defined by

$$|u_1\rangle = \begin{bmatrix} 1 \\ 0 \end{bmatrix}_1 \quad ; \quad |d_1\rangle = \begin{bmatrix} 0 \\ 1 \end{bmatrix}_1 \quad : \quad |u_2\rangle = \begin{bmatrix} 1 \\ 0 \end{bmatrix}_2 \quad ; \quad |d_2\rangle = \begin{bmatrix} 0 \\ 1 \end{bmatrix}_2$$

A general expression for a *separable* product state for the pair of spins is the following (dropping the subscript on state vector since notation does not need it)

$$|\Psi_S\rangle = \left[a|u_1\rangle + b|d_1\rangle \right] \left[\alpha|u_2\rangle + \beta|d_2\rangle \right] \quad ; \quad |a|^2 + |b|^2 = 1 = |\alpha|^2 + |\beta|^2$$

In contrast, an example of an *entangled state* for the two spins, using the rules of tensor product of vectors given in Sect. 2.6, is given by

$$|\Psi_E\rangle = a|u_1\rangle|d_2\rangle + b|d_1\rangle |u_2\rangle \quad ; \quad |a|^2 + |b|^2 = 1 \tag{5.25}$$

$$= a \begin{pmatrix} 1 \\ 0 \end{pmatrix} \otimes \begin{pmatrix} 0 \\ 1 \end{pmatrix} + b \begin{pmatrix} 0 \\ 1 \end{pmatrix} \otimes \begin{pmatrix} 1 \\ 0 \end{pmatrix} = \begin{pmatrix} 0 \\ a \\ b \\ 0 \end{pmatrix}$$

The entangled state vector $|\Psi_E\rangle$ has been studied extensively and plays a central role in the EPR paradox as well as in empirical tests of Bell's theorem [1].

The proof that Eq. 5.25 is an entangled state requires the evaluation of the reduced density matrix. The density matrix is given by a tensor (outer) product and, using the rules given in Sect. 2.6, yields

$$\rho_E = |\Psi_E\rangle\langle\Psi_E|$$
$$= |a|^2|u_1\rangle\langle u_1| \otimes |d_2\rangle\langle d_2| + |b|^2|d_1\rangle\langle d_1|\otimes)|u_2\rangle\langle u_2| + \text{off-diagonal} \tag{5.26}$$
$$= \begin{pmatrix} 0 & 0 & 0 & 0 \\ 0 & |a|^2 & ab^* & 0 \\ 0 & a^*b & |b|^2 & 0 \\ 0 & 0 & 0 & 0 \end{pmatrix}$$

The *reduced density matrix* is defined by taking the partial trace over the degree of freedom of the second spin; under the partial trace, the off-diagonal terms in Eq. 5.26 are all zero. The result is the following

$$\rho_{ER} = tr_2(\rho_E)$$
$$= |a|^2 |u_1\rangle\langle u_1| + |b|^2 |d_1\rangle\langle d_1| \qquad (5.27)$$
$$= \begin{bmatrix} |a|^2 & 0 \\ 0 & |b|^2 \end{bmatrix}$$

Taking the trace of the reduced matrix over the degree of freedom of the first spin yields, from Eq. 5.27, the following

$$tr\rho_{ER} = |a|^2 + |b|^2 = 1 \quad : \quad \text{Normalization}$$
$$tr(\rho_{ER}^2) = |a|^4 + |b|^4 = 1 - 2|ab|^2 < 1 \qquad (5.28)$$

Note if either a or b is zero, Eq. 5.28 shows that there is no entanglement, as indeed is the case since the state vector given in Eq. 5.25 becomes a product state and is separable.

Hence, we conclude from Eq. 5.28 that, since the reduced density matrix $tr(\rho_{ER}^2) <$ 1, the state vector given in Eq. 5.25 is entangled, namely

$$|\Psi_E\rangle = a|u_1\rangle|d_2\rangle + b|d_1\rangle |u_2\rangle : \quad \text{Entangled}$$

5.10 Quantum Entropy

Entropy is a measure of the ignorance regarding a system. The concept of entropy in statistical physics has a natural analog for quantum systems S and, following von Neumann, is defined as follows

$$S = -\text{Tr}(\rho \ln \rho) = -\sum_{i=1}^{N} p_i \ln p_i \qquad (5.29)$$
$$\rho = U\text{diag}(p_1, p_2, \ldots, p_N)U^\dagger \ ; \ \ UU^\dagger = \mathbb{I}$$

Consider a pure state with $p_1 = 1$ and $p_i = 0$; $i \neq 1$; then

$$\rho = |\psi\rangle\langle\psi| \ \Rightarrow \ S = -\text{Tr}(\rho \ln \rho) = 0$$

A pure state yields zero entropy since, as expected, there is no ignorance in knowing the state of the system. In contrast to a pure state, if one has no information about a system, then one expects that entropy should be a maximum.

The entropy of a mixed state, from Eqs. 5.9 and 5.29, is the following

$$S = -\text{Tr}(\rho_M \ln \rho_M) = -\sum_{i=1}^{N} p_i \ln p_i$$
$$\rho_M = V\text{diag}(p_1, p_2, \ldots, p_N)V^\dagger \ ; \ \ VV^\dagger = \mathbb{I}$$

To find the density matrix that yields a maximum value of entropy S, we maximize S with respect to all the p_i's, with the constraint that $\sum_{i=1}^{N} p_i = 1$; using Lagrange multiplier λ yields the maximization problem

$$L = S + \lambda[\sum_{i=1}^{N} p_i - 1]$$

$$0 = \frac{\partial L}{\partial p_I} = -k_B(\ln p_I + 1) + \lambda \quad \Rightarrow \quad p_I = \text{constant}$$

$$0 = \frac{\partial L}{\partial \lambda} = \sum_{i=1}^{N} p_i - 1 \quad \Rightarrow \quad p_I = \frac{1}{N}$$

The result above shows that maximum entropy state is one for which all the states are equally likely. The fact that all states are equally likely is precisely what one expects for a system about which one is totally ignorant.

The density matrix is proportional to the identity operator $\mathbb{I}$ since, due to the completeness equation, we have

$$\sum_{i=1}^{N} |\psi_i\rangle\langle\psi_i| = \mathbb{I}$$

Hence, for a N-state maximally uncertain system

$$\rho_{\text{max}} = \frac{1}{N}\mathbb{I} \quad \Rightarrow \quad \text{Tr}(\rho_{\text{max}}) = 1$$

$$S_{\text{max}} = -\text{Tr}(\rho \ln \rho) = \frac{1}{N} \ln(N)\text{Tr}(\mathbb{I})$$

$$\Rightarrow S = \ln(N) \quad : \quad \text{maximum entropy} \tag{5.30}$$

5.11   Maximally Entangled States

For a state vector with two degrees of freedom consider the following entangled state in the Schmidt representation [1]

$$|\Psi_E\rangle = \sum_{i=1}^{N} c_i|\psi_i^I\rangle|\psi_i^{II}\rangle \; ; \quad \sum_{i=1}^{N} |c_i|^2 = 1$$

$$\langle\psi_i^I|\psi_j^I\rangle = \delta_{i-j} = \langle\psi_i^{II}|\psi_j^{II}\rangle$$

that yields, from Eq. 5.23, the reduced density matrix for the entangled state as follows

$$\rho_{ER} = \text{Tr}_2\Big(|\Psi_E\rangle\langle\Psi_E|\Big) = \sum_{i=1}^{N} |c_i|^2 |\psi_i^I\rangle\langle\psi_i^I|$$

The maximally entangled state has the maximum entropy and hence yields

$$\rho_{ER}\Big|_{\text{Maximal}} = \left[\sum_{i=1}^{N} |c_i|^2 |\psi_i^I\rangle\langle\psi_i^I|\right]_{\text{Maximal}} = \frac{1}{N}\mathbb{I} \tag{5.31}$$

Since the completeness of the eigenfunctions of a Hermitian operator gives a resolution of the identity operator,[5] one obtains

$$|c_i|^2 = \frac{1}{N} \;\Rightarrow\; c_i = \frac{1}{\sqrt{N}}e^{i\phi_i}$$

and yields the *maximally entangled state* given by[6]

$$|\Psi_E\rangle = \frac{1}{\sqrt{N}}\sum_{i=1}^{N} e^{i\phi_i}|\psi_i^I\rangle|\psi_i^{II}\rangle \tag{5.32}$$

An example of a pair of maximally entangled spins is given by the following density matrix of a non-separable system

$$\rho_{NS} = \frac{1}{4}\mathbb{I}\otimes\mathbb{I} \;;\; \text{Tr}(\rho_{NS}) = 1 \;;\; \mathbb{I} = \begin{bmatrix} 1 & 0 \\ 0 & 1 \end{bmatrix}$$

The reduced density matrix shows that the pair of spins is entangled since

$$\rho_{NS,R} = \text{Tr}_2(\rho_{NS}) = \frac{1}{2}\mathbb{I} \;;\; \text{Tr}(\rho_{NS,R}^2) = \frac{1}{2} \;:\; \text{Maximally entangled}$$

5.11.1 An Entangled State of Two Binary Degrees of Freedom

An entangled state for two spin degrees of freedom, from Eq. 5.25, is given by

$$|\Psi_E\rangle = a|u_1\rangle|d_2\rangle + b|d_1\rangle|u_2\rangle \;;\; |a|^2 + |b|^2 = 1 \tag{5.33}$$

---

[5] Namely $\sum_{i=1}^{N} |\psi_i^I\rangle\langle\psi_i^I| = \mathbb{I}$.

[6] The maximally entangled state is the same whether the partial trace is performed over quantum system I or system II.

with the reduced density matrix, from Eq. 5.27, given by

$$
\begin{aligned}
\rho_{ER} &= \mathrm{Tr}_2(\rho_E) \\
&= |a|^2 |u_1\rangle\langle u_1| + |b|^2 |d_1\rangle\langle d_1| \\
&= \begin{bmatrix} |a|^2 & 0 \\ 0 & |b|^2 \end{bmatrix} \\
&\Rightarrow p_1 = |a|^2 \; ; \; p_2 = |b|^2
\end{aligned}
\tag{5.34}
$$

Hence, from Eq. 5.29, the entropy of this state is given by

$$
\begin{aligned}
S &= -\mathrm{Tr}(\rho_{ER} \ln \rho_{ER}) = -p_1 \ln p_1 - p_2 \ln p_2 \tag{5.35} \\
&= -|a|^2 \ln(|a|^2) - |b|^2 \ln(|b|^2) \tag{5.36}
\end{aligned}
$$

For the following special case, and from Eq. 5.30

$$
\begin{aligned}
|a| &= \frac{1}{\sqrt{2}} = |b| \\
&\Rightarrow S = \ln(2) : \text{Maximum entropy}
\end{aligned}
\tag{5.37}
$$

Hence, a *maximally entangled* state of two spins is given by

$$
|\Psi_E\rangle = \frac{1}{\sqrt{2}}\left[e^{i\phi}|u_1\rangle|d_2\rangle + |d_1\rangle\,|u_2\rangle \right]
\tag{5.38}
$$

5.12 Pure and Mixed Density Matrix

The density matrix, introduced in Sect. 5.4, is a Hermitian operator closely related to the state vector; recall from Eq. 5.6 that the pure density matrix for a state vector $|\chi\rangle$ is defined by

$$
\rho_P = |\chi\rangle\langle\chi|
\tag{5.39}
$$

The measurement of the expectation value of observable $\mathcal{O}$ can be expressed in terms of the density matrix of a pure state ρ_p as follows

$$
E_\chi[\mathcal{O}] \equiv \langle\chi|\mathcal{O}|\chi\rangle = \mathrm{Tr}(\mathcal{O}|\chi\rangle\langle\chi|) = \mathrm{Tr}(\mathcal{O}\rho_P)
\tag{5.40}
$$

The expectation value of an operator $\mathcal{O}$ with eigenvectors $\mathcal{O}|\psi_i\rangle = \lambda_i|\psi_i\rangle$ can be rewritten in terms of the mixed density matrix ρ_M as follows

$$E_\psi[\mathcal{O}] = \langle \psi | \hat{O} | \psi \rangle = \mathrm{Tr}(\sum_i |c_i|^2 \mathcal{O} |\psi_i\rangle\langle\psi_i|) = \mathrm{Tr}(\mathcal{O}\rho_M)$$

$$\Rightarrow \rho_M = \sum_i p_i |\psi_i\rangle\langle\psi_i| \; ; \; p_i = |c_i|^2$$

The mixed density matrix ρ_M can be used for evaluating the expectation value of any function of the operator $\mathcal{O}$. However, if one uses ρ_M for evaluating the expectation value of *another* operator Q that does not commute with $\mathcal{O}$, namely $[\mathcal{O}, Q] \neq 0$, then there are *unavoidable* errors. The magnitude of these errors is set by the Heisenberg Uncertainty Principle and is discussed in Belal [1].

Consider a quantum mechanical system to be in thermal equilibrium with a heat bath at temperature T. The system now has a quantum mechanical indeterminacy as well as classical uncertainty due to thermal randomness. The behavior of the quantum system is described by the canonical ensemble's probability distribution of energy eigenstates—given by the Boltzmann distribution.

Let H be the quantum mechanical Hamiltonian with the following spectral decomposition in terms of the energy eigenfunctions $|\psi_i\rangle$ and eigenvalues E_i

$$H = \sum_i E_i |\psi_i\rangle\langle\psi_i|$$

A quantum system with *thermal uncertainty* is described by the density matrix ρ_T given by

$$\rho_T = \frac{1}{Z} e^{-H/k_B T} = \frac{1}{Z} \sum_i e^{-E_i/k_B T} |\psi_i\rangle\langle\psi_i| \; ; \; Z = \mathrm{Tr}\, e^{-H/k_B T} \qquad (5.41)$$

$$\Rightarrow \rho_T = \sum_i p_i |\psi_i\rangle\langle\psi_i| \; ; \; \mathrm{Tr}(\rho_T) = \sum_i p_i = 1 \; ; \; p_i = \frac{1}{Z} e^{-E_i/k_B T}$$

where k_B is the Boltzmann constant. The thermal density matrix ρ_T for the canonical ensemble is a mixed state since

$$\mathrm{Tr}(\rho_T^2) = \sum_i p_i^2 < 1$$

The reason that ρ_T is a mixed state is that thermal randomness leads to a *classical uncertainty* in the state of the system; this in turn entails that all the quantum state vectors $|\psi_i\rangle$ must be decoherent since there are no quantum correlations between the different quantum states—unlike the case for a pure density matrix ρ_p that has off-diagonal terms as given in Eq. 5.8.

The expectation value of an operator $\mathcal{O}$, for which $[\mathcal{O}, H] \neq 0$, and that is in equilibrium with a heat bath is given by

$$E_T[\mathcal{O}] = \mathrm{Tr}(\mathcal{O}\rho_T) = \sum_i p_i \langle \psi_i | \mathcal{O} | \psi_i \rangle$$

$$= \sum_i p_i \alpha_i \; ; \; \alpha_i = \langle \psi_i | \mathcal{O} | \psi_i \rangle$$

The thermal density matrix ρ_T encodes both thermal and quantum uncertainty, reflected in the probability p_i that the quantum system is in eigenstate E_i and the expectation value α_i of the operator in this eigenstate.

References

1. Belal E (2013) Baaquie. The theoretical foundations of quantum mechanics. Springer, UK
2. Feynman RP (2007) The character of physical law. Penguin Books, USA
3. Baaquie BE (2020) Mathematical methods and quantum mathematics for economics and finance. Springer, Singapore
4. Nielsen MA, Chang IL (2000) Quantum computation and quantum information. Cambridge University Press, UK

Chapter 6
Binary Degrees of Freedom and Qubits

6.1 Introduction

In this chapter, we review the notations used in quantum mechanics and discuss some fundamental ideas of quantum mechanics within the context of quantum computers. The underlying principles of quantum mechanics and the quantum theory of measurement have been discussed in Chap. 4. A few highlights of the chapter are reviewed in the context of qubits.

Consider n-binary degrees of freedom with computational basis states given by $|x_0, x_1, \ldots, x_{n-1}\rangle$. A quantum algorithm provides an answer to the input qubits and which is expressed in the final output state. The answer being sought has to be extracted by measurements performed on the final output state. Measurements are carried out using an experimental device and represented by projection operators that represent the physical construction of irreversible measurement gates

$$|x_0, x_1, \ldots, x_{n-1}\rangle\langle x_0, x_1, \ldots, x_{n-1}| \; ; \; x_i = 0, 1 \; : \; i = 1, 2, \ldots, n$$

The quantum system carrying the qubit $|q\rangle = |q(x_0, x_1, \cdots, x_{n-1})\rangle$ is subjected to a measurement in which *all the measurements gates* are applied on the system, and the outcome of the measurement is that some definite and determinate measurement gate register a change. This change is a signal that the gate has been triggered by the measurement; which measurement gates register a signal is completely random.

A list is made of the frequency with which the various measurement gates register a signal. The state vector

$$|q\rangle \; \Rightarrow \; q(x_0, x_1, \ldots, x_{n-1}) = \langle x_0, x_1, \ldots, x_{n-1}|q\rangle$$

determines the probability, which is given by the normalized frequency distribution of the various measurement gates registering a signal and is equal to

$$|q(x_0, x_1, \ldots, x_{n-1})|^2$$

© The Author(s), under exclusive license to Springer Nature Singapore Pte Ltd. 2023
B. E. Baaquie and L.-C. Kwek, *Quantum Computers*,
https://doi.org/10.1007/978-981-19-7517-2_6

6.2 Degrees of Freedom and Qubits

This section is a review of the main results of Chap. 4. The quantum gates, circuits and qubits, based on the discussion above, are summarized with some repetitions for greater clarity:

- The classical bit taking values of 0 and 1 is identified as the underlying *binary degree of freedom* of the quantum system of a quantum computer.
- Recall, as discussed in Sect. 4.2, that the degree of freedom is intrinsically indeterminate, with the binary degree of freedom *simultaneously* taking both the values 0 and 1.
- The quantum state vector (wave function) of a quantum computer is a complex-valued function of the degree of freedom that yields the likelihood of observing the different determinate values of the degree of freedom.
- The quantum state of a single binary degree of freedom is called a *qubit* and given in Eq. 4.3. A qubit is a quantum *superposition* of the 0 and 1 binary degrees of freedom.
- The qubit encodes the likelihood of observing 0 or 1 if a measurement is performed on it.
- The single classical bit is replaced by a single qubit.
- All quantum gates are reversible (unitary) transformations.
- In a quantum circuit, a single qubit is represented by a single horizontal line and gates by various symbols acting on the qubits.
- The quantum circuit is defined by a sequence of unitary gates applied to the input string state vector of n-binary degrees of freedom to obtain the final output state vector.
- Once an input string is specified, the computational processes are carried out by transforming the initial string through a series of gates (more on this later)—using the laws of quantum mechanics.
- During the computational process, the classical bit is superseded by the qubits, denoted by $|\psi\rangle$, which consists of a *superposition* of the basis states of the computational basis.

In principle, for a quantum computer, the input and output string state vector can be expressed in many ways. But, for simplicity, we work in the computational basis in which the input and output strings are expressed in terms of the n-binary degrees of freedom $\{0, 1\}^n$ given in Eq. 2.26. For a quantum computer, the input is a determinate string and the output is also a determinate string. A fundamental difference between a classical and quantum computer is that the output string of a quantum computer is random and uncertain—and can only be obtained after performing a quantum measurement. The output string state is obtained as the average of repeated measurements of the same algorithm run many times—with identical preparation. The central role of measurement in quantum mechanics is discussed in Sect. 4.5.

As given in Eq. 4.4, the state vector $|\psi\rangle$ of the n-binary degrees of freedom can be represented in the computational basis $|x\rangle, x = 0, 1, 2, \ldots, 2^n - 1$ by the following expansion

$$|\psi\rangle = \sum_{x=0}^{2^n-1} \alpha_x |x\rangle \; ; \; \sum_{x=0}^{2^n-1} |\alpha_x|^2 = 1 \tag{6.1}$$

The state vector $|\psi\rangle$ is called the qubits of the n-binary degrees of freedom and is the **superposition** of the computational basis states.

The qubits come into play in the following manner. Once the initial state is provided—and until the output is ready—the transformations inside a quantum computer are all performed on the qubits: the process of quantum computation *cannot* be observed as it will destroy the process. Suppose there are many qubits required for the algorithm; the quantum computer will simultaneously update all the qubits—sometimes called quantum parallelism. During the entire computational process, all the qubits are in an indeterminate state that consists of the superposition of the qubit states.

Once the computation is completed, the output string is in an indeterminate state. The process of measurement *collapses* the indeterminate output state—with a certain probability—to a specific and determinate output string. The quantum mechanical measurement can, in principle, result in a whole range of possible outcomes, with the output being in a specific determinate string having only a certain likelihood of occurrence.

6.3 Single Qubit

We consider the special cases of $n = 1$ and $n = 2$ as they play a key role in quantum algorithms. For $n = 1$, the computational basis is given by

$$|0\rangle \; ; \; |1\rangle$$

From Eq. 6.1, the fundamental state vector, corresponding to the classical 1-bit, is given by

$$|\psi\rangle \equiv |q\rangle = \alpha|0\rangle + \beta|1\rangle \; ; \; |\alpha|^2 + |\beta|^2 = 1$$

The special notation $|q\rangle$ is used for the state vector of a single binary degree of freedom and is called the *qubit*. The crux of the difference between a classical and a quantum computer is the generalization of the definition of a bit to a *qubit*. Deutsch's 1985 paper showed that the fundamental object for quantum algorithms is not the classical bit, but instead the qubit.

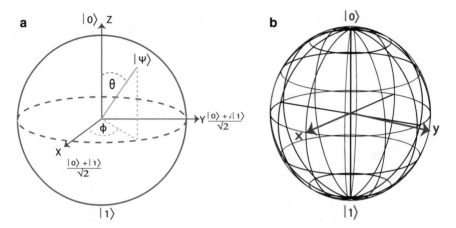

In column vector notation, a qubit is given by the following

$$|q\rangle = \alpha|0\rangle + \beta|1\rangle = \alpha\begin{bmatrix} 1 \\ 0 \end{bmatrix} + \beta\begin{bmatrix} 0 \\ 1 \end{bmatrix} \tag{6.2}$$

In general, $\alpha = e^{i\chi}a$, $\beta = e^{i\phi}b$ are in general complex numbers, with a, b real. The qubit can only be measured up to a global phase, and hence, we can set $\chi = 0$ and yields

$$|q\rangle = a|0\rangle + e^{i\phi}b|1\rangle \; ; \; a^2 + b^2 = 1$$

To summarize our earlier discussion on qubits, the qubits are quantum mechanical state vectors that are a function of the binary degree of freedom 0, 1 that is *essentially indeterminate*, having no intrinsic value: the qubit is a quantum state vector that is the *superposition* of the single bit states. Metaphorically speaking, the qubit has the probability of being in the two distinct states given by a^2, b^2, respectively. The superposed state of qubit is forbidden in a classical computer, and it has an interpretation that requires the quantum theory of measurement.

For the probabilistic interpretation of quantum mechanics, the norm of a qubit is equal to 1, which implies $\langle q|q\rangle = 1$; hence

$$\langle q|q\rangle = 1 \implies a^2 + b^2 = 1 \implies a = \cos\left(\frac{\theta}{2}\right) \; ; \; b = \sin\left(\frac{\theta}{2}\right)$$

and yields the general single qubit

$$|q(\theta, \phi)\rangle = \cos\left(\frac{\theta}{2}\right)|0\rangle + e^{i\phi}\sin\left(\frac{\theta}{2}\right)|1\rangle = \cos\left(\frac{\theta}{2}\right)\begin{bmatrix}1\\0\end{bmatrix} + e^{i\phi}\sin\left(\frac{\theta}{2}\right)\begin{bmatrix}0\\1\end{bmatrix}$$
(6.3)

The single qubit is a great enhancement of the 1-bit. The angles θ, ϕ parametrize a two-dimensional sphere, called the Bloch sphere, for the qubit and shown in Fig. 6.1. Every point on the surface of Bloch sphere is a possible (pure) state vector of the qubit.

6.3.1 Density Matrix

There is another way to visualize the qubit: by expressing it in terms of a density matrix. For any pure state $|q\rangle$, one can associate a density matrix $\rho \equiv |q\rangle\langle q|$ with the state and is given by

$$
\begin{aligned}
|\rho\rangle = |q\rangle\langle q| &= \begin{bmatrix}\cos(\frac{\theta}{2})\\e^{i\phi}\sin(\frac{\theta}{2})\end{bmatrix}\begin{bmatrix}\cos(\frac{\theta}{2}) & e^{-i\phi}\sin(\frac{\theta}{2})\end{bmatrix}\\
&= \begin{bmatrix}\cos^2(\frac{\theta}{2}) & e^{-i\phi}\cos(\frac{\theta}{2})\sin(\frac{\theta}{2})\\e^{i\phi}\cos(\frac{\theta}{2})\sin(\frac{\theta}{2}) & \sin^2(\frac{\theta}{2})\end{bmatrix}\\
&= \frac{1}{2}\begin{bmatrix}1+\cos\theta & e^{-i\phi}\sin\theta\\e^{i\phi}\sin\theta & 1-\cos\theta\end{bmatrix}\\
&= \frac{1}{2}\left(\mathbb{I} + \sin\theta\cos\phi\begin{bmatrix}0 & 1\\1 & 0\end{bmatrix} + \sin\theta\sin\phi\begin{bmatrix}0 & -i\\i & 0\end{bmatrix} + \cos\theta\begin{bmatrix}1 & 0\\0 & -1\end{bmatrix}\right)\\
&= \frac{1}{2}\left(\mathbb{I} + \sin\theta\cos\phi\,\sigma_x + \sin\theta\sin\phi\,\sigma_y + \cos\theta\,\sigma_z\right)
\end{aligned}
$$

where σ_x, σ_y and σ_z are Pauli's spin matrices given by

$$\sigma_x = \begin{bmatrix}0 & 1\\1 & 0\end{bmatrix}, \quad \sigma_y = \begin{bmatrix}0 & -i\\i & 0\end{bmatrix}, \quad \sigma_z = \begin{bmatrix}1 & 0\\0 & -1\end{bmatrix}.$$
(6.4)

One can rewrite the density matrix as follows:

$$\rho = |q\rangle\langle q| = \tfrac{1}{2}(\mathbb{I} + \mathbf{n}\cdot\boldsymbol{\sigma}) \quad \mathbf{n}\cdot\mathbf{n} = 1$$

where the unit vector $\mathbf{n}$ is given by

$$\mathbf{n} = \begin{bmatrix}n_x\\n_y\\n_z\end{bmatrix} = \begin{bmatrix}\sin\theta\cos\phi\\\sin\theta\sin\phi\\\cos\theta\end{bmatrix}, \quad \sigma = \begin{bmatrix}\sigma_x\\\sigma_y\\\sigma_z\end{bmatrix}.$$

Note that σ is a vector of Pauli matrices. With this parametrization, we see that the vector $\mathbf{n}$ can be represented by a point on a unit sphere (Bloch sphere) with azimuthal angle θ and polar angle ϕ.

Fully distinguishable and distinct states lie on antipodal points of the sphere. Two special distinct states of a single 1-bit are simply *two points* on the Bloch sphere, which in our parametrization are the North Pole $|0\rangle$ and the South Pole $|1\rangle$. Some special cases of the qubit q are shown in Fig. 6.1.

- Classical bits:

$$|q(0,0)\rangle = |0\rangle \;\; ; \;\; |q(\pi,0)\rangle = |1\rangle$$

- Superposed states:

$$|q(\pi/2,0)\rangle = \frac{1}{\sqrt{2}}[|0\rangle + |1\rangle] \;\; ; \;\; |q(\pi/2,\pi/2)\rangle = \frac{1}{\sqrt{2}}[|0\rangle + i|1\rangle]$$

6.4 Bell Entangled Qubits

The case of $n = 2$ consists of two binary degrees of freedom $\{0,1\} \otimes \{0,1\}$. The computational basis yields the following basis states

$$|00\rangle \;\; ; \;\; |01\rangle \;\; ; \;\; |10\rangle \;\; ; \;\; |11\rangle$$

A general expansion of the two qubits state vector is given by

$$|\psi\rangle = \alpha|00\rangle + \beta|01\rangle + \gamma|10\rangle + \delta|11\rangle \tag{6.5}$$
$$|\alpha|^2 + |\beta|^2 + |\gamma|^2 + |\delta|^2 = 1$$

A single degree of freedom $\{0,1\}$ yields superposed states but has no entangled states since for entangled quantum states, one needs *at least two degrees of freedom*. Hence, for $n = 2$, we have a set of entangled states, called the Bell states, that in many cases are more convenient to use as basis states in place of the computational basis states. The Bell states are given by

$$|B_1\rangle = \frac{1}{\sqrt{2}}(|00\rangle + |11\rangle) \;\; ; \;\; |B_2\rangle = \frac{1}{\sqrt{2}}(|00\rangle - |11\rangle)$$
$$|B_3\rangle = \frac{1}{\sqrt{2}}(|01\rangle + |10\rangle) \;\; ; \;\; |B_4\rangle = \frac{1}{\sqrt{2}}(|01\rangle - |10\rangle)$$

A succinct notation for the Bell states is given by

$$|xy\rangle_B = \frac{1}{\sqrt{2}}(|0y\rangle + (-1)^x|1\bar{y}\rangle) \; ; \; x, y = 0, 1$$

where $\bar{y}$ is NOT y.

6.5 Bell States: Maximally Entangled

Consider the basis states $|u_1\rangle, |d_1\rangle$ and $|u_2\rangle, |d_2\rangle$ being defined by

$$|0\rangle_1 = |u_1\rangle = \begin{bmatrix} 1 \\ 0 \end{bmatrix}_1 \; ; \; |1\rangle_1 = |d_1\rangle = \begin{bmatrix} 0 \\ 1 \end{bmatrix}_1$$

and

$$|0\rangle_2 = |u_2\rangle = \begin{bmatrix} 1 \\ 0 \end{bmatrix}_2 \; ; \; |1\rangle_2 = |d_2\rangle = \begin{bmatrix} 0 \\ 1 \end{bmatrix}_2$$

Case I

Another class of maximally entangled states is given in general by Eq. 5.32

$$|\Psi_E\rangle = \frac{1}{\sqrt{N}} \sum_{i=1}^{N} e^{i\phi_i} |\psi_i^I\rangle |\psi_i^{II}\rangle \tag{6.6}$$

For $N = 2$, denoting two-dimensional column vectors by u_1, d_1, u_2, d_2, we have the following

$$\begin{bmatrix} \psi_1^I \\ \psi_2^I \end{bmatrix} = \begin{bmatrix} u_1 \\ d_1 \end{bmatrix} \; ; \; \begin{bmatrix} \psi_1^{II} \\ \psi_2^{II} \end{bmatrix} = \begin{bmatrix} u_2 \\ d_2 \end{bmatrix}$$

Hence, ignoring an overall phase, from Eq. 6.6 we have the following entangled state vectors for the two binary degrees of freedom

$$|\Psi_E\rangle = \frac{1}{\sqrt{2}}(|u_1 u_2\rangle + e^{i\phi}|d_1 d_2\rangle) \tag{6.7}$$

Choosing $\phi = 0$ and $\phi = \pi$, Eq. 6.7 yields, respectively

$$|B_1\rangle = \frac{1}{\sqrt{2}}(|00\rangle + |11\rangle) \; ; \; |B_2\rangle = \frac{1}{\sqrt{2}}(|00\rangle - |11\rangle)$$

Case II

Recall from Eq. 5.38 that a *maximally entangled* state of two binary degrees of freedom is given by[1]

$$|\Psi_E\rangle = \frac{1}{\sqrt{2}}\Big[|u_1\rangle|d_2\rangle + e^{i\phi}|d_1\rangle\,|u_2\rangle\Big] \tag{6.8}$$

Choosing $\phi = 0$ and $\phi = \pi$ Eq. 6.8 yields, respectively

$$|B_3\rangle = \frac{1}{\sqrt{2}}(|01\rangle + |10\rangle) \;\; ; \;\; |B_4\rangle = \frac{1}{\sqrt{2}}(|01\rangle - |10\rangle)$$

Hence, we see that all the Bell states are maximally entangled.

---

[1] The phase can be moved due to the overall phase being irrelevant.

Chapter 7
Quantum Gates and Circuits

A quantum circuit, like the classical case, represents qubits by lines and gates by various symbols. At the hardware level, a quantum circuit, like a classical computer, is a physical object—composed of tiny capacitors with Josephson junctions creating superconducting qubits that are manipulated using resonators, magnetic and electric fields and other interactions. The information transmitted along the circuit is no longer high and low voltages of a classical bit but the quantum mechanical state (wave) function.

The mathematical structure of the qubits and quantum gates is fundamentally different from the classical case. For a classical computer, gates are in general irreversible, and hence, the classical gates cannot be mapped directly to quantum computers. All quantum gates are *unitary transformations*, being reversible (unitary) operations that are applied to one or more qubits; for a reversible computation, the number of input qubits must equal the number of output qubits.

The general scheme of quantum computation is that the qubit $|x\rangle$ is transformed by some gates and results in taking values $|x \oplus f(x)\rangle$. Since the initial qubit state is transformed to the final qubit state by reversible operations, the algorithms of a quantum computer must necessarily be reversible. For a quantum computer, at least two qubits are required for a reversible computation so that no information is lost in each step of the algorithm. Hence, to preserve the information contained in the initial qubit $|x\rangle$, which serves as the initial state for the quantum algorithm, a second ancillary qubit $|y\rangle$ is introduced that carries the result of the computation [1].

Consider a binary function f that is defined by the following mapping ($|x\rangle \otimes |y\rangle \equiv |x\rangle|y\rangle$)

$$f : |x\rangle|y\rangle \rightarrow |x\rangle|y \oplus f(x)\rangle$$

where recall that $y \oplus f(x)$ is binary addition. A quantum gate $\mathcal{O}_f$ implements the function by the following operation

$$\mathcal{O}_f\big(|x\rangle|y\rangle\big) = |x\rangle|y \oplus f(x)\rangle$$

B. E. Baaquie and L.-C. Kwek, *Quantum Computers*,
https://doi.org/10.1007/978-981-19-7517-2_7

139

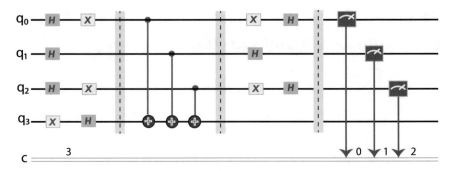

Fig. 7.1 A general circuit of a quantum computer. Published with permission of © Belal E. Baaquie and L. C. Kwek. All Rights Reserved

The *main difference* between the classical and quantum circuit diagram is that (a) all quantum gates are reversible (as are some, but not all, classical gates) and (b) the input and output n-qubit strings states are, in principle, the *superposition* of the binary degrees of freedom.

If the state $|x\rangle$ needs n qubits and the function needs ancillary m qubits, then the computation will need $n + m$ qubits. The dual-register architecture of the input qubit and the ancillary qubit comes into play in many non-classical algorithms.

A typical circuit of the quantum computer, shown in Fig. 7.1, consists of (a) the initial and ancillary states, (b) a number of quantum gates, and (c) a special irreversible quantum gate that terminates the algorithm by performing a measurement.

The quantum circuit shown in Fig. 7.1 has the following components:

1. The horizontal axis denotes time.
2. Each line represents a qubit. The input qubits are indicated by $q_0, \ldots, q_3$.
3. The double line labeled c records the results of measurements.
4. The quantum gates are unitary matrices.
5. Gates denoted by symbols X, H act on only a single qubit.
6. The vertical line with a dot on top and a cross on the bottom are gates that simultaneously act on the indicated two qubits.
7. In principle, quantum gates can simultaneously act on n-qubit strings.
8. The vertical dashed lines are not part of the circuit and are used to indicate the difference between gates that act on a single and on two qubits.
9. The measurement gates are irreversible (non-unitary) projection operators, indicated by a counter and pointer in Fig. 7.1.
10. Qubits are terminated with a measurement; in Fig. 7.1, a measurement is performed only on q_0, q_1, q_2 but not on q_3.
11. The result of the measurement is recorded in a counter, denoted by double line labeled by c, with classical binary bits—denoted by 0, 1, 2, 3—reserved for each qubit in the counter c, as shown in Fig. 7.1.
12. The qubit q_3 is not measured so the binary bit 3 is shown on the left of the double line.

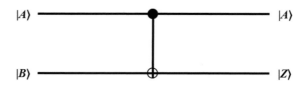

7.1 Quantum Gates

There are three Pauli gates: X, Y and Z; from Eq. 6.4, one changes the notation of the Pauli matrices to the following

$$X = \begin{bmatrix} 0 & 1 \\ 1 & 0 \end{bmatrix}, \ Y = \begin{bmatrix} 0 & -i \\ i & 0 \end{bmatrix}, \ Z = \begin{bmatrix} 1 & 0 \\ 0 & -1 \end{bmatrix}. \tag{7.1}$$

The Pauli gates act on a single qubit and can be used for creating superposed qubit states. The Pauli X gate acts like the classical NOT gate on the basis states; i.e., it changes the state $|0\rangle$ to $|1\rangle$ and vice versa.

$$X|0\rangle = |1\rangle; \quad X|1\rangle = |0\rangle$$

The CNOT gate is the same as the XOR classical gate, and discussed in Sect. 3.4 and given in Fig. 7.2. Two gates in Fig. 7.1 are Pauli X gates, which is the same as the classical NOT gate.

Unitary rotation matrices, acting on a single qubit, are built from the Pauli gates. Using the fact that

$$X^2 = Y^2 = Z^2 = \mathbb{I}$$

yields

$$R_x(\theta) = \exp i\theta X = \cos(\theta)\mathbb{I} + i \sin \theta X$$
$$R_y(\theta) = \exp i\theta Y = \cos(\theta)\mathbb{I} + i \sin \theta Y$$
$$R_z(\theta) = \exp i\theta Z = \cos(\theta)\mathbb{I} + i \sin \theta Z \tag{7.2}$$

The unitary Hadamard gate H is given by

$$H = \frac{1}{\sqrt{2}} \begin{bmatrix} 1 & 1 \\ 1 & -1 \end{bmatrix} = H^\dagger$$

Hence

$$H\begin{bmatrix} 1 \\ 0 \end{bmatrix} = \frac{1}{\sqrt{2}} \begin{bmatrix} 1 \\ 1 \end{bmatrix}; \quad H\begin{bmatrix} 0 \\ 1 \end{bmatrix} = \frac{1}{\sqrt{2}} \begin{bmatrix} 1 \\ -1 \end{bmatrix} \tag{7.3}$$

The Hadamard gate can be used to change the NOT or Pauli-X gate into a Pauli-Z gate since

$$HXH = Z$$

The Hadamard gate is unitary since

$$HH^\dagger = H^2 = \begin{bmatrix} 1 & 0 \\ 0 & 1 \end{bmatrix} = \mathbb{I} \Rightarrow H^2|0\rangle = |0\rangle : H^2|1\rangle = |1\rangle$$

From Eq. 7.3

$$H|0\rangle = \frac{1}{\sqrt{2}}[|0\rangle + |1\rangle]; \quad H|1\rangle = \frac{1}{\sqrt{2}}[|0\rangle - |1\rangle] \tag{7.4}$$

In Eq. 7.4, the Hadamard gate has created a superposed state for the single qubit, something forbidden for a classical gate acting on a classical bit. Equation 7.4 can be rewritten as follows

$$H|x\rangle = \frac{1}{\sqrt{2}} \sum_{y=0}^{1} (-)^{xy} |y\rangle; \quad x = 0, 1 \tag{7.5}$$

Hence, for $|x\rangle = |x_1 x_2 \ldots x_n\rangle$, Eq. 7.5 yields

$$H|x\rangle = \frac{1}{\sqrt{2^n}} \sum_{y=0}^{1} (-)^{xy} |y\rangle; \quad xy = x_1 y_1 \oplus x_2 y_2 \cdots \oplus x_n y_n; \quad x_i, y_i = 0, 1 \tag{7.6}$$

The Hadamard gate H is used extensively in many quantum algorithms. Most computations start with the initial state $|0\rangle$; the Hadamard gate is applied to the input state $|0\rangle$ to render it into a superposed state of the two basis states with equal amplitudes, since

$$H|0\rangle = \frac{1}{\sqrt{2}}[|0\rangle + |1\rangle] \tag{7.7}$$

Consider the Hadamard gate acting on 2-qubits $|q_1 q_2\rangle$; using the tensor product notation

$$H^{\otimes 2}|q_1 q_2\rangle = (H \otimes H)(|q_1\rangle \otimes |q_2\rangle) = H|q_1\rangle \otimes H|q_2\rangle$$
$$\Rightarrow H^{\otimes 2}|00\rangle = \frac{1}{\sqrt{2^2}} \, [|0\rangle + |1\rangle][|0\rangle + |1\rangle]$$
$$= \frac{1}{\sqrt{2^2}} \, [|00\rangle + |01\rangle + |10\rangle + |11\rangle] \tag{7.8}$$

To create the computational basis states, the Hadamard gate is often used. Equation 7.8 can be generalized to n-degrees of freedom. Consider the basis state

$$\otimes^{\otimes n}|0\rangle = \underbrace{|0000\ldots 00\rangle}_{n\text{-fold tensor product}}$$

Hence, similar to Eq. 7.8

$$H^{\otimes n}\left[\otimes^{\otimes n}|0\rangle\right] = \frac{1}{\sqrt{2^n}}\,[|0\rangle + |1\rangle][|0\rangle + |1\rangle]\cdots[|0\rangle + |1\rangle]$$

$$= \frac{1}{\sqrt{2^n}}\sum_{x=0}^{2^n-1}|x\rangle \tag{7.9}$$

7.2 Superposed and Entangled Qubits

Consider the general single qubit

$$|q\rangle = a|0\rangle + b|1\rangle; \quad a^2 + b^2 = 1$$

We can construct this state starting from the initial state of $|0\rangle$ in the following manner. Consider the rotation matrix

$$R_y(\theta)|0\rangle = (\cos\theta\,\mathbb{I} + i\sin\theta\,Y)|0\rangle = \cos\theta|0\rangle + \sin\theta|1\rangle \tag{7.10}$$

Hence, choosing

$$a = \cos\theta; \quad b = \sin\theta$$

yields

$$|q\rangle = R_y(\theta)|0\rangle$$

The quantum circuit diagram is shown in Fig. 7.3.

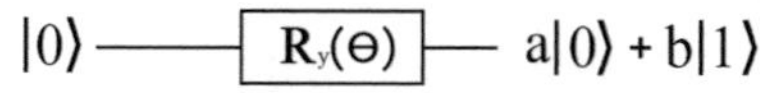

$$|0\rangle \longrightarrow \boxed{R_y(\theta)} \longrightarrow a|0\rangle + b|1\rangle$$

Fig. 7.3 Preparation of a general single qubit. Published with permission of © Belal E. Baaquie and L. C. Kwek. All Rights Reserved

The most general two-qubit state vector, from Eq. 6.5, is given by

$$
\begin{aligned}
|\psi\rangle &= \alpha|00\rangle + \beta|01\rangle + \gamma|10\rangle + \delta|11\rangle \\
&= \sqrt{\alpha^2 + \beta^2}\,|0\rangle \left(\frac{\alpha|0\rangle + \beta|1\rangle}{\sqrt{\alpha^2 + \beta^2}} \right) + \sqrt{\gamma^2 + \delta^2}\,|1\rangle \left(\frac{\gamma|0\rangle + \delta|1\rangle}{\sqrt{\gamma^2 + \delta^2}} \right) \quad (7.11) \\
&\quad |\alpha|^2 + |\beta|^2 + |\gamma|^2 + |\delta|^2 = 1
\end{aligned}
$$

Choose U so that

$$
U \frac{1}{\sqrt{\alpha^2 + \beta^2}} \begin{pmatrix} \alpha \\ \beta \end{pmatrix} = \frac{1}{\sqrt{\gamma^2 + \delta^2}} \begin{pmatrix} \gamma \\ \delta \end{pmatrix} \implies UU^\dagger = \mathbb{I}
$$

such that

$$
U \left(\frac{1}{\sqrt{\alpha^2 + \beta^2}} (\alpha|0\rangle + \beta|1\rangle) \right) = \frac{1}{\sqrt{\gamma^2 + \delta^2}} (\gamma|0\rangle + \delta|1\rangle)
$$

Such a U exists since both states

$$
|\phi_1\rangle = \sqrt{\alpha^2 + \beta^2}\,|0\rangle + \sqrt{\gamma^2 + \delta^2}\,|1\rangle; \quad |\phi_2\rangle = \frac{1}{\sqrt{\alpha^2 + \beta^2}} (\alpha|0\rangle + \beta|1\rangle)
$$

can be regarded as points on the Bloch sphere and U is simply the unitary (rotation) matrix that transforms (rotates) the vector $|\phi_1\rangle$ to $|\phi_2\rangle$.

Hence, from Eq. (7.11),

$$
\begin{aligned}
|\psi\rangle &= \mathcal{U}\left[|\phi_1\rangle \otimes |\phi_2\rangle \right] \quad (7.12) \\
&= \mathcal{U}\left[\left(\sqrt{\alpha^2 + \beta^2}\,|0\rangle + \sqrt{\gamma^2 + \delta^2}\,|1\rangle \right) \otimes |\phi_2\rangle \right] \\
&= \sqrt{\alpha^2 + \beta^2}\,|0\rangle|\phi_2\rangle + \sqrt{\gamma^2 + \delta^2}\,|1\rangle U|\phi_2\rangle \\
&= |0\rangle(\alpha|0\rangle + \beta|1\rangle) + |1\rangle(\gamma|0 + \delta|1\rangle) \\
&= \alpha|00\rangle + \beta|01\rangle + \gamma|10\rangle + \delta|11\rangle \quad (7.13)
\end{aligned}
$$

where the Control-U gate is given by

$$
\mathcal{U} = |0\rangle\langle 0| \otimes \mathbb{I} + |1\rangle\langle 1| \otimes U
$$

One prepares the one-qubit states using the result of Eq. 7.10. Prepare the initial state $|\psi_i\rangle$ given by

$$
|\psi_i\rangle = (|0\rangle + |1\rangle) \otimes (\alpha|0\rangle + \beta|1\rangle)
$$

Hence, from Eq. 7.12, the 2-qubit state is given by

Fig. 7.4 Preparation of a
general 2-qubits state.
Published with permission of
© Belal E. Baaquie and L. C.
Kwek. All Rights Reserved

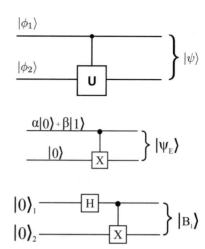

Fig. 7.5 Preparation of a
general 2-qubits entangled
state and the special case of
the Bell state $|B_1\rangle$.
Published with permission of
© Belal E. Baaquie and L. C.
Kwek. All Rights Reserved

$$|\psi\rangle = \mathcal{U}|\psi_i\rangle$$

The quantum circuit for preparing the 2-qubit state is given in Fig. 7.4.

Entangled states are a major resource in quantum algorithms and to express them in a circuit diagram, one needs to group two or more qubits together. Consider the entangled state

$$|\psi_E\rangle = \alpha|00\rangle + \delta|11\rangle$$

Prepare initial unentangled (factorized) of two 1-qubits state given by

$$|\psi_i\rangle = (\alpha|0\rangle + \delta|1\rangle)|0\rangle = \alpha|00\rangle + \delta|10\rangle$$

Consider the CNOT gate

$$\mathcal{X} = |0\rangle\langle 0| \otimes \mathbb{I} + |1\rangle\langle 1| \otimes X$$

Applying the CNOT gate to the initial state yields the entangled state

$$\mathcal{X}|\psi_i\rangle = \alpha|0\rangle|0\rangle + \delta|1\rangle X|0\rangle = \alpha|00\rangle + \delta|11\rangle = |\psi_E\rangle$$

The circuit diagram is given in Fig. 7.5.

Note that in the derivation above, by setting $\alpha = \delta = 1/\sqrt{2}$, we have a mapping of Bell states, given in Eq. 6.6 into computational basis. Consider the following Bell state

$$|B_1\rangle = \frac{1}{\sqrt{2^2}}(|00\rangle + |11\rangle) = \frac{1}{\sqrt{2^2}}\mathcal{X}\Big[(|0\rangle + |1\rangle)|0\rangle\Big]$$

$$\Rightarrow |B_1\rangle = \frac{1}{\sqrt{2}}\mathcal{X}\Big[H|0\rangle \otimes |0\rangle\Big] = \frac{1}{\sqrt{2}}\mathcal{X}(H \otimes \mathbb{I})|00\rangle$$

since the Hadamard gate yields,

$$H|0\rangle = \frac{1}{\sqrt{2}}(|0\rangle + |1\rangle)$$

All the other Bell states can similarly be mapped into the computational basis.

7.3   Two- and Three-Qubit Quantum Gates

An example of a two-qubit gate is the Control-Not (CNOT) gate 7.2, given by

$$\text{CNOT} = \begin{bmatrix} 1 & 0 & 0 & 0 \\ 0 & 1 & 0 & 0 \\ 0 & 0 & 0 & 1 \\ 0 & 0 & 1 & 0 \end{bmatrix}. \tag{7.14}$$

This gate acts on two qubits and if the first qubit is in state $|0\rangle$, it does nothing to the second qubit, but if the first qubit is in state $|1\rangle$, it applies the X gate to the second qubit. Another way of thinking of the CNOT gate is

$$\text{CNOT} = |0\rangle\langle 0| \otimes \mathbb{I} + |1\rangle\langle 1| \otimes X \tag{7.15}$$

where $\mathbb{I}$ is the identity matrix.

Another important quantum gate that arises from the development of reversible computations is the Control-Control-NOT (CCNOT) gate and is a quantum version of the classical Toffoli gate. The circuit diagram of the CCNOT gate is given in Fig. 3.8 and reproduced below in for clarity Fig. 7.6. By analogy with the Toffoli gate, the CCNOT gate is given by the following.

Fig. 7.6 CCNOT quantum gate.

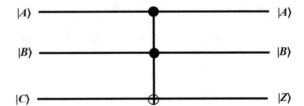

$$\text{CCNOT} = \Big(|00\rangle\langle 00| + |00\rangle\langle 00| + |00\rangle\langle 00|\Big) \otimes \mathbb{I} + |11\rangle\langle 11| \otimes X$$

$$\text{CCNOT} = \begin{pmatrix} 1 & 0 & 0 & 0 & 0 & 0 & 0 & 0 \\ 0 & 1 & 0 & 0 & 0 & 0 & 0 & 0 \\ 0 & 0 & 1 & 0 & 0 & 0 & 0 & 0 \\ 0 & 0 & 0 & 1 & 0 & 0 & 0 & 0 \\ 0 & 0 & 0 & 0 & 1 & 0 & 0 & 0 \\ 0 & 0 & 0 & 0 & 0 & 1 & 0 & 0 \\ 0 & 0 & 0 & 0 & 0 & 0 & 0 & 1 \\ 0 & 0 & 0 & 0 & 0 & 0 & 1 & 0 \end{pmatrix} \tag{7.16}$$

In the computational basis, CCNOT gate is described by the matrix given in Eq. 7.16. The Toffoli gate is a universal gate for classical algorithms. For quantum algorithms, a generalization of the Toffoli gate proposed by Deutsch is the Toffoli gate together with a matrix built from one qubit and is given by the following unitary matrix

$$\text{CCNOT}_\varrho = \begin{pmatrix} 1 & 0 & 0 & 0 & 0 & 0 & 0 & 0 \\ 0 & 1 & 0 & 0 & 0 & 0 & 0 & 0 \\ 0 & 0 & 1 & 0 & 0 & 0 & 0 & 0 \\ 0 & 0 & 0 & 1 & 0 & 0 & 0 & 0 \\ 0 & 0 & 0 & 0 & 1 & 0 & 0 & 0 \\ 0 & 0 & 0 & 0 & 0 & 1 & 0 & 0 \\ 0 & 0 & 0 & 0 & 0 & 0 & i\sin\theta & \cos\theta \\ 0 & 0 & 0 & 0 & 0 & 0 & \cos\theta & i\sin\theta \end{pmatrix} \tag{7.17}$$

7.4 Arithmetic Addition of Binary Qubits

The logic of adding two numbers on a quantum computer is the same as that for a classical computer [2, 3]. The quantum adder does not possess any quantum advantage over the classical adder. Nor does it save any resources. We discuss the quantum full adder here mainly for pedagogical purpose, in particular, to illustrate the fact that quantum gates are linear operators on the underlying bits as well the interconnection of the quantum adder with the classical full adder.

The addition of two quantum qubits follows the same logical truth tables and gates as the classical full adder. In particular, the reversible classical gate $\mathcal{F}$ obtained in Sect. 3.8 and given by the matrix in Eq. 3.36 is also the quantum gate. The main and significant difference is that quantum adder can also add incoming qubits that are the superposition of the binary degrees of freedom.

Let the state vector $|\psi\rangle$ be the result of the addition of two qubits $|A\rangle$ and $|B\rangle$; then, similar to Eq. 3.33, we have

$$|\psi\rangle = \mathcal{F}|ABC_{\text{in}}0\rangle \equiv \mathcal{F}\Big(|A\rangle \otimes |B\rangle \otimes |C_{\text{in}}\rangle \otimes |0\rangle\Big) \tag{7.18}$$

where for the quantum case the incoming qubits are given by

$$|A\rangle = \alpha_1|0\rangle + \beta_1|1\rangle$$
$$|B\rangle = \alpha_2|0\rangle + \beta_2|1\rangle \tag{7.19}$$

Since $\mathcal{F}$ is a linear operator (matrix) acting on a linear vector space, we have from Eq. 7.18—explicitly writing out the incoming state vectors—the following

$$\begin{aligned}
|\psi\rangle &= \mathcal{F}\Big((\alpha_1|0\rangle + \beta_1|1\rangle) \otimes (\alpha_2|0\rangle + \beta_2|1\rangle) \otimes |C_{in}\rangle \otimes |0\rangle\Big) \\
&= \alpha_1\alpha_2\mathcal{F}(|00C_{in}0\rangle) + \alpha_1\beta_2\mathcal{F}(|01C_{in}0\rangle) \\
&\quad + \beta_1\alpha_2\mathcal{F}(|10C_{in}0\rangle) + \beta_1\beta_2\mathcal{F}(|11C_{in}0\rangle)
\end{aligned} \tag{7.20}$$

We analyze the two cases for the quantum adder.

Case I: $C_{in} = 0$
Using the result for the classical full-adder, such as in the examples given in Eq. 3.35, we have the following

$$\begin{aligned}
|\psi\rangle &= \mathcal{F}\Big((\alpha_1|0\rangle + \beta_1|1\rangle) \otimes (\alpha_2|0\rangle + \beta_2|1\rangle) \otimes |0\rangle \otimes |0\rangle\Big) \\
&= \alpha_1\alpha_2\mathcal{F}(|0000\rangle) + \alpha_1\beta_2\mathcal{F}(|0100\rangle) \\
&\quad + \beta_1\alpha_2\mathcal{F}(|1000\rangle) + \beta_1\beta_2\mathcal{F}(|1100\rangle) \\
|\psi\rangle &= \alpha_1\alpha_2|0000\rangle + \alpha_1\beta_2|0101\rangle + \beta_1\alpha_2|1001\rangle + \beta_1\beta_2|1110\rangle
\end{aligned} \tag{7.21}$$

Case II: $C_{in} = 1$
Using the result for the classical full-adder, such as in the examples given in Eq. 3.35, we have the following

$$\begin{aligned}
|\psi\rangle &= \mathcal{F}\Big((\alpha_1|0\rangle + \beta_1|1\rangle) \otimes (\alpha_2|0\rangle + \beta_2|1\rangle) \otimes |1\rangle \otimes |0\rangle\Big) \\
&= \alpha_1\alpha_2\mathcal{F}(|0010\rangle) + \alpha_1\beta_2\mathcal{F}(|0110\rangle) \\
&\quad + \beta_1\alpha_2\mathcal{F}(|1010\rangle) + \beta_1\beta_2\mathcal{F}(|1110\rangle) \\
|\psi\rangle &= \alpha_1\alpha_2|0001\rangle + \alpha_1\beta_2|0110\rangle + \beta_1\alpha_2|1010\rangle + \beta_1\beta_2|1111\rangle
\end{aligned} \tag{7.22}$$

7.5   Quantum Measurements of Qubits

The qubits and gates discussed in the earlier sections prepare the final quantum state from the initial input quantum state. The process of computation needs one more crucial step, which is the process of quantum measurement. In the chapter on quantum mechanics, in particular, in Sects. 4.5 and 4.10, the Born rule and the generalized Born rule have been discussed. These rules are studied in the context of qubits, and a detailed discussion is given of measurements performed on 1 qubit and 2 qubits.

For a single qubit given by

$$|q\rangle = \alpha|0\rangle + \beta|1\rangle$$

the density matrix is given by

$$\rho = |q\rangle\langle q| = |\alpha|^2|0\rangle\langle 0| + \alpha\beta^*|0\rangle\langle 1| + \alpha^*\beta|1\rangle\langle 0| + |\beta|^2|1\rangle\langle 1|$$

Recall from Sect. 4.5 one needs to decide which value of binary degree of freedom is the device going to study statistically. The final result for a measurement is given by the arithmetic mean of, in principle, an infinite number of observations. In practice, a sufficiently large number of observations is sufficient. One needs to keep in mind the statistical nature of quantum measurements, and that in principle one can never directly observe the (binary) degree of freedom [4].

The device provides projection operators for the measurement given by

$$\mathcal{M}_0 = |0\rangle\langle 0|; \quad \mathcal{M}_1 = |1\rangle\langle 1|$$

The projection operators are non-Unitary irreversible gates that cause the qubit to collapse to an eignestates one of the projection operators. The measurement gate is given in Fig. 7.7.

The measurement requires the density matrix

$$\rho = |q\rangle\langle q|$$

The measurement of a single qubit is given in Fig. 7.8.

- Consider the following Born measurement

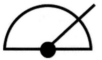

Fig. 7.7 Irreversible measurement gate. Published with permission of © Belal E. Baaquie and L. C. Kwek. All Rights Reserved

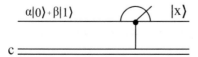

Fig. 7.8 Measurement of a single qubit; counter C is for recording the result of the measurement. Published with permission of © Belal E. Baaquie and L. C. Kwek. All Rights Reserved

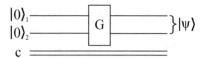

Fig. 7.9 Preparation of a general 2-qubit state using the reversible gate G. C is a counter for recording the outcome of the measurement. Published with permission of © Belal E. Baaquie and L. C. Kwek. All Rights Reserved

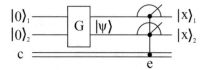

Fig. 7.10 Simultaneous measurement of both the degree of freedom of a 2-qubit. C indicates a recording counter that records the result of the measurement. Published with permission of © Belal E. Baaquie and L. C. Kwek. All Rights Reserved

$$|q\rangle \;\rightarrow\; \text{Measurement} \;\rightarrow\; |x\rangle; \quad x = 0 \text{ or } 1$$

- The probability of the qubit collapsing to the state $|x\rangle = |0\rangle$ is given by the projection operator $\mathcal{M}_0$

$$\text{Tr}(\mathcal{M}_0\rho) = |\alpha|^2$$

- The probability of the qubit collapsing to the state $|x\rangle = |1\rangle$ is given by the projection operator M_1

$$\text{Tr}(M_1\rho) = |\beta|^2$$

Figure 7.10 shows the preparation of $|\psi\rangle$ using the gate G, and the result of the measurement of both qubits is indicated by e.

The most general two qubits state vector $|\psi\rangle$ can be prepared as shown in Fig. 7.9 using a reversible gate G. From Eq. 6.5, it is given by

$$|\psi\rangle = \alpha|00\rangle + \beta|01\rangle + \gamma|10\rangle + \delta|11\rangle; \quad \rho = |\psi\rangle\langle\psi|$$

Similar to the single qubit case, we have the following projection operators required for carrying out a quantum measurement on a 2-qubits string.

$$\mathcal{M}_{00} = |0\rangle\langle0| \otimes |0\rangle\langle0|; \quad \mathcal{M}_{01} = |0\rangle\langle0| \otimes |1\rangle\langle1|$$

and

$$\mathcal{M}_{10} = |1\rangle\langle1| \otimes |0\rangle\langle0|; \quad \mathcal{M}_{11} = |1\rangle\langle1| \otimes |1\rangle\langle1|$$

The simultaneous measurement of both the degree of freedom, shown in Fig. 7.10, yields the following result

$$|\psi\rangle \rightarrow \text{Measurement} \rightarrow |x_1\rangle|x_2\rangle; \quad x_1, x_2 = 0 \text{ or } 1$$

The outcome of the measurement is indicated by e in Fig. 7.10. The probability of a specific result $|x_1\rangle|x_2\rangle$ is given by

$$p(ij) = \text{Tr}(\mathcal{M}_{ij}\rho); \quad x_1, x_2 = 0, 1 \tag{7.23}$$

and yields the following probability of occurrence of the state $|x_1\rangle|x_2\rangle$

$$p(00) = |\alpha|^2; \, p(01) = |\beta|^2; \, p(10) = |\gamma|^2; \, p(11) = |\delta|^2$$

7.5.1 Partial Measurement

The generalized Born measurement for the two-qubit case—including the consistency of the generalized measurement discussed in Sect. 4.10—is worked in detail for pedagogical purpose. A measurement for quantum algorithms that is a partial Born measurement, discussed in Sect. 4.10, is one in which only a few of the qubits are measured. To exemplify this measurement, it is sufficient to consider the two-qubit case with a measurement being carried out on only one of the qubits, followed by another measurement of the second qubit.

Consider measurement processes shown in Fig. 7.11. To express the result of these measurements, for clarity we label the first and second qubits and rewrite the state vector $|\psi\rangle$ as follows

$$|\psi\rangle = |0\rangle_1\left(\alpha|0\rangle_2 + \beta|0\rangle_2\right) + |1\rangle_1\left(\gamma|0\rangle_2 + \delta|0\rangle_2\right) \tag{7.24}$$
$$= p_1(0)|0\rangle_1|\Phi_1(0)\rangle + p_1(1)|1\rangle_1|\Phi_1(1)\rangle \tag{7.25}$$

where

$$p_1(0) = |\alpha|^2 + |\beta|^2; \quad |\Phi_1(0)\rangle = \frac{1}{\sqrt{p_1(0)}}\left(\alpha|0\rangle_2 + \beta|0\rangle_2\right) \tag{7.26}$$

$$p_1(1) = |\gamma|^2 + |\delta|^2; \quad |\Phi_1(1)\rangle = \frac{1}{\sqrt{p_1(1)}}\left(\gamma|0\rangle_2 + \delta|0\rangle_2\right) \tag{7.27}$$

The measurement of the first qubit yields the following

$$|\psi\rangle \rightarrow |x_1\rangle|\Phi_1(x)\rangle \tag{7.28}$$

with

- Probability of state $|0\rangle|\Phi_1(0)\rangle$ occurring being given by $p_1(0)$
- Probability of state $|1\rangle|\Phi_1(1)\rangle$ occurring being given by $p_1(1)$

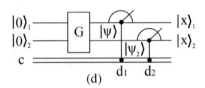

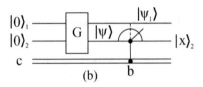

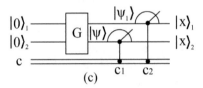

Another measurement carried out on state $|x_1\rangle|\Phi_1(x)\rangle$, as shown in Fig. 7.11, yields

$$|x_1\rangle|\Phi_1(x)\rangle \;\rightarrow\; |x_1\rangle|x_2\rangle \qquad (7.29)$$

with

- Probability of state $|0\rangle_1|0\rangle_1$ occurring being given by

$$p_1(0)\frac{|\alpha|^2}{p_1(0)} = |\alpha|^2 = p(00)$$

- Probability of state $|0\rangle_1|1\rangle_1$ occurring being given by

$$p_1(0)\frac{|\beta|^2}{p_1(0)} = |\beta|^2 = p(01)$$

- Probability of state $|1\rangle_1|0\rangle_1$ occurring being given by

$$p_1(1)\frac{|\gamma|^2}{p_1(1)} = |\gamma|^2 = p(10)$$

- Probability of state $|1\rangle_1|1\rangle_1$ occurring being given by

$$p_1(1)\frac{|\delta|^2}{p_1(1)} = |\delta|^2 = p(11)$$

In summary, the two successive measurements are given by the following

$$|\psi\rangle \;\rightarrow\; |x_1\rangle|\Phi_1(x)\rangle \;\rightarrow\; |x_1\rangle|x_2\rangle$$

The result above shows that the result of performing successive measurements on the two qubit yields the same result as the one obtained by simultaneously observing the two qubits

$$|\psi\rangle \;\rightarrow\; |x_1\rangle|x_2\rangle$$

with the result given in Eq. 7.23.

One can also choose to perform the successive measurements by first measuring the second degree of freedom $|x_2\rangle$ and then the first degree of freedom, as shown in Fig. 7.12. One then has, in a similar notation as above

$$|\psi\rangle \;\rightarrow\; |\Phi_2(x)\rangle|x_2\rangle \;\rightarrow\; |x_1\rangle|x_2\rangle$$

Similar to the above derivation, it can be shown that reversing the order of measuring the qubits yields the same as a single measurement of both the degrees of freedom in this case.

As mentioned in Sect. 4.11, the degrees of freedom are independent of each other. Hence, the order of measuring the qubits of the degrees of freedom does not matter. We have verified that measuring a particular qubit does not depend on measurements carried out on the other qubit.

However, in a more general case involving measurements of entangled states the order in which the measurements are made matters. If one uses entangled state vectors as the basis states, such as the Bell states discussed in Sect. 6.4 for two binary degrees of freedom, then one needs to keep track of the order of the measurements. However, in almost all applications measurements are made using the computational basis, and the quantum circuit is measured for non-entangled states; for these states, as mentioned above, the order in which the various individual qubits are measured does not matter.

References

1. Nielsen MA, Chuang IL (2012) Quantum computation and quantum information. Cambridge University Press, UK
2. Barbosa GA (2006) Quantum half-adder. Phys Rev A 73(5):052321
3. Bomble L, Lauvergnat D, Remacle F, Desouter-Lecomte M (2009) Controlled full adder or subtractor by vibrational quantum computing. Phys Rev A 80(2):022332
4. Baaquie BE (2013) The theoretical foundations of quantum mechanics. Springer, UK

Chapter 8
Phase Estimation and quantum Fourier Transform (qFT)

8.1   Introduction

Phase estimation and the quantum Fourier transform (qFT) are the inverse of each other.[1] From a pedagogical point of view, to start the discussion with phase estimation provides greater clarity. The reason being, as will become clear in the sections below, that phase estimation requires the removal of phase factors from a qubit, whereas the qFT requires the addition of phase factors. Removing a phase factor from a qubit has a transparent representation in terms of the quantum gates, and hence, we start with phase estimation; the quantum Fourier transform is the inverse of the transformation for phase estimation [1–4].

8.2   Eigenvalue of Unitary Operator

Consider a unitary operator U that has an eigenvector $|u\rangle$ such that

$$U|u\rangle = e^{2\pi i\omega}|u\rangle$$

Both the unitary operator U and its eigenvector $|u\rangle$ are known, and we need to determine the unknown phase ω. The following procedure is followed. Apply the operator U on $|u\rangle 2^n$ to yield

$$U^y|u\rangle = e^{2\pi i\omega y}|u\rangle; \quad y = 0, 1, 2, \ldots, 2^n - 1$$

Form the state vector $|\psi\rangle$ given by

$$|\psi\rangle = \frac{1}{\sqrt{2^n}} \sum_{y=0}^{2^n-1} \exp\{2\pi i\omega y\}|y\rangle : |y\rangle = |y_1 \ldots y_n\rangle; \quad y_i = 0, 1 \qquad (8.1)$$

© The Author(s), under exclusive license to Springer Nature Singapore Pte Ltd. 2023   155
B. E. Baaquie and L.-C. Kwek, *Quantum Computers*,
https://doi.org/10.1007/978-981-19-7517-2_8

where one notes that the labeling for the ket vector $|y\rangle$ is taken from Eq. 2.4.

The quantum circuits for U^y and $|\psi\rangle$ are given in [5], where it is shown how the action of $U^y|u\rangle$ is employed to extract the phase $\exp\{2\pi i\omega y\}$ to form the state vector $|\psi\rangle$. The problem of phase estimation is to start with state vector $|\psi\rangle$ and determine the phase ω.

8.3   Phase Estimation

Consider a real number ω with $0 < \omega < 1$. If one has a system of n qubits, ω can be expressed approximately as

$$\omega \simeq \frac{\alpha_1}{2} + \frac{\alpha_2}{2^2} + \frac{\alpha_3}{2^3} + \cdots + \frac{\alpha_{n-1}}{2^{n-1}} + \cdots + \frac{\alpha_n}{2^n}$$

where $\alpha_i = 0$ or 1 for each i. The binomial expression given above is exact only if $\omega = x/2^n$, where x is an integer. As an example, the real number $\omega' \dots$ can be expressed approximately (for $n = 12$) as

$$\omega' \simeq \frac{1}{2} + \frac{1}{2^3} + \frac{1}{2^6} + \frac{1}{2^9} + \frac{1}{2^{12}} = 0.101001001001$$

For an arbitrary value of $0 < \omega < 1$, the approximate value of ω can be made arbitrarily accurate by increasing the value of n. The accuracy of the phase estimation algorithm for ω depends on the number of qubits that are being used to estimate it. For arbitrary ω, let $x/2^n$ be an integer multiple of $1/2^n$ closest to ω. The phase estimation algorithm returns the value x with probability at least $4/\pi^2$ [1].

In general, for ω close to $x/2^n$, we can write ω as

$$\omega \simeq \frac{1}{2}x_1 + \frac{1}{2^2}x_2 + \cdots + \frac{1}{2^n}x_n = \sum_{i=1}^{n} x_i 2^{-i}; \quad x_i = 0, 1$$

In binary notation, ω is written as follows

$$\omega = 0.x_1 x_2 \dots x_n \tag{8.2}$$

Since we have n qubits, the computational basis $|x_0 x_1 \dots x_{n-1}\rangle$ has 2^n basis states and is labeled by $y = 0, 1, 2, \dots, 2^{n-1}$. To estimate ω, consider the state vector given in Eq. 8.21

$$|\psi\rangle = \frac{1}{\sqrt{2^n}} \sum_{y=0}^{2^n-1} \exp\{2\pi i\omega y\}|y\rangle : |y\rangle = |y_1 \dots y_n\rangle; \quad y_i = 0, 1 \tag{8.3}$$

Furthermore

$$y = \sum_{i=1}^{n} y_i 2^{n-i} = 2^{n-1} y_1 + 2^{n-2} y_2 + \cdots + 2 y_{n-1} + 2^0 y_n$$

We have the following

$$y\omega = \left(\sum_{i=1}^{n} y_i 2^{n-i} \right) \times 0.x_1 x_2 \dots x_n \tag{8.4}$$

Recall from Eq. 2.7

$$2^k x = x_1 x_2 \dots x_k . x_{k+1} x_{k+2} \dots x_m \dots$$
$$= x_1 x_2 \dots x_k + 0.x_{k+1} x_{k+2} \dots x_m \dots \tag{8.5}$$

Hence

$$2^{n-i} \times 0.x_1 x_2 \dots x_n = x_1 x_2 \dots x_{n-i} + 0.x_{n-i+1} \dots x_n$$

and yields

$$2^{n-i} y_i \times 0.x_1 x_2 \dots x_n = \text{integer} + y_i \times 0.x_{n-i+1} \dots x_n$$
$$\Rightarrow y\omega = \sum_{i=1}^{n} y_i \times 0.x_{n-i+1} \dots x_n + \text{integer}; \quad y_i = 0, 1 \tag{8.6}$$

For $n = 3$, Eq. 8.6 yields

$$\omega y = y_1 (0.x_3) + y_2 (0.x_2 x_3) + y_3 (0.x_1 x_2 x_3); \quad y_1, y_2, y_3 = 0, 1$$

From Eqs. 8.3 and 8.6, the sum over the qubits factorizes and yields

$$|\psi\rangle = \sum_{y=0}^{2^n - 1} \exp\{2\pi i \sum_{i=1}^{n} y_i \times 0.x_{n-i+1} \dots x_n\} |y\rangle$$
$$= \frac{1}{\sqrt{2^n}} \prod_{i=1}^{n} \left[\sum_{y_i=0}^{1} \exp\{2\pi i y_i \times 0.x_{n-i+1} \dots x_n\} |y_i\rangle \right]$$
$$= \frac{1}{\sqrt{2^n}} \left[\sum_{y_1=0}^{1} \exp\{2\pi i y_1 (0.x_n)\} |y_1\rangle \right] \left[\sum_{y_2=0}^{1} \exp\{2\pi i y_2 (0.x_{n-1} x_n)\} |y_2\rangle \right] \dots$$
$$\times \left[\sum_{y_n=0}^{1} \exp\{2\pi i y_n (0.x_1 x_2 \dots x_n)\} |y_n\rangle \right]$$
$$= \frac{1}{\sqrt{2^n}} \left[|0\rangle + e^{2\pi i (0.x_n)} |1\rangle \right]_1 \left[|0\rangle + e^{2\pi i (0.x_{n-1} x_n)} |1\rangle \right]_2 \cdots \left[|0\rangle + e^{2\pi i (0.x_1 x_2 \dots x_n)} |1\rangle \right]_n \tag{8.7}$$

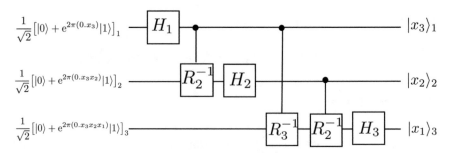

Fig. 8.1 Three-qubit phase estimation. Gates are placed on the target qubit and the heavy dots specify the control qubit. Published with permission of © Belal E. Baaquie and L. C. Kwek. All Rights Reserved

8.3.1 Phase Estimation for n = 3

To analyze the general equation given in Eq. 8.7, we start with the case of $n = 3$ as this has all the features of the general case.

For $n = 3$, Eq. 8.7 yields the following

$$|\psi_3\rangle = \frac{1}{\sqrt{2^3}} \Big[|0\rangle + e^{2\pi i(0.x_3)}|1\rangle\Big]_1 \Big[|0\rangle + e^{2\pi i(0.x_2 x_3)}|1\rangle\Big]_2 \Big[|0\rangle + e^{2\pi i(0.x_1 x_2 x_3)}|1\rangle\Big]_3$$
(8.8)

Note that for $n = 3$ all unitary gates are $2^3 \times 2^3$ matrices acting on a 2^3-dimensional state space and $|\psi_3\rangle$, which is a 2^3-dimensional complex-valued state vector. The quantum circuit for the $n = 3$ phase estimation is given in Fig. 8.1.

Note that

$$|0\rangle + e^{2\pi i(0.x_3)}|1\rangle = |0\rangle + \exp\Big\{2\pi i \frac{x_3}{2}\Big\}|1\rangle = \begin{cases} |0\rangle + |1\rangle : x_3 = 0 \\ |0\rangle - |1\rangle : x_3 = 1 \end{cases}$$

The Hadamard gate yields the following

$$|0\rangle = \frac{1}{\sqrt{2}}H\Big[|0\rangle + |1\rangle\Big]; \quad |1\rangle = \frac{1}{\sqrt{2}}H\Big[|0\rangle - |1\rangle\Big]$$

and hence

$$H\Big[\frac{|0\rangle + \exp\{2\pi i x/2\}|1\rangle}{\sqrt{2}}\Big] = |x\rangle = \begin{cases} |0\rangle : x = 0 \\ |1\rangle : x = 1 \end{cases}$$
(8.9)

Define

$$H_i = \underbrace{\mathbb{I} \otimes \cdots H \cdots \otimes \mathbb{I}}_{i\text{th position}} = H_i^\dagger = H_i^{-1}; \quad H_1 = H \otimes \mathbb{I} \otimes \mathbb{I}$$

Hence

$$H_1|\psi_3\rangle = \frac{1}{\sqrt{2^2}}|x_3\rangle_1\left[|0\rangle + e^{2\pi i(0.x_2x_3)}|1\rangle\right]_2\left[|0\rangle + e^{2\pi i(0.x_1x_2x_3)}|1\rangle\right]_3 \quad (8.10)$$

The term referring to qubit 2 in Eq. 8.10 cannot be simplified using the Hadamard gate since the binary number $0.x_2x_3$ in the exponential needs to simplified. A conditional gate depending on the first qubit $|x_3\rangle$ can be used to transform $0.x_2x_3$ to $0.x_2$ as follows. Consider the following (the notation of $\mathcal{U}_{12}^\dagger$ is used for consistency with the one used in the literature). $\mathcal{U}_{12}^\dagger$ leaves the third qubit unchanged and hence

$$\mathcal{U}_{12}^\dagger = |0\rangle\langle0| \otimes \mathbb{I} \otimes \mathbb{I} + |1\rangle\langle1| \otimes \mathcal{R}_2^{-1} \otimes \mathbb{I} \equiv U_{12}^\dagger \otimes \mathbb{I} \quad (8.11)$$

$$\text{where } \mathcal{R}_2^{-1} = \begin{bmatrix} 1 & 0 \\ 0 & \exp\{-2\pi i/2^2\} \end{bmatrix} \quad (8.12)$$

We have the following

$$\mathcal{U}_{12}^\dagger H_1|\psi_3\rangle = \frac{1}{\sqrt{2^2}}\left\{U_{12}^\dagger\left(|x_3\rangle_1\left[|0\rangle + e^{2\pi i(0.x_2x_3)}|1\rangle\right]_2\right)\right\}$$
$$\times\left[|0\rangle + e^{2\pi i(0.x_1x_2x_3)}|1\rangle\right]_3 \quad (8.13)$$

Hence

$$U_{12}^\dagger\left(|x_3\rangle_1\left[|0\rangle + e^{2\pi i(0.x_2x_3)}|1\rangle\right]_2\right)$$
$$= \left(|0\rangle\langle0| \otimes \mathbb{I} + |1\rangle\langle1| \otimes \mathcal{R}_2^{-1}\right)\left\{|x_3\rangle_1\left[|0\rangle + e^{2\pi i(0.x_2x_3)}|1\rangle\right]_2\right\}$$
$$= \left(|0\rangle_1\langle0|x_3\rangle \otimes \mathbb{I} + |1\rangle_1\langle1|x_3\rangle \otimes \mathcal{R}_2^{-1}\right)\left[|0\rangle + e^{2\pi i(0.x_2x_3)}|1\rangle\right]_2$$
$$= \begin{cases} |0\rangle_1\left[|0\rangle + e^{2\pi i(0.x_2)}|1\rangle\right]_2 & : x_3 = 0 \\ |1\rangle_1\mathcal{R}_2^{-1}\left[|0\rangle + e^{2\pi i(0.x_21)}|1\rangle\right]_2 & : x_3 = 1 \end{cases} \quad (8.14)$$

Note that, for $x_3 = 1$

$$\mathcal{R}_2^{-1}\left[|0\rangle + e^{2\pi i(0.x_21)}|1\rangle\right]_2 = \left[|0\rangle + e^{2\pi i(0.x_21 - 0.01)}|1\rangle\right]_2 = \left[|0\rangle + e^{2\pi i(0.x_2)}|1\rangle\right]_2$$

and hence, from Eq. 8.14

$$U_{12}^\dagger\left(|x_3\rangle_1\left[|0\rangle + e^{2\pi i(0.x_2x_3)}|1\rangle\right]_2\right) = |x_3\rangle_1\left[|0\rangle + e^{2\pi i(0.x_2)}|1\rangle\right]_2$$

This yields

$$\mathcal{U}_{12}^{\dagger}H_1|\psi_3\rangle = \frac{1}{\sqrt{2^2}}|x_3\rangle_1 \Big[|0\rangle + e^{2\pi i(0.x_2)}|1\rangle\Big]_2 \Big[|0\rangle + e^{2\pi i(0.x_1x_2x_3)}|1\rangle\Big]_3 \quad (8.15)$$

Hence, as shown in Fig. 8.1, applying the Hadamard transformation to the second qubit yields from above

$$H_2\mathcal{U}_{12}^{\dagger}H_1|\psi_3\rangle = \frac{1}{\sqrt{2}}|x_3\rangle_1|x_2\rangle_2\Big[|0\rangle + e^{2\pi i(0.x_1x_2x_3)}|1\rangle\Big]_3 \quad (8.16)$$

Reducing $0.x_1x_2x_3$ to $0.x_1$ needs two steps, since the gates are conditional on the values of x_2, x_3. First step is to remove dependence on x_3 using R_3^{-1} and the second step is to remove x_2 using R_2^{-1}.

Define unitary gate by $\mathcal{U}_{13}$ that acts on third qubit $\Big[|0\rangle + e^{2\pi i(0.x_1x_2x_3)}|1\rangle\Big]_3$ conditioned on $|x_3\rangle_1$; it has to leave the qubit $|x_2\rangle_2$ unchanged. Hence,

$$\mathcal{U}_{13}^{\dagger} = |0\rangle\langle 0| \otimes \mathbb{I} \otimes \mathbb{I} + |1\rangle\langle 1| \otimes \mathbb{I} \otimes R_3^{-1} \quad (8.17)$$

$$\text{where } R_3^{-1} = \begin{bmatrix} 1 & 0 \\ 0 & \exp\{-2\pi i/2^3\} \end{bmatrix} \quad (8.18)$$

In general

$$R_k^{-1} = \begin{bmatrix} 1 & 0 \\ 0 & \exp\{-2\pi i/2^k\} \end{bmatrix} \quad (8.19)$$

Furthermore, similar to Eq. 8.16

$$\mathcal{U}_{13}^{\dagger}H_2\mathcal{U}_{12}^{\dagger}H_1|\psi_3\rangle = \frac{1}{\sqrt{2}}|x_3\rangle_1|x_2\rangle_2\Big[|0\rangle + e^{2\pi i(0.x_1x_2)}|1\rangle\Big]_3 \quad (8.20)$$

Define unitary gate

$$\mathcal{U}_{23}^{\dagger} = \mathbb{I} \otimes |0\rangle\langle 0| \otimes \mathbb{I} + \mathbb{I} \otimes |1\rangle\langle 1| \otimes R_2^{-1} \quad (8.21)$$

which yields from Eq. 8.22

$$\mathcal{U}_{23}^{\dagger}\mathcal{U}_{13}^{\dagger}H_2\mathcal{U}_{12}^{\dagger}H_1|\psi_3\rangle = \tfrac{1}{\sqrt{2}}|x_3\rangle_1|x_2\rangle_2\Big[|0\rangle + e^{2\pi i(0.x_1)}|1\rangle\Big]_3 \quad (8.22)$$

Applying the Hadamard gate H_3 gate to Eq. 8.22 yields the following final result given in circuit in Fig. 8.1

$$H_3\mathcal{U}_{23}^{\dagger}\mathcal{U}_{13}^{\dagger}H_2\mathcal{U}_{12}^{\dagger}H_1|\psi_3\rangle = |x_3\rangle_1|x_2\rangle_2|x_1\rangle_3 \quad (8.23)$$

Using the three bit swap gate S_3 to reverse the order of the qubits yields

$$S_3\left(|x_3\rangle_1|x_2\rangle_2|x_1\rangle_3\right) = |x_1\rangle_1|x_2\rangle_2|x_3\rangle_3 = |x_1 x_2 x_3\rangle = |\omega\rangle$$

where the phase ω is given, as expected, by

$$\omega = \frac{1}{2}x_1 + \frac{1}{2^2}x_2 + \frac{1}{2^3}x_3$$

In summary, the phase estimation is given by the following

$$S_3 H_3 \mathcal{U}_{23}^\dagger \mathcal{U}_{13}^\dagger H_2 \mathcal{U}_{12}^\dagger H_1 |\psi_3\rangle \equiv \mathcal{P}_3 |\psi_3\rangle = |x_1 x_2 x_3\rangle = |\omega\rangle \quad (8.24)$$

Note that, since $\mathcal{P}_3$ is unitary, from Eq. 8.24

$$\mathcal{P}_3^{-1} = \mathcal{P}_3^\dagger = H_1 \mathcal{U}_{12} H_2 \mathcal{U}_{13} \mathcal{U}_{23} H_3 S_3 \quad (8.25)$$

and

$$(\mathcal{R}_k^{-1})^\dagger = \mathcal{R}_k = \begin{bmatrix} 1 & 0 \\ 0 & \exp\{2\pi i/2^k\} \end{bmatrix}; \quad \mathcal{R}_k \mathcal{R}_k^\dagger = \mathbb{I}$$

The general case of $|\omega\rangle = |x_1 x_2 x_3 \ldots x_n\rangle$ follows the same steps and yields an estimation of

$$\omega = \frac{1}{2}x_1 + \frac{1}{2^2}x_2 + \frac{1}{2^3}x_3 + \cdots + \frac{1}{2^n}x_n$$

with the generalization of Eq. 8.24 given by

$$\mathcal{P}|\psi\rangle = |x_1 x_2 x_3 \ldots x_n\rangle = |\omega\rangle \;\Rightarrow\; |\psi\rangle = \mathcal{P}^\dagger|\omega\rangle \quad (8.26)$$

where $\mathcal{P}$ is unitary gate given by a $2^n \times 2^n$ matrix and $|\psi_n\rangle$ is an 2^n-dimensional complex-valued state vector. The result Eq. 8.26 can be written in the following more formal manner

$$|\omega\rangle = \mathcal{P}|\psi\rangle = \mathcal{P}\left(\frac{1}{\sqrt{2^n}} \sum_{y=0}^{2^n-1} \exp\{2\pi i \omega y\}|y\rangle\right) \quad (8.27)$$

The following is a summary of the algorithm for ascertaining the unknown phase ω.

- The phase ω is expressed as the state vector $|\omega\rangle$ that is to be determined.
- Start with the right-hand side of Eq. 8.27 that is known and given by

$$|\psi\rangle = \frac{1}{\sqrt{2^n}} \sum_{y=0}^{2^n-1} \exp\{2\pi i \omega y\}|y\rangle$$

- Apply the quantum gate $\mathcal{P}$ to $|\psi\rangle$ and obtain

$$\mathcal{P}\left(\frac{1}{\sqrt{2^n}}\sum_{y=0}^{2^n-1}\exp\{2\pi i\omega y\}|y\rangle\right)$$

- The final result is given by

$$\mathcal{P}\left(\frac{1}{\sqrt{2^n}}\sum_{y=0}^{2^n-1}\exp\{2\pi i\omega y\}|y\rangle\right) = |\omega\rangle$$

8.4 quantum Fourier Transform

The quantum Fourier transform is obtained by the inverting the steps of the algorithm required for phase estimation.

Consider a binary number $x > 1$ with the binary expansion given in Eq. 2.2

$$x = x_1 2^{n-1} + x_2 2^{n-2} + \cdots + x_j 2^{n-j} + \cdots + x_n 2^0; \quad x_j = 0, 1$$
$$\omega = \frac{x}{2^n} = x_1\frac{1}{2} + x_2\frac{1}{2^2} + \cdots + x_n\frac{1}{2^n}$$
$$\Rightarrow \omega \equiv 0.x_1 x_2 \ldots x_n$$
$$y = y_n 2^0 + y_{n-1} 2^1 + \cdots + y_j 2^{n-j} + \cdots + y_1 2^{n-1}; \quad y_j = 0, 1$$

and hence for xy being ordinary multiplication

$$\frac{xy}{2^n} = \frac{x}{2^n}y = \omega y$$

Note due to the choice of the binary expansion, we have two equivalent notations for the computational basis

$$|x\rangle = |x_1 x_2 \ldots x_n\rangle; \quad |y\rangle = |y_1 y_2 \ldots y_n\rangle; \quad x, y = 0, 1, \ldots, 2^n - 1$$

A quantum Fourier transform is a unitary transformation that maps the computational basis $|x\rangle$ to another equivalent basis $|y\rangle$. The discrete Fourier transform is a unitary transformation $\mathcal{F}$ of the computational basis $|x\rangle$ of the n-qubit to a new computational basis denoted by $|y\rangle$. Hence, we have the definition of a quantum Fourier Transform (qFT) denoted by $\mathcal{F}$ and given by

$$\mathcal{F}\left(|x\rangle\right) = \frac{1}{\sqrt{2^n}}\sum_{y=0}^{2^n-1}\exp\left\{i\frac{2\pi xy}{2^n}\right\}|y\rangle \qquad (8.28)$$

Note there is a periodicity of the qFT in the definition given in Eq. 8.28

$$\mathcal{F}\Big(|x\rangle\Big) = \mathcal{F}\Big(|x + 2^n\rangle\Big)$$

The periodicity is consistent with the definition of the integer $0 \leq x \leq 2^n - 1$; for values of x outside this range, since there are in total only n-qubits, the qFT $\mathcal{F}\Big(|x\rangle\Big)$ is defined only for x modulo 2^n. All values of x outside the range $0 \leq x \leq 2^n - 1$ are mapped back into the range $0 \leq x \leq 2^n - 1$.

The definitions given for binary numbers yield the following

$$\mathcal{F}\Big(|x\rangle\Big) = \frac{1}{\sqrt{2^n}} \sum_{y=0}^{2^n-1} \exp\left\{i\frac{2\pi xy}{2^n}\right\} |y\rangle = \frac{1}{\sqrt{2^n}} \sum_{y=0}^{2^n-1} \exp\{2\pi i \omega y\} |y\rangle$$

$$= \frac{1}{\sqrt{2^n}} \Big[|0\rangle + e^{2\pi i (0.x_n)}|1\rangle\Big]_1 \Big[|0\rangle + e^{2\pi i (0.x_{n-1}x_n)}|1\rangle\Big]_2$$

$$\dots \Big[|0\rangle + e^{2\pi i (0.x_1 x_2 \dots x_n)}|1\rangle\Big]_n \tag{8.29}$$

The expansion of the right-hand side of Eq. 8.29 is given in Eq. 8.7.

The discussion on the quantum Fourier transform is based on the derivation of phase estimation. Note that $\mathcal{P}$ is a unitary gate since it is a product of unitary matrices; hence its inverse exists and is given by

$$\mathcal{F} = \mathcal{P}^{-1} = \mathcal{P}^\dagger \tag{8.30}$$

The quantum Fourier transform is *unitary* since

$$\mathcal{F} \cdot \mathcal{F}^\dagger = \mathbb{I}_{2^n \times 2^n}$$

Equations 8.28 and 8.3 yield

$$\mathcal{P}^\dagger |x_1 x_2 x_3 \dots x_n\rangle = |\psi_n\rangle = \frac{1}{\sqrt{2^n}} \sum_{y=0}^{2^n-1} \exp\{2\pi i xy/2^n\} |y\rangle$$

Hence, Eqs. 8.26 and 8.29 yield

$$\mathcal{F}|x\rangle = \frac{1}{\sqrt{2^n}} \sum_{y=0}^{2^n-1} \exp\{2\pi i xy/2^n\} |y\rangle$$

Similar to the phase estimation algorithm, the quantum Fourier transform algorithm is given by the following. Note that the role of what is known and what is to be determined is interchanged in going from phase estimation to Fourier transform, with one being the inverse of the other.

- The quantum Fourier transform

$$\frac{1}{\sqrt{2^n}} \sum_{y=0}^{2^n-1} \exp\{2\pi ixy/2^n\}|y\rangle$$

  is to be determined.
- Start with the value of $|x\rangle$.
- Apply the quantum gate $\mathcal{F}$ to $|x\rangle$ and obtain

$$\mathcal{F}|x\rangle$$

  The quantum gate $\mathcal{F}$ adds *phase factors* to the qubits, whereas the quantum gate for phase estimation $\mathcal{P}$ *removes the phase factors*.
- The final result is given by

$$\mathcal{F}|x\rangle = \frac{1}{\sqrt{2^n}} \sum_{y=0}^{2^n-1} \exp\{2\pi ixy/2^n\}|y\rangle$$

8.4.1   quantum Fourier Transform for n = 3

To show the close relation between the Fourier transform to phase estimation is studied for the case of $n = 3$. The quantum Fourier transform for $n = 3$, from Eq. 8.8, is given by

$$\mathcal{F}_3|x\rangle_3 = \frac{1}{\sqrt{2^3}} \Big[|0\rangle + e^{2\pi i(0.x_3)}|1\rangle\Big]_1 \Big[|0\rangle + e^{2\pi i(0.x_2 x_3)}|1\rangle\Big]_2 \Big[|0\rangle + e^{2\pi i(0.x_1 x_2 x_3)}|1\rangle\Big]_3 \tag{8.31}$$

From Eqs. 8.25 and 8.30

$$\mathcal{F}_3 = \mathcal{P}_3^\dagger = H_1 \mathcal{U}_{12} H_2 \mathcal{U}_{13} \mathcal{U}_{23} H_3 \mathcal{S}_3 \tag{8.32}$$

An explicit derivation is given of Eq. 8.31 to verify that the quantum Fourier transform is, in fact, determined by gate $\mathcal{F}_3$ given in Eq. 8.32. The quantum Fourier transform is built from two types of gates, one is the Hadamard gate and the other is a two-qubit controlled rotation R_k. Consider the Hadamard transform on a single qubit $|x\rangle$; since $H^2 = \mathbb{I}$, from Eq. 8.9

$$H|x\rangle = \frac{1}{\sqrt{2}} \Big[|0\rangle + \exp\Big\{2\pi i \frac{x}{2}\Big\}|1\rangle\Big]; \quad x = 0, 1 \tag{8.33}$$

Hence, for H_i acting on the computational basis

$$H_i|x_1x_2\ldots x_n\rangle = \frac{1}{\sqrt{2}}|x_1x_2\ldots\rangle\left[|0\rangle + \exp\left\{\frac{2\pi i}{2}x_i\right\}|1\rangle\right]_i|\ldots x_{n-1}x_n\rangle \quad (8.34)$$

From Eq. 8.32, the $n = 3$ quantum Fourier transform is given by

$$\mathcal{F}_3|x_1x_2x_3\rangle = H_1\mathcal{U}_{12}H_2\mathcal{U}_{13}\mathcal{U}_{23}H_3\mathcal{S}_3|x_1x_2x_3\rangle$$
$$= H_1\mathcal{U}_{12}H_2\mathcal{U}_{13}\mathcal{U}_{23}H_3\left(|x_3\rangle_1|x_2\rangle_2|x_1\rangle_3\right) \quad (8.35)$$

The subscript on the ket vectors indicates its position in the tensor product.

The unitary gates are given below and are obtained by taking the Hermitian conjugation of the earlier results using the identity

$$(A \otimes B)^\dagger = A^\dagger \otimes B^\dagger$$

Taking the Hermitian conjugation of Eq. 8.13 yields

$$\mathcal{U}_{12} = |0\rangle\langle 0| \otimes \mathbb{I} \otimes \mathbb{I} + |1\rangle\langle 1| \otimes \mathcal{R}_2 \otimes \mathbb{I} \quad (8.36)$$

Similarly, taking the Hermitian conjugation of Eq. 8.17 yields

$$\mathcal{U}_{13} = |0\rangle\langle 0| \otimes \mathbb{I} \otimes \mathbb{I} + |1\rangle\langle 1| \otimes \mathbb{I} \otimes \mathcal{R}_3 \quad (8.37)$$

and lastly from Eq. 8.21

$$\mathcal{U}_{23} = \mathbb{I} \otimes |0\rangle\langle 0| \otimes \mathbb{I} + \mathbb{I} \otimes |1\rangle\langle 1| \otimes \mathcal{R}_2 \quad (8.38)$$

where

$$\mathcal{R}_k = \begin{bmatrix} 1 & 0 \\ 0 & \exp\{2\pi i/2^k\} \end{bmatrix}$$

The quantum circuit for the $n = 3$ quantum Fourier transform is given in Fig. 8.2 and mathematically expressed in Eq. 8.32. The quantum circuit given in Fig. 8.2 is the inverse of the one for phase estimation given in Fig. 8.1: the *order* of all the gates have been *reversed* and all the gates replaced by their Hermitian conjugation; since the gates are unitary, Hermitian conjugation results in the gate being replaced by their *inverses*.

Comparing the circuits for the phase factor and quantum Fourier transform, given in Figs. 8.1 and 8.2, respectively, note that the state vectors with and without the phase factors have been interchanged for the two circuits.

The gate H_3 acts on the third qubit $|x_1\rangle_3$ gates $\mathcal{U}_{13}$ and $\mathcal{U}_{23}$ act on $|x_1\rangle_3$ conditioned by $|x_3\rangle_1$ and $|x_2\rangle_2$, respectively. To illustrate the workings of the unitary gates consider, from Eq. 8.35, the following fragment

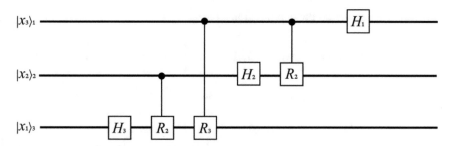

Fig. 8.2 Quantum Fourier transform for three qubits. Published with permission of © Belal E. Baaquie and L. C. Kwek. All Rights Reserved

$$\mathcal{U}_{23} H_3 \left(|x_3\rangle_1 |x_2\rangle_2 |x_1\rangle_3 \right)$$

$$= \mathcal{U}_{23} \left(|x_3\rangle_1 |x_2\rangle_2 \left[|0\rangle + \exp\left\{ 2\pi i \frac{x_1}{2} \right\} |1\rangle \right]_3 \right)$$

$$= \left\{ \mathbb{I} \otimes |0\rangle\langle 0| \otimes \mathbb{I} + \mathbb{I} \otimes |1\rangle\langle 1| \otimes \mathcal{R}_2 \right\} \left(|x_3\rangle_1 |x_2\rangle_2 \left[|0\rangle + \exp\{ 2\pi i (0.x_1) \} |1\rangle \right]_3 \right)$$

$$= |x_3\rangle_1 \left\{ |0\rangle_2 \delta(x_2) \left[|0\rangle + \exp\{ 2\pi i (0.x_1) \} |1\rangle \right]_3 \right.$$

$$\left. + |1\rangle_2 \delta(x_2 - 1) \left[|0\rangle + \exp\{ 2\pi i (0.x_1 + 0.01) \} |1\rangle \right]_3 \right\}$$

$$= |x_3\rangle_1 |x_2\rangle_2 \left[|0\rangle + e^{2\pi i (0.x_1 x_2)} |1\rangle \right]_3 \tag{8.39}$$

since

$$|0\rangle_2 \delta(x_2) \left[|0\rangle + \exp\{ 2\pi i (0.x_1) \} |1\rangle \right]_3$$

$$+ |1\rangle_2 \delta(x_2 - 1) \left[|0\rangle + \exp\{ 2\pi i (0.x_1 + 0.01) \} |1\rangle \right]_3$$

$$= \begin{cases} |0\rangle_2 \left[|0\rangle + e^{2\pi i (0.x_1)} |1\rangle \right]_3 & : x_2 = 0 \\ |1\rangle_2 \left[|0\rangle + e^{2\pi i (0.x_1 x_2)} |1\rangle \right]_3 & : x_2 = 1 \end{cases}$$

$$= |x_2\rangle_2 \left[|0\rangle + e^{2\pi i (0.x_1 x_2)} |1\rangle \right]_3 \tag{8.40}$$

The other gates in Eq. 8.35 have similar action on the qubits and yield

$$\mathcal{F}_3|x_1x_2x_3\rangle = H_1\mathcal{U}_{12}H_2\mathcal{U}_{13}\mathcal{U}_{23}H_3\big(|x_3\rangle_1|x_2\rangle_2|x_1\rangle_3\big)$$

$$= \frac{1}{\sqrt{2}}H_1\mathcal{U}_{12}H_2\mathcal{U}_{13}\mathcal{U}_{23}\Big(|x_3\rangle_1|x_2\rangle_2\big[|0\rangle + \exp\Big\{2\pi i\frac{x_1}{2}\Big\}|1\rangle\big]_3\Big)$$

$$= \frac{1}{\sqrt{2}}H_1\mathcal{U}_{12}H_2\Big(|x_3\rangle_1|x_2\rangle_2\big[|0\rangle + e^{2\pi i(0.x_1x_2x_3)}|1\rangle\big]_3\Big)$$

$$= \frac{1}{\sqrt{2^2}}H_1\Big(|x_3\rangle_1\big[|0\rangle + e^{2\pi i(0.x_2x_3)}|1\rangle\big]_2\big[|0\rangle + e^{2\pi i(0.x_1x_2x_3)}|1\rangle\big]_3\Big)$$

$$= \frac{1}{\sqrt{2}}\big[|0\rangle + e^{2\pi i(0.x_3)}|1\rangle\big]_1 \cdot \frac{1}{\sqrt{2}}\big[|0\rangle + e^{2\pi i(0.x_2x_3)}|1\rangle\big]_2$$

$$\times \frac{1}{\sqrt{2}}\big[|0\rangle + e^{2\pi i(0.x_1x_2x_3)}|1\rangle\big]_3 \tag{8.41}$$

Equation 8.41 is the result given earlier in Eq. 8.31, verifying that the unitary gate $\mathcal{F}_3$ yields the quantum Fourier transform. Note the order of the factors in Eq. 8.41 is correct since the swap gate $\mathcal{S}_3$ was used to reorder the qubits before applying the quantum gates.

8.5 Quantum Circuit of qFT

For the general case, we need to obtain the n-qubit quantum Fourier transformed state starting from the input state $|x_1x_2\ldots x_n\rangle$ to output state $|y_1y_2\ldots y_n\rangle$ is given by the unitary transformation (gate) $\mathcal{F}$ and, from Eq. 8.29 yields

$$\mathcal{F}(|x\rangle) = \mathcal{F}\big(|x\rangle\big) = \frac{1}{\sqrt{2^n}}\sum_{y=0}^{2^n-1}\exp\Big\{i\frac{2\pi xy}{2^n}\Big\}|y\rangle$$

The construction of a qFT using a quantum circuit is shown in Fig. 8.3 and discussed in detail in [1]. As was illustrated for the $n = 3$ case, the quantum Fourier transform needs two gates, one is the Hadamard gate and the other is a two-qubit controlled rotation $\mathcal{U}_{jk}$.

The two-qubit controlled rotation $\mathcal{U}_{ij}$ acts on $|x_i\rangle$, $|x_j\rangle$, where the first qubit $|x_i\rangle$, $i = 1,\ldots,n$ is the target qubit and the second qubit $|x_j\rangle$ is the control qubit. Similar to the case of $n = 3$ (in 2×2 block diagonal notation), for the conditioning qubit $|x_j\rangle$ and target qubit $|x_i\rangle$, we have

$$\mathcal{U}_{ij} = \underbrace{\mathbb{I}\otimes\mathbb{I}\cdots|0\rangle\langle 0|\otimes\mathbb{I}\cdots\mathbb{I}\otimes\mathbb{I}}_{\text{control qubit: }|0\rangle\langle 0|\ j\text{th position}}\oplus\underbrace{\mathbb{I}\otimes\mathbb{I}\cdots|1\rangle\langle 1|\otimes\mathbb{I}\cdots\otimes\mathcal{R}_k\cdots\otimes\mathbb{I}}_{\mathcal{R}_k\text{ acting on }i\text{th target qubit}} \tag{8.42}$$

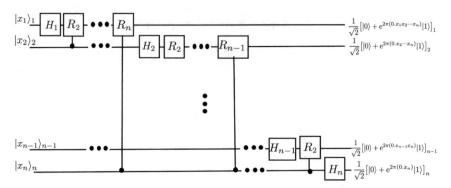

Fig. 8.3 Circuit diagram for quantum Fourier transform. Published with permission of © Belal E. Baaquie and L. C. Kwek. All Rights Reserved

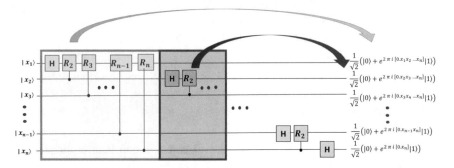

Fig. 8.4 Gates for the quantum Fourier transform. Published with permission of © Belal E. Baaquie and L. C. Kwek. All Rights Reserved

where $k = j - i$ and as mentioned before $\mathcal{R}_k$ is a 2×2 diagonal matrix given by

$$\mathcal{R}_k = \begin{bmatrix} 1 & 0 \\ 0 & \exp\{2\pi i/2^k\} \end{bmatrix} \tag{8.43}$$

Hence, as shown in the circuit diagram given in Fig. 8.3, the quantum Fourier transform is given by

$$\mathcal{F}(|x\rangle) = H_n \mathcal{U}_{n,n-1} H_{n-1} \ldots H_3 \mathcal{U}_{3,n-1} \ldots \mathcal{U}_{3,4} H_2 \mathcal{U}_{2,n} \ldots \mathcal{U}_{2,3} H_1 |x\rangle \tag{8.44}$$

Note a general feature of quantum circuit diagrams exemplified in Eq. 8.42 is that the order of the gates (operators) given in the circuit diagrams is *reversed* when these operators act on the state vector [3].

A graphical representation of some features of the algorithm that generates the quantum Fourier transform is shown in Fig. 8.4.

Hence, similar to the derivation for the $n = 3$ case, applying the gates as shown in Fig. 8.3 yields

$$\mathcal{F}|x_1 x_2 \ldots x_n\rangle = \frac{1}{\sqrt{2}}\Big[|0\rangle + e^{2\pi i (0.x_1 \ldots x_n)}|1\rangle\Big]_n \cdot \frac{1}{\sqrt{2}}\Big[|0\rangle + e^{2\pi i (0.x_2 \ldots x_n)}|1\rangle\Big]_n$$

$$\ldots \frac{1}{\sqrt{2}}\Big[|0\rangle + e^{2\pi i (0.x_{n-1} x_n)}|1\rangle\Big]_2 \cdot \frac{1}{\sqrt{2}}\Big[|0\rangle + e^{2\pi i (0.x_n)}|1\rangle\Big]_1 \quad (8.45)$$

Comparing Eqs. 8.29 and 8.45, we have obtained the quantum Fourier transform of the n-qubit $|x\rangle$ upto to a cyclical *reordering* of the n-qubits, which can be realized by the Swap gate. For the $n = 3$, the swap was implemented before applying the quantum gates and hence the final result did not need a reordering.

If the number of qubits is large, the angle of rotation $x/2^n$ becomes small for many of the qubits. An approximate qFT can be defined for which $x/2^n > L$, with the terms with rotation smaller than L being ignored. This approximation greatly reduces the number of operations and is useful for many applications.

References

1. Phillip K, Raymond L, Michele M (2007) Introduction to quantum computing. Cambridge University Press, USA
2. David M (2008) Quantum computing explained. MIT Press, USA
3. Mermin ND (2007) Quantum computer science. Cambridge University Press, UK
4. Nielsen MA, Chuang IL (2012) Quantum computation and quantum information. Cambridge University Press, UK
5. Nielsen MA, Chuang I (2002) Quantum computation and quantum information

Part II
Quantum Algorithms

Chapter 9
Deutsch Algorithm

The Deutsch algorithm illustrates, using a very special example, that a quantum computer, *in principle*, is more efficient than a classical computer.

Consider a binary function f of a single bit. There are four possible binary valued functions and are given below.

Table 9.1 shows that the four functions can be grouped as follows: a *constant function*, that consists of f_0, f_1—a constant function gives the same output as the input—and a *balanced function*, consisting of f_3, f_4, which gives an equal number of 0's and 1's as output.

The question posed by Deutsch is the following: how many times does the function f have to be measured in order to determine whether it is a constant or a balanced function? The answer for a classical computer is that the function f has to be called *at least two times* and both $f(0)$, $f(1)$ need to be evaluated. One then has the following result classical result, requiring two measurements for each case:

- $f(0) = f(1)$, implying the function is a constant since
  either $f_0(0) = 0 = f_0(1)$ or $f_1(0) = 0 = f_1(1)$
- $f(0) \neq f(1)$, in which case it is a balanced function.

Table 9.1 Binary function of one bit

Function	$x = 0$	$x = 1$
Constant		
f_0	0	0
f_1	1	1
Balanced		
f_3	0	1
f_4	1	0

© The Author(s), under exclusive license to Springer Nature Singapore Pte Ltd. 2023
B. E. Baaquie and L.-C. Kwek, *Quantum Computers*,
https://doi.org/10.1007/978-981-19-7517-2_9

The Deutsch algorithm shows that the quantum computer needs to call the function *only once* to decide whether is a constant or a balanced function, in contrast to a classical algorithm that needs to call it at least twice. However, unlike the classical case where both the values $f(0)$, $f(1)$ are know, the Deutsch algorithm cannot determine the individual values $f(0)$, $f(1)$, only if it is a constant or balanced. This is a general feature of quantum algorithms, that a very specific question needs to be asked to see its quantum advantage over a classical algorithm.

9.1   The Deutsch Quantum Circuit

The quantum circuit for the Deutsch algorithm is given in Fig. 9.1. Note at the heart of the Deutsch algorithm is a gate called the oracle U_f that, as shown in Fig. 9.1b, reads input x and produces output $f(x)$ and carries out the following unitary transformation

$$U_f|x\rangle|y\rangle = |x\rangle|y \oplus f(x)\rangle \tag{9.1}$$

where

$$|x\rangle : \text{input state};\quad |y\rangle : \text{ancillary state}$$

The oracle is unitary since, using binary addition $f(x) \oplus f(x) = 0$ yields

$$U_f^2|x\rangle|y\rangle = U_f|x\rangle|y \oplus f(x)\rangle = |x\rangle|y\rangle$$

The qubit $|x\rangle|y \oplus f(x)\rangle$ obtained by the application of the oracle is an *entangled* quantum state and forms a necessary part of the quantum circuit. Entangled states cannot be factorized into a tensor product of the ingredient states; in the case of Eq. 9.1, as it stands, the ancillary state $|y\rangle$ cannot be factored from the input qubit $|x\rangle$. However, it will be shown that due to the binary nature of the ancillary state $|y\rangle$,

Fig. 9.1 a Quantum circuit for the Deutsch algorithm. **b** The action of the U_f gate, also called the oracle. Published with permission of © Belal E. Baaquie and L. C. Kwek. All Rights Reserved

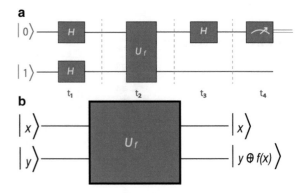

a transformation is possible that removes the entanglement and leads to a simpler quantum algorithm.

In fact for all quantum algorithms considered in this book, the oracle always behaves like the expression given in Eq. 9.1, with the only difference being in the nature of the qubit and the function f.

The oracle needs to know the function f; in the algorithm, we do not know whether the function is constant or balanced until a measurement is made, but the oracle knows the properties of the function during the process of the quantum computation.

The initial input and ancillary qubits, at time t_0, are given by $|\psi\rangle = |0\rangle|1\rangle$. The following steps are taken sequentially in time, as shown in Fig. 9.1.

- $|\psi(t_0)\rangle = |0\rangle|1\rangle$
- $|\psi(t_1)\rangle = H^{\otimes 2}|\psi(t_0)\rangle = H^{\otimes 2}|0\rangle|1\rangle = (H \otimes \mathbf{I})(\mathbf{I} \otimes H)|0\rangle|1\rangle$
- $|\psi(t_2)\rangle = U_f|\psi(t_1)\rangle$
- $|\psi(t_3)\rangle = (H \otimes \mathbf{I})|\psi(t_2)\rangle$
- Measurement at time t_4 of is only of the input qubit: $\text{Tr}(|0\rangle\langle 0|\psi(t_3)\rangle)$.

The steps leading to the result are the following. From Eq. 7.4

$$|\psi(t_1)\rangle = H|0\rangle H|1\rangle = \frac{1}{2}\Big[|0\rangle + |1\rangle\Big]\Big[|0\rangle - |1\rangle\Big]$$
$$= \frac{1}{2}\Big\{|0\rangle\big[|0\rangle - |1\rangle\big] + |1\rangle\big[|0\rangle - |1\rangle\big]\Big\} \tag{9.2}$$

The oracle is the gate U_f defined in Eq. 9.1 and shown in Fig. 9.1, is given by

$$U_f\Big(|x\rangle|y\rangle\Big) = |x\rangle|y \oplus f(x)\rangle$$

Hence, using binary addition $0 \oplus f = f$, yields from Eq. 9.2

$$|\psi(t_2)\rangle$$
$$= U_f|\psi(t_1)\rangle = \frac{1}{2}U_f\Big\{\Big[|0\rangle\big[|0\rangle - |1\rangle\big] + |1\rangle\big[|0\rangle - |1\rangle\big]\Big]\Big\}$$
$$= \frac{1}{2}\Big[|0\rangle\Big\{|0 \oplus f(0)\rangle - |1 \oplus f(0)\rangle\Big\} + |1\rangle\Big\{|1 \oplus f(1)\rangle - |1 \oplus f(1)\rangle\Big\}\Big] \tag{9.3}$$

Case (i): $f(0) = f(1) = f$. From Eq. 9.3

$$|\psi(t_2)\rangle = \frac{1}{2}\Big[|0\rangle\Big\{|f\rangle - |1 \oplus f\rangle\Big\} + |1\rangle\Big\{|f\rangle - |1 \oplus f\rangle\Big\}\Big]$$
$$= \frac{1}{2}\Big[|0\rangle + |1\rangle\Big]\Big[|f\rangle - |1 \oplus f\rangle\Big] \tag{9.4}$$

Note the following important identity, which is used in many of the quantum algorithms discussed later.

$$|f\rangle - |1 \oplus f\rangle = \begin{cases} \Big[|0\rangle - |1\rangle\Big] & : f = 0 \\ -\Big[|0\rangle - |1\rangle\Big] & : f = 1 \end{cases}$$

$$\Rightarrow |f\rangle - |1 \oplus f\rangle = (-1)^f \Big[|0\rangle - |1\rangle\Big] \tag{9.5}$$

Equation 9.5 shows that the input qubit has been dis-entangled from the ancillary qubit due to the binary nature of the qubits and leads to many further simplifications.

Hence, from Eq. 9.5, we have

$$|\psi(t_2)\rangle = (-1)^f \frac{1}{2}\Big[|0\rangle + |1\rangle\Big]\Big[|0\rangle - |1\rangle\Big] : f(0) = f(1) \tag{9.6}$$

Applying the Hadamard gate to the input qubit yields

$$|\psi(t_3)\rangle = (H \otimes \mathbf{I})|\psi(t_2)\rangle = (-1)^f |0\rangle \Big[\frac{|0\rangle - |1\rangle}{\sqrt{2}}\Big] : f(0) = f(1)$$

Case (ii): $f(0) \neq f(1)$. From Eq. 9.3, using the binary identities

$$f(0) = 1 \oplus f(1); \quad 1 \oplus f(0) = f(1)$$

yields

$$\begin{aligned} |\psi(t_2)\rangle &= \frac{1}{2}\Big[|0\rangle\big\{|f(0)\rangle - |1 \oplus f(0)\rangle\big\} + |1\rangle\big\{|f(1)\rangle - |1 \oplus f(1)\rangle\big\}\Big] \\ &= \frac{1}{2}\Big[|0\rangle\big\{|f(0)\rangle - |f(1)\rangle\big\} + |1\rangle\big\{|f(1)\rangle - |f(0)\rangle\big\}\Big] \\ &= -\frac{1}{2}\Big[|0\rangle - |1\rangle\Big]\Big[|f(1)\rangle - |1 \oplus f(1)\rangle\Big] \end{aligned} \tag{9.7}$$

From Eq. 9.5

$$\begin{aligned} |\psi(t_2)\rangle &= -(-1)^{f(1)}\frac{1}{2}\Big[|0\rangle - |1\rangle\Big]\Big[|0\rangle - |1\rangle\Big] \\ &= (-1)^{f(0)}\frac{1}{2}\Big[|0\rangle - |1\rangle\Big]\Big[|0\rangle - |1\rangle\Big] : f(0) \neq f(1) \end{aligned} \tag{9.8}$$

Applying the Hadamard gate to the input qubit yields

$$|\psi(t_3)\rangle = (H \otimes \mathbf{I})|\psi(t_2)\rangle = (-1)^{f(0)}|1\rangle\Big[\frac{|0\rangle - |1\rangle}{\sqrt{2}}\Big] : f(0) \neq f(1)$$

Note the fact the ancillary qubit for both cases is unchanged by the quantum circuit. This is the feature of the ancillary for many quantum algorithms, and the ancillary qubit can be thought of as a 'catalyst', coming out unchanged in the final state. The ancillary qubit is not observed in the final state as it carries no information about the function f.

Summarizing the results, we have

$$|\psi(t_3)\rangle = \begin{cases} (-1)^f |0\rangle \left[\frac{|0\rangle - |1\rangle}{\sqrt{2}}\right] & : f(0) = f(1) = f \\ (-1)^{f(0)} |1\rangle \left[\frac{|0\rangle - |1\rangle}{\sqrt{2}}\right] & : f(0) \neq f(1) \end{cases} \tag{9.9}$$

Note the sign $(-1)^f = \pm$ on the final state is irrelevant since, as mentioned before, the quantum state is only measured up to a phase.

Measurement of the single binary degree of freedom is executed by applying the measurement gates, of which there are two gates for the single qubit, to obtain the final state of the quantum circuit. The measurement gates are

$$M_0 = |0\rangle\langle 0| \otimes \mathbf{I}; \quad M_1 = |1\rangle\langle 1| \otimes \mathbf{I}$$

A single measurement is performed, ignoring the ancillary qubit, which is one of the simplest examples of a generalized Born measurement; the result of the measurement, from Sect. 4.10, is given by

$$|\psi(t_3)\rangle \rightarrow \text{Measurement} \rightarrow |x\rangle \frac{1}{\sqrt{2}}\Big[|0\rangle - |1\rangle\Big]; \quad x = 0, 1$$

The probability for the two possible outcomes is given by

$$\text{Tr}\Big(M_x|\psi(t_3)\rangle\langle\psi(t_3)|\Big) = \begin{cases} 1 \text{ if } x = 0 & : f(0) = f(1) \\ 1 \text{ if } x = 1 & : f(0) \neq f(1) \end{cases}$$

Hence, a single measurement for input degree of freedom yields the desired result: if the result of the measurement for outcome $|0\rangle$ has the probability 1, then the function is a constant, and if not, then the function is balanced . Note that, as mentioned before, the result of the measurement does not yield any information of what are the values of $f(0)$, $f(1)$, but instead only if the functions are constant or balanced.

We rewrite the result of the Deutsch algorithm to make a connection with its generalization given by Deutsch–Jozsa algorithm. Note that

$$f(0) \oplus f(1) = 0 : \text{constant}; \quad f(0) \oplus f(1) = 1 : \text{balanced}$$

We can hence rewrite Eqs. 9.6 and 9.8 as follows

$$|\psi(t_2)\rangle = (-1)^{f(0)} \frac{1}{2}\Big[|0\rangle + (-1)^{f(0)\oplus f(1)}|1\rangle\Big]\Big[|0\rangle - |1\rangle\Big] \tag{9.10}$$

Ignoring the auxiliary qubit and applying the Hadamard transformation on the input qubit yields

$$H\Big[|0\rangle + (-1)^{f(0)\oplus f(1)}|1\rangle\Big] = \Big[H|0\rangle + (-1)^{f(0)\oplus f(1)}H|1\rangle\Big]$$

$$= \frac{1}{\sqrt{2}}\Big[|0\rangle + |1\rangle + (-1)^{f(0)\oplus f(1)}(|0\rangle - |1\rangle)\Big]$$

$$= \frac{1}{\sqrt{2}}\Big[\{1 + (-1)^{f(0)\oplus f(1)}\}|0\rangle + \{1 - (-1)^{f(0)\oplus f(1)}|1\rangle\}\Big] \qquad (9.11)$$

Hence, from Eqs. 9.10 and 9.11 we obtain

$$|\psi(t_3)\rangle = (-1)^{f(0)}\frac{1}{2}\Big[\{1 + (-1)^{f(0)\oplus f(1)}\}|0\rangle + \{1 - (-1)^{f(0)\oplus f(1)}|1\rangle\}\Big]$$
$$\times\Big[\frac{|0\rangle - |1\rangle}{\sqrt{2}}\Big] \qquad (9.12)$$

and yields the expected result given in Eq. 9.9

$$|\psi(t_3)\rangle = \begin{cases} (-1)^{f(0)}|0\rangle\Big[\frac{|0\rangle - |1\rangle}{\sqrt{2}}\Big] & : f(0) = f(1) = f \\ (-1)^{f(0)}|1\rangle\Big[\frac{|0\rangle - |1\rangle}{\sqrt{2}}\Big] & : f(0) \neq f(1) \end{cases}$$

The expression given in Eq. 9.12 yields the following

$$\begin{cases} \frac{1}{4}|(-1)^{f(0)} + (-1)^{f(1)}|^2 & : \text{Probability to observe } |0\rangle \\ \frac{1}{4}|(-1)^{f(0)} - (-1)^{f(1)}|^2 & : \text{Probability to observe } |1\rangle \end{cases} \qquad (9.13)$$

The result given in Eq. 9.13 is the special case of the result obtained for the Deutsch–Jozsa algorithm and given in Eq. 10.5.

Chapter 10
Deutsch–Jozsa Algorithm

The Deutsch–Jozsa algorithm generalizes the Deutsch algorithm to the case of n-degrees of freedom. The derivation of the more complex quantum Fourier transform algorithm can be adapted to yield the result. The main reason for analyzing the Deutsch–Jozsa algorithm is to illustrate the cancelation of amplitudes that arises from representing the n-qubits as the superposition of the computational basis states.

Consider n qubits together with a single ancillary qubit. On applying the Hadamard gate to the qubit $|0\rangle^{\otimes n}$, the result is all possible combinations of the qubit computational basis with equal amplitude. There are 2^n terms in the expansion, and let $x = \{0, 1, 2, \ldots, 2^n - 1\}$ be a labeling of the states. The notation is based on the mapping of x to the binary basis, as given in Eq. 2.2, and yields the Dirac ket vector

$$x = |x_1 x_2 \ldots x_n\rangle; \quad x_i = 0, 1 : i = 1, 2, \ldots, n$$

The binary expansion of x in terms of the x_i yields the following N values for x

$$0 \leq x \leq N - 1; \quad N = 2^n$$

The circuit shown in Fig. 10.1 has the following components.

$$|\psi(t_0)\rangle = |0\rangle^{\otimes n}|1\rangle$$

and

$$|\psi(t_1)\rangle = \left(H^{\otimes n}|0\rangle^{\otimes n}\right)H|1\rangle = \frac{1}{\sqrt{N}}\sum_{x=0}^{N-1}|x\rangle\frac{1}{\sqrt{2}}\left[|0\rangle - |1\rangle\right]; \quad N = 2^n$$

© The Author(s), under exclusive license to Springer Nature Singapore Pte Ltd. 2023
B. E. Baaquie and L.-C. Kwek, *Quantum Computers*,
https://doi.org/10.1007/978-981-19-7517-2_10

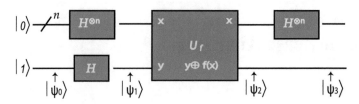

Fig. 10.1 The Deutsch–Jozsa algorithm. Published with permission of © Belal E. Baaquie and L. C. Kwek. All Rights Reserved

Hence

$$|\psi(t_2)\rangle = U_f|\psi(t_1)\rangle = \frac{1}{\sqrt{N}}\sum_{x=0}^{N-1}|x\rangle\frac{1}{\sqrt{2}}\Big[|0 \oplus f(x)\rangle - |1 \oplus f(x)\rangle\Big]$$

Recall from Eq. 9.5

$$\frac{1}{\sqrt{2}}[|f\rangle - |1 \oplus f\rangle] = (-1)^f \frac{1}{\sqrt{2}}\Big[|0\rangle - |1\rangle\Big] = (-1)^f H|1\rangle$$

and hence

$$|\psi(t_2)\rangle = U_f|\psi(t_1)\rangle = \frac{1}{\sqrt{N}}\sum_{x=0}^{N-1}(-1)^{f(x)}|x\rangle H|1\rangle \tag{10.1}$$

All the computational basis states in Eq. 10.1 have the same likelihood of occurrence so a measurement carried out on $|\psi(t_2)\rangle$ will yield no information. Another Hadamard transformation $H^{\otimes n}$ needs to done on n-qubit state $|\psi(t_2)\rangle$ so that the algorithm can leverage on the 'magic' of constructive and destructive interference that occurs for superposed states and yield the sought for result.

The binary function $f(x)$ is a mapping from $\{0, 1\}^{\otimes n} \rightarrow \{0, 1\}$. Hence,

$$f(x) = \text{constant} : \text{constant function}$$

If $f(x)$ is balanced, then

$$f(x) = \text{equal number of zeroes and ones} : \text{balanced function}$$

To write the Hadamard transformation for the n-qubit, consider first the action of a single Hadamard H gate on a single qubit. Recall from Eq. 7.4

$$H|0\rangle = \frac{1}{\sqrt{2}}(|0\rangle + |1\rangle); \quad H|1\rangle = \frac{1}{\sqrt{2}}(|0\rangle - |1\rangle)$$

$$\Rightarrow H|x\rangle = \frac{1}{\sqrt{2}}\sum_{y=0}^{1}(-1)^{xy}|y\rangle \tag{10.2}$$

Hence, for the n-qubit, Eq. 10.2 yields

$$H^{\otimes n}|x_1, x_2, \ldots, x_n\rangle = \frac{1}{\sqrt{2^n}}\sum_{y_1,y_2,\ldots,y_n=0}^{1}(-1)^{x_1 y_1 \oplus x_2 y_2 \oplus \cdots \oplus x_n y_n}|y_1, y_2, \ldots, y_n\rangle$$

$$\Rightarrow H^{\otimes n}|x\rangle = \frac{1}{\sqrt{2^n}}\sum_{y=0}^{1}(-1)^{xy}|y\rangle; \quad xy = x_1 y_1 \oplus x_2 y_2 \oplus \cdots \oplus x_n y_n \tag{10.3}$$

Equation 10.3 is written in abbreviated vector notation with $|x\rangle$ (and $|y\rangle$ as well) stands for the label of n qubits

$$|x\rangle = |x\rangle^{\otimes n} = |x_1 x_2 \ldots x_n\rangle; \quad x_i = 0, 1 : i = 1, 2, \ldots, n$$

Applying $H^{\otimes n}$ to n-qubit state $|\psi(t_2)\rangle$ given in Eq. 10.1 yields

$$|\psi(t_3)\rangle = \left(H^{\otimes n} \otimes \mathbb{I}\right)|\psi(t_2)\rangle = \frac{1}{N}\sum_{y}\left[\sum_{x=0}^{N-1}(-1)^{xy\oplus f(x)}\right]|y\rangle H|1\rangle \tag{10.4}$$

Performing a generalized Born measurement of only the input qubits (and not of the auxiliary qubit) yields, using the results of Sect. 4.10, the following

$$|\psi(t_3)\rangle \rightarrow |y\rangle H|1\rangle$$

The measurement of all the qubits results in $|y\rangle$, which is one of the possible determinate value of the qubits. The probability $P(y)$ for obtaining the determinate value $|y\rangle$ as the final state is given by

$$P(y) = \frac{1}{N^2}\left|\sum_{x=0}^{N-1}(-1)^{xy\oplus f(x)}\right|^2$$

Similar to the Deutsch algorithm, to check whether the binary function is balanced or a constant, we need to find the probability that the final state of the input qubits is given by

$$|y\rangle = |000\ldots000\rangle$$

In other words, what is probability $P(0)$ that the state $|\psi(t_3)\rangle$ collapses to the state $|0\rangle^{\otimes n}$?

Let $\mathcal{M}_0 = |0\rangle^{\otimes n \,\otimes n}\langle 0| \otimes \mathbf{I}$; then

$$P(0) = \mathrm{Tr}\Big(\mathcal{M}_0 |\psi(t_3)\rangle\langle\psi(t_3)|\Big) = \frac{1}{N^2}\left| \sum_{x=0}^{N-1}(-1)^{xy\oplus f(x)}\right|^2_{\Big|_{y=0}}$$

$$= \frac{1}{N^2}\left| \sum_{x=0}^{N-1}(-1)^{f(x)}\right|^2 \tag{10.5}$$

$$= \begin{cases} 1 & : f(x) \;:\; \text{constant} \\ 0 & : f(x) \;:\; \text{balanced} \end{cases}$$

Recall the sum over x is over all basis states. The Deutsch–Jozsa algorithm states that the quantum computation yields a final state which, with one observation, can unambiguously decide whether the function is constant or balanced. As noted earlier, Eq. 9.13 is the special case of Eq. 10.5 for $x = 0, 1$.

If the function $f(x)$ is a *constant*, then *every term* in the sum in Eq. 10.5 is either $+1$ (for $f(0) = 0$) or -1 (for $f(0) = 1$); hence, there is *constructive interference* of all the terms and yields

$$\frac{1}{N^2}(\pm 1)^2\left| \sum_{x=0}^{N-1}1\right|^2 = 1 : \text{constant function}$$

leading to probability 1 (certainty) that the input state $|0\rangle^{\otimes n}$ will be observed when the degrees of freedom for the output state are measured.

If the function is *balanced*, then there are equal number of $+1$ and -1 terms in Eq. 10.5 and yields

$$\frac{1}{N^2}|\sum_{x=0}^{N-1}(-1)^{f(x)}|^2 = 0 : \text{balanced function}$$

since all the terms exactly cancel to zero. In other words, there is *destructive interference*, with the probability of the output qubit being in the state $|0\rangle^{\otimes n}$ is zero.

To know precisely what are the values $f(x)$ for the balanced case needs more information, including the complete definition of the function $f(x)$. For the purpose of the Deutsch–Jozsa algorithm, the detailed knowledge of $f(x)$ is not necessary.

To determine whether the function $f(x)$ is constant or balanced, a classical computer would have to call the function $2^{n-1} + 1$ times, whereas the quantum computer can obtain the result with only one measurement. This is an enormous and exponential improvement. However, this example is not very useful as it does not have much utility in practice.

If a classical randomized algorithm is used, the number of times the function needs to be called is much less than 2^n; but if one wants to be certain about the result, $2^{n-1} + 1$ calls of the function are needed for the classical case.

The constructive or destructive interference of the superposed quantum state given in Eq. 10.4, and which results in Eq. 10.5, is essentially a quantum effect that has no analog in classical algorithms. Quantum superposition and quantum entanglement—two resources unique to quantum algorithms—are, as mentioned earlier, at the root of the advantage quantum computers and algorithms have over the classical case.

Chapter 11
Grover's Algorithm

Grover's algorithm and Shor's algorithm for factorizing (large) primes are the two masterpieces of quantum computing. Although Grover's algorithm is usually considered in the context of an efficient search for entries in a large database , its potential uses are pervasive, apparently occurring even in the biological process of protein synthesis and, in material science, where electrons perform a Grover search in the search for defects in two-dimensional lattices.[1]

Consider the following problem. Suppose one has a database of one million entries. Each entry is, and to a number (the register) and a corresponding object. Suppose one would like to find the register of a particular object. How many steps will the computer need to find the number (register) associated with the object?

The classical computer will need to compare each number with the object associated with it. For a database with size N, this will take $N/2$ steps on the average, since sometimes the correspondence may be found after a few steps, and sometimes it may need N steps. The Grover algorithm is far more efficient, requiring only $\sqrt{N}$ steps. It can be shown that no classical computer can be as efficient as the Grover algorithm. For a database with one million entries, the Grover algorithm requires only 1000 searches. This is a big saving, in terms of time and resources.

A natural question arises in the Grover search algorithm, which is, if we know that a particular qubit satisfies the criterion of the sought for result, then are not we assuming that we know the answer? The answer is 'No', since *there is a vast difference between knowing the solution and being able to recognize the solution* [1].

To illustrate the difference between knowing and recognizing a solution, consider a telephone book with the names arranged alphabetically. Suppose now one is given the phone number and one needs to find the name for that number. Since one does not have the answer, one needs to check the number with, in principle, every name in the phone book. Once one finds the name that matches the number, recognizing

---

[1] https://www.technologynetworks.com/informatics/news/important-quantum-algorithm-may-be-a-property-of-nature-324062.

© The Author(s), under exclusive license to Springer Nature Singapore Pte Ltd. 2023
B. E. Baaquie and L.-C. Kwek, *Quantum Computers*,
https://doi.org/10.1007/978-981-19-7517-2_11

the name is automatic. Hence, recognizing the solution is different from knowing the solution. For a telephone book with a million names, doing the search using a classical computer will take roughly a million steps whereas the Grover algorithm can complete the search in a thousand steps.

Another example of the difference between knowing and recognizing a solution is the factorization of primes. Consider a search algorithm for finding the primes of a large integer $m = pq$, where p, q are primes: One needs an algorithm to find p, q. If the solution to the factorization is *known*, then one knows the values of p, q.

If one does not have the solution, then one needs to find p, q, $q < p$. For a classical algorithm, one starts by generating numbers, with x ranging from $1, \ldots, \sqrt{m}$ since the smaller prime q is less than $\sqrt{m}$. For each number x, the code carries out the relatively easy operation m/x; if x divides m, it is the number q that we are looking for. The larger prime is then determined by $p = m/x$.

Hence, *recognizing* the solution means determining whether x can divide m. Dividing m by x is far easier that directly factoring m into the prime numbers p, q. The process of checking, for each x, whether x divides m allows the search algorithm to 'mark' the solution being sought. The advantage of the Grover algorithm over the classical case is discussed later in Sect. 11.8.

11.1 Phase Inversion and Amplitude Amplification

The main steps in the Grover algorithm are outlined with a simple example using intuitive reasoning to illustrate the more complex operations for the general case. The essential idea is encapsulated in the phase inversion and the amplitude amplification.

The steps of the Grover algorithm are illustrated in Fig. 11.1. Let $|x\rangle$ be the basis state for $n = 2$ qubits. The Grover algorithm's objective is to identify a specific (marked) basis state that is given to be $|\xi\rangle = |10\rangle$. It has the following four steps:

1. Create an initial state in which all the basis states are equally superposed and, for $n = 2$, is given by

$$|\psi(t_1)\rangle = \frac{1}{2^{n/2}} \sum_{i=0}^{2^n-1} |x_i\rangle = \frac{1}{2}\Big[|00\rangle + |01\rangle + |10\rangle + |11\rangle\Big]$$

   At this stage all the basis states $|x_i\rangle$ have the same amplitude and shown in Fig. 11.1a.
2. Construct the **oracle** gate to identify and mark the state being sought by *flipping* the basis state $|\xi\rangle$, which is done by inverting the phase of the marked state from 1 to -1, shown in Fig. 11.1b, and given by

$$|\psi(t_1)\rangle \rightarrow |\psi(t_2)\rangle = \frac{1}{2}\Big[|00\rangle + |01\rangle - |10\rangle + |11\rangle\Big]$$

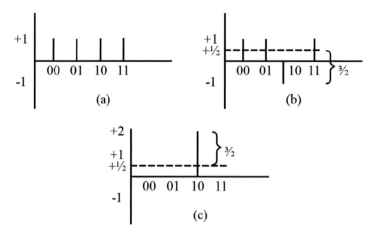

Fig. 11.1 Phase inversion and amplification in Grover's algorithm. Published with permission of © Belal E. Baaquie and L. C. Kwek. All Rights Reserved

3. The flipped basis state $|\xi\rangle$ is then amplified using what is called the Grover **diffusion gate**, which is not required in this example.
4. The mean of all the amplitudes after flipping the marked state $|\xi\rangle$ is $+1/2$. All the basis states are reflected about their mean, which in effect, reduces the amplitude of all the basis states except $|\xi\rangle$, which is further amplified, and shown in Fig. 11.1c and given by

$$|\psi(t_2)\rangle \rightarrow W|\psi(t_2)\rangle = |10\rangle = |\xi\rangle$$

5. The Grover algorithm stops after these three steps since the output is equal to the marked state with probability one (certainty).
6. For the general case, the process is continued until sufficient accuracy is attained so that a measurement, with high likelihood, only detects the marked state $|\xi\rangle$.

One can see from the steps enumerated above that the Grover algorithm first performs a phase inversion to identify the qubit being sought. In general, the algorithm amplifies the amplitude for the marked basis state and then does another inversion to attenuate the amplitude of all the basis states, except the marked state.

11.2 Grover's Quantum Circuit

We now discuss the general case for n-qubits. Let $|x\rangle$ signify the basis states that are being scanned, and for each $|x\rangle$ the oracle U_f looks up the name associated with the number. Let ξ correspond to the number that is associated with the name that one is looking for. The search function f is defined as follows

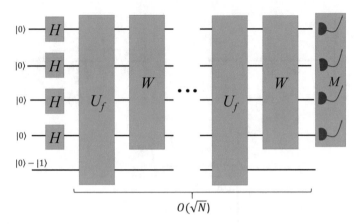

Fig. 11.2 Quantum circuit for the Grover algorithm. M in the figure stands for measurement.
Published with permission of © Belal E. Baaquie and L. C. Kwek. All Rights Reserved

$$f(x) = \begin{cases} 0 & : x \neq \xi \\ 1 & : x = \xi \end{cases} \tag{11.1}$$

The oracle , similar to the Deutsch algorithm, carries out the following transformation

$$U_f |x\rangle |y\rangle = |x\rangle |y \oplus f(x)\rangle \ ; \ x, y = 0, 1, 2, \ldots, 2^n - 1 \tag{11.2}$$

where, for $N = 2^n$, we are using the following notation

$$|x\rangle \equiv \underbrace{|x_1\rangle \otimes |x_2\rangle \cdots \otimes |x_n\rangle}_{n \text{ times}} \ ; \ |y\rangle \equiv \underbrace{|y_1\rangle \otimes |y_2\rangle \cdots \otimes |y_n\rangle}_{n \text{ times}} \ : \ N = 2^n$$

The oracle U_f looks up for every string of $|x\rangle$, say from the phone book that is supplied to it, whether the string corresponds to the name one is looking for. If it finds that one of the string ξ exactly matches the name, it gives an output of 1, otherwise the output is 0. The algorithm provides no information of what is the value of the string ξ until a measurement is performed.

The transformation is applied, in Grover's case, $\sqrt{N}$ times, we see that at every time step, all the possible values of the string are updated, with only the amplitude of $x = \xi$ qubit increasing. It is only at the end of the computation, when a measurement is performed, that the result for ξ emerges. The Grover algorithm is an inverse function algorithm: given $f(x)$, the algorithm finds ξ.

In summary, the output of the process of quantum computation for the Grover algorithm produces, with a high likelihood, the qubit $|\xi\rangle$.

The Grover quantum circuit, shown in Fig. 11.2, has the following states and gates.

- The input state is the n-fold tensor product of the qubit $|0\rangle$ time the ancillary qubit, and given by

$$|\psi(t_0)\rangle = |0\rangle^{\otimes n} H|1\rangle = \underbrace{|0\rangle \otimes |0\rangle \cdots \otimes |0\rangle}_{n \text{ times}} H|1\rangle$$

- Apply the Hadamard operator on the initial state to yield

$$|\psi(t_1)\rangle = \left(H^{\otimes n}|0\rangle^{\otimes n}\right)H|1\rangle$$

$$= \underbrace{H|0\rangle \otimes H|0\rangle \cdots \otimes H|0\rangle}_{n \text{ times}} H|1\rangle = \frac{1}{\sqrt{N}} \sum_{x=0}^{N-1} |x\rangle H|1\rangle$$

- The oracle is defined similar to the case of the Deutsch–Jozsa algorithm by

$$U_f|x\rangle|y\rangle = |x\rangle|y \oplus f(x)\rangle = \left(\mathcal{O}_f|x\rangle\right)|y\rangle$$

where $|y\rangle = H|1\rangle$ is the auxiliary qubit.
- Apply the oracle U_f to obtain

$$|\psi(t_2)\rangle = U_f|\psi(t_1)\rangle = U_f\left(H^{\otimes n}|0\rangle^{\otimes n} H|1\rangle\right) = \left(\mathcal{O}_f\left(H^{\otimes n}|0\rangle^{\otimes n}\right)\right)H|1\rangle$$

The oracle marks the state being sought by a phase inversion.
- Apply the diffusion operator W to obtain

$$|\psi(t_3)\rangle = W|\psi(t_2)\rangle = WU_f|\psi(t_1)\rangle$$

to increase the amplitude of the marked state.
- The application of the oracle U_f and the diffusion operator W is repeated $\sqrt{N}$ times, each time increasing the amplitude of the marked qubit.
- All the qubits, except for the auxiliary qubit, are subjected to a measurement by applying projection operators $M(x), x = 1, 2, \ldots N$ on all the degrees of freedom $|x\rangle$
- The probability of the detectors registering a specific determinate value of qubit x^{Obs} for the outcome is given by $\text{Tr}\left(M(x^{Obs})|\psi(t_3)\rangle\langle\psi(t_3)|\right)$.

The oracle operator U_f has enough information to 'mark' the string state $|\xi\rangle$; once this state has been identified by a phase inversion, the diffusion operator W is applied to enhance the probability of this state being observed. The process is repeated $\sqrt{N}$ times so that one is almost certain to observe the state $|\xi\rangle$ when the measurement is performed.

11.3 Grover Algorithm: Two-Qubit

As a warm-up to the general case, we provide a more explicit derivation of the two-qubit case. Applying the Hadamard transformation yields the input state vector

$$\psi(t_1)\rangle = H^{\otimes 2}|0\rangle^{\otimes 2} H|1\rangle = \frac{1}{2}\Big[|00\rangle + |01\rangle + |10\rangle + |11\rangle\Big]\frac{1}{\sqrt{2}}\Big[|0\rangle - |1\rangle\Big] \quad (11.3)$$

Let the sought after qubit be $|\xi\rangle = |10\rangle$ and hence $f(\xi) = 1$; let U_f be the oracle gate. Equations 9.5, 11.1 and 11.2 yield

$$U_f|x\rangle|y\rangle = |x\rangle|y \oplus f(x)\rangle$$

Hence

$$U_f|10\rangle\Big[|0\rangle - |1\rangle\Big] = |10\rangle|0 \oplus 1\rangle - |10\rangle|1 \oplus 1\rangle = -|10\rangle\Big[|0\rangle - |1\rangle\Big]$$

with the rest of the qubits remaining unchanged; note the ancillary qubit $|0\rangle - |1\rangle$ has been restored to its initial value, leaving a net negative sign to the input state $|\xi\rangle = |10\rangle$. Hence, the action of the oracle U_f yields

$$|\psi(t_2)\rangle = U_f|\psi(t_1)\rangle = \frac{1}{2}\Big[|00\rangle + |01\rangle - |10\rangle + |11\rangle\Big]\frac{1}{\sqrt{2}}\Big[|0\rangle - |1\rangle\Big]$$

The ancillary qubit $\Big[|0\rangle - |1\rangle\Big]/\sqrt{2}$ does not play any further role and is dropped. The state vector having no reference to the ancillary qubit is given by the following

$$|\psi(t_2)\rangle \rightarrow \frac{1}{2}\Big[|00\rangle + |01\rangle - |10\rangle + |11\rangle\Big] \quad (11.4)$$

At this stage, if a measurement is carried out on $|\psi(t_2)\rangle$, from Eq. 4.18, all the components of $|\psi(t_2)\rangle$ would have equal likelihood of occurrence and one would not be able to detect the sought for qubit $|10\rangle$.

There is a unitary transformation (reversible gate), denoted by W, that *amplifies the amplitude* for $|10\rangle$ and is the key to the Grover algorithm. The Grover *amplification* works by *reflecting* a sequence of numbers about their mean. If a number is above the mean, it is reflected to a value below the mean. And if it is less, it is reflected to a value above the mean.

For the state vector given in Eq. 11.4, its coefficients are 1, 1, −1, 1, with their average being 1/2. The coefficients with +1 are 1/2 above the average and hence are reflected to 0; the coefficient −1 is reflected to 2 since it is below the average and is reflected to the value of 1/2 + 3/2 = 2. Hence, after the reflection we have, as shown in Fig. 11.3

Fig. 11.3 Amplification by reflection and diffusion gate in Grover's algorithm. Published with permission of © Belal E. Baaquie and L. C. Kwek. All Rights Reserved

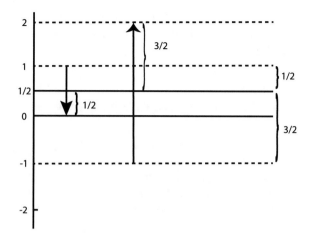

$$1, 1, -1, 1 \rightarrow \text{ reflection } \rightarrow 0, 0, 2, 0 \tag{11.5}$$

The reversible gate W makes the Grover reflection and yields the result

$$|\psi(t_3)\rangle = W|\psi(t_2)\rangle = |10\rangle = |\xi\rangle \tag{11.6}$$

For the two-qubit case, the Grover search algorithm yields the result with *one updating*. The gate W for the two qubits is given by Chris [2]

$$W = \frac{1}{2} \begin{bmatrix} -1 & 1 & 1 & 1 \\ 1 & -1 & 1 & 1 \\ 1 & 1 & -1 & 1 \\ 1 & 1 & 1 & -1 \end{bmatrix} \; ; \; |10\rangle = \begin{bmatrix} 0 \\ 1 \end{bmatrix} \otimes \begin{bmatrix} 1 \\ 0 \end{bmatrix} = \begin{bmatrix} 0 \\ 0 \\ 1 \\ 0 \end{bmatrix}$$

and the reflected state vector is given by

$$|\psi(t_2)\rangle = \frac{1}{2}\Big[|00\rangle + |01\rangle - |10\rangle + |11\rangle\Big] = \frac{1}{2}\begin{bmatrix} 1 \\ 1 \\ -1 \\ 1 \end{bmatrix}$$

It can be verified that W is reversible since $W^T W = \mathbf{I}$. The matrices and input qubits above yield Eq. 11.6, as expected

$$W|\psi(t_2)\rangle = \frac{1}{4}\begin{bmatrix} -1 & 1 & 1 & 1 \\ 1 & -1 & 1 & 1 \\ 1 & 1 & -1 & 1 \\ 1 & 1 & 1 & -1 \end{bmatrix}\begin{bmatrix} 1 \\ 1 \\ -1 \\ 1 \end{bmatrix} = \begin{bmatrix} 0 \\ 0 \\ 1 \\ 0 \end{bmatrix} = |10\rangle = |\xi\rangle$$

The state $|\xi\rangle$ is normalized to unity, as required by quantum mechanics. The amplitude for all the other qubit basis states is zero, as expected from the reflections given in Eq. 11.5.

11.4 Grover Algorithm: Phase Inversion

To generalize the two-qubit result to the case of n qubits, a considerable bit of algebra is needed. Consider n qubits together with a single ancillary qubit. Similar to Eq. 11.3, on applying the Hamamard gate to the qubit $|0\rangle^{\otimes n}$, the result is all possible combinations of the qubit computational basis with equal amplitude. There are 2^n number of terms in the expansion, and let $x = \{0, 1, \ldots, 2^n - 1\}$ be a labeling of the states.

$$|\psi(t_1)\rangle = \left(H^{\otimes n}|0\rangle^{\otimes n}\right)H|1\rangle = \frac{1}{\sqrt{N}}\sum_{x=0}^{N-1}|x\rangle\frac{1}{\sqrt{2}}\Big[|0\rangle - |1\rangle\Big] \; ; \; N = 2^n$$

Hence

$$|\psi(t_2)\rangle = U_f|\psi(t_1)\rangle = \frac{1}{\sqrt{N}}\sum_{x=0}^{N-1}|x\rangle\frac{1}{\sqrt{2}}\Big[|0 \oplus f(x)\rangle - |1 \oplus f(x)\rangle\Big]$$

Recall from Eq. 9.5

$$\frac{1}{\sqrt{2}}[|f\rangle - |1 \oplus f\rangle] = (-1)^f\frac{1}{\sqrt{2}}\Big[|0\rangle - |1\rangle\Big] = (-1)^f H|1\rangle$$

and hence

$$|\psi(t_2)\rangle = U_f|\psi(t_1)\rangle = \frac{1}{\sqrt{N}}\sum_{x=0}^{N-1}(-1)^{f(x)}|x\rangle H|1\rangle \tag{11.7}$$

From Eq. 11.7, define operator $\mathcal{O}$ and qubit state $|\phi\rangle$ by

$$|\psi(t_2)\rangle = \left(\mathcal{O}\frac{1}{\sqrt{N}}\sum_{x=0}^{N-1}|x\rangle\right)H|1\rangle \tag{11.8}$$

$$|\phi\rangle = \frac{1}{\sqrt{N}}\sum_{x=0}^{N-1}|x\rangle = \frac{1}{\sqrt{N}}\sum_{x\neq\xi}^{N}|x\rangle + \frac{1}{\sqrt{N}}|\xi\rangle \; ; \; \langle\phi|\xi\rangle = \frac{1}{\sqrt{N}} \tag{11.9}$$

$$\mathcal{O}|\phi\rangle \equiv \mathcal{O}\left[\frac{1}{\sqrt{N}}\sum_{x=0}^{N-1}|x\rangle\right] = \frac{1}{\sqrt{N}}\sum_{x=0}^{N-1}(-1)^{f(x)}|x\rangle \tag{11.10}$$

Equations 11.7 and 11.8 show that the oracle U_f can be completely replaced by the operator $\mathcal{O}$ acting only on the input qubits. For this reason, the Grover algorithm does not depend on the ancillary qubit $H|1\rangle$ and henceforth it will be dropped.

Note from Eq. 11.9 that the sought for qubit $|\xi\rangle$ has almost no overlap with the input qubit $|\phi\rangle$; hence on observing the input degrees of freedom, the likelihood of observing $|\xi\rangle$ is negligible. To increase the amplitude of the $|\xi\rangle$ qubit being the outcome of the measurement, the reflection process used for two qubits needs to be generalized for n- qubits.

Note from Eq. 11.10 that

$$\mathcal{O}\sum_{x=0}^{N-1}|x\rangle = \sum_{x=0}^{N-1}(-1)^{f(x)}|x\rangle = -|\xi\rangle + \sum_{x\neq\xi}^{N-1}|x\rangle$$

The explicit form of operator $\mathcal{O}$ is given by

$$\mathcal{O} = \mathbf{I} - 2|\xi\rangle\langle\xi| \tag{11.11}$$

and yields, as expected

$$\mathcal{O}\sum_{x=0}^{N-1}|x\rangle = \sum_{x=0}^{N-1}\left(\mathbf{I} - 2|\xi\rangle\langle\xi|\right)|x\rangle = \sum_{x=0}^{N-1}|x\rangle - 2|\xi\rangle$$

Hence

$$\frac{1}{\sqrt{N}}\mathcal{O}|\left(\sum_{x=0}^{N-1}|x\rangle\right) = |\phi\rangle - \frac{2}{\sqrt{N}}|\xi\rangle \;\; ; \;\; \mathcal{O}|\xi\rangle = -|\xi\rangle$$

11.5 Grover Diffusion Gate W

The oracle U_f has marked the qubit that we are looking for. As in the two-qubit case, the marked qubit $|\xi\rangle$ in the state function $|\psi(t_2)\rangle$ has to be *amplified* so that on measurement of the state function we observe qubit $|\xi\rangle$ with high likelihood [3].

The amplification is carried out by what is known as the diffusion operator W, which is the generalization of the reflection operator used for the two-qubit case. The *diffusion gate* W is defined by

$$W = 2|\phi\rangle\langle\phi| - \mathbf{I}$$
$$\Rightarrow W|\phi\rangle = |\phi\rangle \;\; ; \;\; W|\xi\rangle = \frac{2}{\sqrt{N}}|\phi\rangle - |\xi\rangle$$

In terms of the diffusion gate, the equation for the Grover algorithm is given by the following

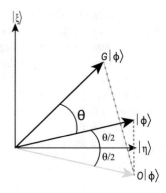

$$|\psi(t_3)\rangle = (W \otimes \mathbf{I})U_f\left(H^{\otimes n}|0\rangle^{\otimes n}H|1\rangle\right) = (W \otimes \mathbf{I})U_f\left(|\phi\rangle H|1\rangle\right)$$

$$= (W \otimes \mathbf{I})\left(\mathcal{O}|\phi\rangle H|1\rangle\right) = W\mathcal{O}|\phi\rangle H|1\rangle \equiv \left(G|\phi\rangle\right)H|1\rangle \quad (11.12)$$

where the Grover rotation operator G is defined by

$$G = W\mathcal{O} = \left(2|\phi\rangle\langle\phi| - \mathbf{I}\right)\left(\mathbf{I} - 2|\xi\rangle\langle\xi|\right) \quad (11.13)$$

From Eq. 11.12, we see that the recursion equation that needs to be studied is $G|\phi\rangle$. After k-recursions, the final state function for Grover's algorithm is

$$|\psi_k\rangle = G^k|\phi\rangle$$

The solution sought for by the algorithm, as shown below, is achieved after applying operator G $K = \pi\sqrt{N}/4$ times and yields

$$|\psi(t_3)\rangle \rightarrow G^K|\phi\rangle H|1\rangle \ ; \ \ K = \frac{\pi}{4}\sqrt{N}$$

The scheme of computation is given in Fig. 11.4. Every recursion of $G^k|\phi\rangle$ rotates the state $|\phi\rangle$ toward $|\xi\rangle$.

Consider the following decomposition

$$|\phi\rangle = \frac{1}{\sqrt{N}}\sum_{x=0}^{N-1}|x\rangle = \sqrt{\frac{N-1}{N}}\frac{1}{\sqrt{N-1}}\sum_{x\neq\xi}^{N-1}|x\rangle + \frac{1}{\sqrt{N}}|\xi\rangle$$

$$\Rightarrow |\phi\rangle = \sqrt{\frac{N-1}{N}}|\eta\rangle + \frac{1}{\sqrt{N}}|\xi\rangle \ ; \ \ |\eta\rangle = \frac{1}{\sqrt{N-1}}\sum_{x\neq\xi}^{N-1}|x\rangle \quad (11.14)$$

Define

$$\cos\left(\frac{\theta}{2}\right) = \sqrt{\frac{N-1}{N}} \quad \Rightarrow \quad \sin\left(\frac{\theta}{2}\right) = \frac{1}{\sqrt{N}} \tag{11.15}$$

and hence, from Eq. 11.14

$$|\phi\rangle = \cos\left(\frac{\theta}{2}\right)|\eta\rangle + \sin\left(\frac{\theta}{2}\right)|\xi\rangle \; ; \; \langle\eta|\xi\rangle = 0 \; ; \; \langle\eta|\eta\rangle = 1 = \langle\xi|\xi\rangle \tag{11.16}$$

A straightforward calculation using Eq. 11.13 yields

$$G|\phi\rangle = \left(1 - \frac{4}{N}\right)|\phi\rangle + \frac{2}{\sqrt{N}}|\xi\rangle \; ; \; G|\xi\rangle = |\xi\rangle - \frac{2}{\sqrt{N}}|\phi\rangle \tag{11.17}$$

From Eqs. 11.14 and 11.17

$$\sqrt{\frac{N-1}{N}}G|\eta\rangle = G|\phi\rangle - \frac{1}{\sqrt{N}}G|\xi\rangle$$

$$= (1 - \frac{2}{N})\left(\sqrt{\frac{N-1}{N}}|\eta\rangle + \frac{1}{\sqrt{N}}|\xi\rangle\right) + \frac{1}{\sqrt{N}}|\xi\rangle$$

$$\Rightarrow G|\eta\rangle = \frac{N-2}{N}|\eta\rangle + 2\frac{\sqrt{N-1}}{N}|\xi\rangle \tag{11.18}$$

Similarly, from Eqs. 11.14 and 11.17

$$G|\xi\rangle = |\xi\rangle - \frac{2}{\sqrt{N}}\left(\sqrt{\frac{N-1}{N}}|\eta\rangle + \frac{1}{\sqrt{N}}|\xi\rangle\right) = -2\frac{\sqrt{N-1}}{N}|\eta\rangle + \frac{N-2}{N}|\xi\rangle$$

$$\tag{11.19}$$

Equation 11.15 yields the identities

$$\frac{N-2}{N} = \cos(\theta) \; ; \; \frac{2\sqrt{N-1}}{N} = \sin(\theta)$$

and hence, from Eqs. 11.18 and 11.19

$$G|\eta\rangle = \cos(\theta)|\eta\rangle + \sin(\theta)|\xi\rangle$$
$$G|\xi\rangle = -\sin(\theta)|\eta\rangle + \cos(\theta)|\xi\rangle \tag{11.20}$$

Equation 11.20 shows that the action of G, on the two-dimensional space spanned by $|\eta\rangle$, $|\xi\rangle$, is to *rotate* the vectors $|\phi\rangle$ toward $|\xi\rangle$, as shown in Fig. 11.4.

11.6   Grover Recursion Equation

The Grover algorithm consists of doing a sequential application of the oracle and diffusion operators to make the initial state function parallel to the qubit $|\xi\rangle$. Each application of $\mathcal{O}$ marks the sought for state and yields $\mathcal{O}|\phi\rangle$; W then makes a *reflection* of the state $\mathcal{O}|\phi\rangle$ about $|\phi\rangle$ and yields $W\mathcal{O}|\phi\rangle = G|\phi\rangle$, thus bringing the resultant state closer and closer to the qubit $|\xi\rangle$. See Fig. 11.4.

Suppose after k recursions, we obtain [4]

$$|\psi_k\rangle = G^k|\phi\rangle = a_k|\xi\rangle + b_k|\eta\rangle \quad : \quad a_0 = \sin\left(\frac{\theta}{2}\right) \; ; \; b_0 = \cos\left(\frac{\theta}{2}\right) \quad (11.21)$$

Applying G once more yields, from Eq. 11.20

$$|\psi_{k+1}\rangle = G^{k+1}|\phi\rangle$$
$$= \left(\cos(\theta)a_k + \sin(\theta)b_k\right)|\xi\rangle + \left(-\sin(\theta)a_k + \cos(\theta)b_k\right)|\eta\rangle \quad (11.22)$$

Note that

$$|\psi_{k+1}\rangle = G^{k+1}|\phi\rangle = a_{k+1}|\xi\rangle + b_{k+1}|\eta\rangle \quad (11.23)$$

Equations 11.22 and 11.23 provide the following recursion equation

$$\begin{bmatrix} a_{k+1} \\ b_{k+1} \end{bmatrix} = \begin{bmatrix} \cos(\theta) & \sin(\theta) \\ -\sin(\theta) & \cos(\theta) \end{bmatrix} \begin{bmatrix} a_k \\ b_k \end{bmatrix}$$
$$\Rightarrow \begin{bmatrix} a_k \\ b_k \end{bmatrix} = \begin{bmatrix} \cos(\theta) & \sin(\theta) \\ -\sin(\theta) & \cos(\theta) \end{bmatrix}^k \begin{bmatrix} a_0 \\ b_0 \end{bmatrix} \quad (11.24)$$

The matrix identity

$$\exp\left\{\theta\begin{bmatrix} 0 & 1 \\ -1 & 0 \end{bmatrix}\right\} = \begin{bmatrix} \cos(\theta) & \sin(\theta) \\ -\sin(\theta) & \cos(\theta) \end{bmatrix}$$

yields the following

$$\begin{bmatrix} \cos(\theta) & \sin(\theta) \\ -\sin(\theta) & \cos(\theta) \end{bmatrix}^k = \exp\{k\theta\begin{bmatrix} 0 & 1 \\ -1 & 0 \end{bmatrix}\}$$
$$= \begin{bmatrix} \cos(k\theta) & \sin(k\theta) \\ -\sin(k\theta) & \cos(k\theta) \end{bmatrix} \quad (11.25)$$

Equations 11.21 and 11.25 yield

$$\begin{bmatrix} a_k \\ b_k \end{bmatrix} = \begin{bmatrix} \cos(k\theta) & \sin(k\theta) \\ -\sin(k\theta) & \cos(k\theta) \end{bmatrix} \begin{bmatrix} \sin\left(\frac{\theta}{2}\right) \\ \cos\left(\frac{\theta}{2}\right) \end{bmatrix} \qquad (11.26)$$

and we obtain the solution

$$a_k = \sin\left(\frac{\theta}{2} + k\theta\right) \; ; \; b_k = \cos\left(\frac{\theta}{2} + k\theta\right)$$

Hence, from Eq. 11.21, after k-recursions the n-qubit is the following

$$|\psi_k\rangle = G^k|\phi\rangle = \sin\left(\frac{\theta}{2} + k\theta\right)|\xi\rangle + \cos\left(\frac{\theta}{2} + k\theta\right)|\eta\rangle \qquad (11.27)$$

The result obtained in Eq. 11.27 is exact.

The recursion is aimed at making the final state $|\psi_k\rangle$ parallel to $|\xi\rangle$, and for this the amplitude of the $|\xi\rangle$ qubit should be near one. Note from Eq. 11.15, for large N

$$\theta \simeq \frac{2}{\sqrt{N}}$$

and

$$|\psi_k\rangle \simeq \sin\left(\frac{1}{\sqrt{N}} + \frac{2k}{\sqrt{N}}\right)|\xi\rangle + \cos\left(\frac{1}{\sqrt{N}} + \frac{2k}{\sqrt{N}}\right)|\eta\rangle \qquad (11.28)$$

To take the input qubit close to $|\xi\rangle$, one needs to continue the recursion N times until

$$k \to K = \frac{\pi}{4}\sqrt{N} \; ; \; N = 2^n \qquad (11.29)$$

and shown in Fig. 11.5.

Equation 11.28 yields

$$|\psi_K\rangle = \sin\left(\frac{1}{\sqrt{N}} + \frac{\pi}{2}\right)|\xi\rangle + \cos\left(\frac{1}{\sqrt{N}} + \frac{\pi}{2}\right)|\eta\rangle \simeq |\xi\rangle \qquad (11.30)$$

On measuring all the n-output qubits with measurement gates $M = |x\rangle\langle x|$, shown in Fig. 11.5, leads to the result

$$|\psi_K\rangle \to \text{Measurement} \to |x\rangle H|1\rangle \approx |\xi\rangle H|1\rangle \qquad (11.31)$$

The probability of observing the device to give a reading corresponding to the basis state $|x\rangle$ is given by

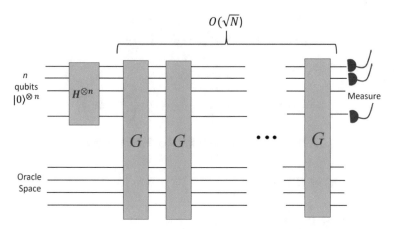

Fig. 11.5 Graphical representation of the Grover algorithm, with $G = W\mathcal{O}$. Published with permission of © Belal E. Baaquie and L. C. Kwek. All Rights Reserved

$$\mathrm{Tr}\Big(|x\rangle\langle x|\psi_K\rangle\langle\psi_K|\Big) = |\langle x|\psi_K\rangle|^2 \simeq \begin{cases} \sin^2\left(\frac{\pi}{2}\right) = 1 & : x = \xi \\ 0 & : x \neq \xi \end{cases} \quad (11.32)$$

Hence, on measuring the degrees of freedom for the output state, one obtains the result of $|\xi\rangle$—with probability almost equal to 1.

In summary, starting from an input n-qubit state in which all the computational basis states are equally likely, after applying the Grover rotation $\pi\sqrt{2^n}/4$ times, on measuring the n-degrees of freedom, the output has probability close to 1 for the sought for state $|\xi\rangle$.

Grover's algorithm can be generalized to seeking a linear combination of M output states as the required outcome. One generalizes the decomposition given in Eq. 11.14 to the following

$$|\phi\rangle = \frac{1}{\sqrt{N}}\sum_{x=0}^{N-1}|x\rangle = \sqrt{\frac{N-M}{N}}\frac{1}{\sqrt{N-M}}\sum_{x=M+1}^{N-M}|x\rangle + \sqrt{\frac{M}{N}}\frac{1}{\sqrt{M}}\sum_{x=1}^{M}|x\rangle$$

$$\Rightarrow |\phi\rangle = \sqrt{\frac{N-M}{N}}|\eta\rangle + \sqrt{\frac{M}{N}}|\xi\rangle \quad (11.33)$$

where

$$|\eta\rangle = \frac{1}{\sqrt{N-M}}\sum_{x=M+1}^{N-M}|x\rangle \ ; \ |\xi\rangle = \frac{1}{\sqrt{M}}\sum_{x=1}^{M}|x\rangle$$

The angle θ_m is defined as

$$\sin\left(\frac{\theta_m}{2}\right) = \sqrt{\frac{M}{N}} \quad \Rightarrow \quad \theta_m \simeq 2\sqrt{\frac{M}{N}}$$

and Eq. 11.33 yields

$$|\phi\rangle = \cos\left(\frac{\theta_m}{2}\right)|\eta\rangle + \sin\left(\frac{\theta_m}{2}\right)|\xi\rangle$$

Similar to Eq. 11.27, after k recursions

$$|\psi_k\rangle = G^k|\phi\rangle = \sin\left(\frac{\theta_m}{2} + k\theta_m\right)|\xi\rangle + \cos\left(\frac{\theta_m}{2} + k\theta_m\right)|\eta\rangle \quad (11.34)$$

Hence, the number of times that the oracle needs to be called, similar to Eq. 11.29, is given by

$$k \to K_m = \frac{\pi}{4}\sqrt{\frac{N}{M}} \quad ; \quad N = 2^n \quad (11.35)$$

11.7 Single Recursion: Two Qubits

To understand the workings of the Grover algorithm, the earlier result for two qubits obtained by the process of reflection is analyzed using the general result obtained for the n-qubit case.

Note for a single recursion ($k=1$), the output qubit from Eq. 11.27 is given by

$$|\psi_1\rangle = G|\phi\rangle = \sin\left(\frac{3\theta}{2}\right)|\xi\rangle + \cos\left(\frac{3\theta}{2}\right)|\eta\rangle \quad (11.36)$$

As before, we can ignore the ancillary qubit and focus on the input qubit. Suppose the qubit we are looking for is $|\xi\rangle = |01\rangle$. As before, the two-qubit input is separated into $|\xi\rangle$, $|\eta\rangle$ as follows

$$|\phi\rangle = \frac{1}{2}\left[|00\rangle + |01\rangle + |10\rangle + |11\rangle\right] = \frac{1}{2}|01\rangle + \frac{\sqrt{3}}{2}\frac{1}{\sqrt{3}}\left[|00\rangle + |10\rangle + |11\rangle\right]$$

$$\Rightarrow |\phi\rangle = \frac{1}{2}|\xi\rangle + \frac{\sqrt{3}}{2}|\eta\rangle \ : \ \langle\xi|\phi\rangle = \frac{1}{2} \ : \ \langle\xi|\eta\rangle = 0 \ ; \ \langle\eta|\eta\rangle = 1 (11.37)$$

Hence, for the two qubits, we have the decomposition shown in Fig. 11.6 and yields

$$|\phi\rangle = \sin\left(\frac{\theta}{2}\right)|\xi\rangle + \cos\left(\frac{\theta}{2}\right)|\eta\rangle \ : \ \sin\left(\frac{\theta}{2}\right) = \frac{1}{2} \ \Rightarrow \ \theta = \frac{\pi}{3} \quad (11.38)$$

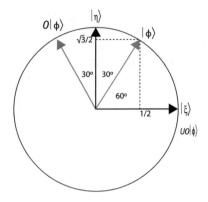

Recall

$$G = \left(2|\phi\rangle\langle\phi| - \mathbf{I}\right)\left(\mathbf{I} - 2|\xi\rangle\langle\xi|\right)$$

Applying operator G to the two-quit bit state given in Eq. 11.38 yields

$$G\phi\rangle = \left(2|\phi\rangle\langle\phi| - \mathbf{I}\right)\left(\mathbf{I} - 2|\xi\rangle\langle\xi|\right)|\phi\rangle = \left(2|\phi\rangle\langle\phi| - \mathbf{I}\right)\left(|\phi\rangle - |\xi\rangle\right)$$
$$\Rightarrow G\phi\rangle = |\xi\rangle = |01\rangle \tag{11.39}$$

and we have recovered the result obtained earlier in Eq. 11.6 for the two qubits using
reflections. The general result, from Eq. 11.36 and for $\theta = \pi/3$, yields

$$G|\phi\rangle = \sin\left(\frac{3\theta}{2}\right)|\xi\rangle + \cos\left(\frac{3\theta}{2}\right)|\eta\rangle = \sin\left(\frac{\pi}{2}\right)|\xi\rangle + \cos\left(\frac{\pi}{2}\right)|\eta\rangle$$
$$\Rightarrow G|\phi\rangle = |\xi\rangle = |01\rangle \tag{11.40}$$

Equation 11.40 confirms the result obtained in Eq. 11.39 that for a two-qubit,
Grover's algorithm needs to be run only once to arrive with certainty at the sought
for qubit $|01\rangle$. In contrast, a classical computer—or classical circuit—for a search in
$N = 2^2 = 4$ numbers would need to the call the oracle, on the average, 2.25 times
[1].

11.8   Discussion

We revisit the factorization of primes to explain the Grover algorithm.[2] To factorize
integer $m = pq$, with primes $p, q, \ q < p$. Since the smaller prime q is less than
$\sqrt{m}$, one needs to generate numbers ranging from $1, \ldots, \sqrt{m}$. To apply the Grover

---

[2] Shor's algorithm is more efficient for factorizing primes, but for pedagogical purposes, the factorization of primes is a good example for explaining Grover's algorithm.

algorithm, the total number of degrees of freedom required for the prime factorization is $N = \sqrt{m}$, with x ranging from $1, \ldots, N$.

For each number x, the oracle carries out the relatively easy integer division given by N/x; if x divides N, it is the marked state ξ that we are looking for. Hence, the oracle for factorizing an integer made out of two primes defines the marked basis state ξ by the following

$$x = \begin{cases} \xi & : \text{if } N/x = \text{integer} \\ x & : \text{if } N/x \neq \text{integer} \end{cases}$$

The oracle (search function) $f(x)$ given in Eq. 11.1 is defined as follows

$$f(x) = \begin{cases} 0 & : x \neq \xi \\ 1 & : x = \xi \end{cases} \tag{11.41}$$

As mentioned earlier, *recognizing* the solution means checking whether if x can divide m—and it does not require knowing the entire factorization of $m = pq$. Recognizing the solution without knowing the full answer allows the search algorithm to 'mark' the solution being sought.

The number of steps required for finding the marked basis state for the Grover algorithm, from Eq. 11.29, is given by

$$\frac{\pi}{4}\sqrt{N}$$

For the classical algorithm, one has to divide $N = \sqrt{m}$ by, in principle, by every number from $1, \ldots, N$ to find which number is the prime, and this requires $\sqrt{m}$ steps. In contrast, for the Grover algorithm, the number of steps required to find the marked state, which is the number we are looking for, is proportional to $\sqrt{N} = m^{1/4}$. This is the famous quadratic acceleration provided by Grover's algorithm.

The advantage of the Grover algorithm over the classical case lies in the fact that, due to the superposition principle, one can *simultaneously* update all the basis states, as given in Eq. 11.7

$$|\psi(t_2)\rangle = U_f|\psi(t_1)\rangle = \frac{1}{\sqrt{N}} \sum_{x=0}^{N-1} (-1)^{f(x)}|x\rangle H|1\rangle$$

In contrast, a classical algorithm cannot update the different allowed state simultaneously since the classical updating needs to do be done one by one.

References

1. Nielsen MA, Chuang IL (2012) Quantum computation and quantum information. Cambridge University Press, UK
2. Chris B (2019) Quantum computing for everyone. MIT Press, USA
3. Bernard Z (2018) A first introduction to quantum computing and information. Springer, UK
4. Rieffel Eleanor G, Polak Wolfgang H (2011) Quantum computing: a gentle introduction. MIT Press, USA

Chapter 12
Simon's Algorithm

At the outset of quantum algorithms in the early nineties, there were not many practical algorithms that show significant advantage of a quantum computer to a classical computer. In 1994, Daniel Simon came up with one of the earliest examples [1].

In Simon's problem, we are given an unknown function f implemented with a black box or oracle such that $f : \{0, 1\}^n \to \{0, 1\}^n$ satisfies the property that for all $x, y \in \{0, 1\}^n$, there exists a nonzero element $a \in \{0, 1\}^n$

$$f(x) = f(y) \text{ if and only if } x \oplus y = a. \tag{12.1}$$

The question is how can we determine a quickly. Note that

$$x \oplus y = a \implies y = x \oplus a$$

Classically, as in the Deutsch–Jozsa algorithm, the worst case scenario to find a with 100 % certainty for a given function f involves checking up to $2^n/2 + 1$ inputs. And like the Deutsch–Jozsa algorithm, if we are lucky, we could classically solve with our first few tries.

12.1 Quantum Algorithm

In the quantum algorithm, we start with the initial state with two registers

$$|\psi_1\rangle = |0\rangle^{\otimes n} |0\rangle^{\otimes n}$$

© The Author(s), under exclusive license to Springer Nature Singapore Pte Ltd. 2023
B. E. Baaquie and L.-C. Kwek, *Quantum Computers*,
https://doi.org/10.1007/978-981-19-7517-2_12

We next apply the Hadamard $H^{\otimes n}$ to the first register so that

$$|\psi_2\rangle = \frac{1}{\sqrt{2^n}} \sum_{i \in \{0,1\}^n} |i\rangle |0\rangle^{\otimes n} \tag{12.2}$$

Note that the Hadamard transformation introduces n-degrees of freedom labeled by i. We then apply the oracle:

$$\mathcal{U}|\psi_2\rangle = |\psi_3\rangle = \frac{1}{\sqrt{2^n}} \sum_{i \in \{0,1\}^n} |i\rangle |f(i)\rangle$$

$$= \frac{1}{\sqrt{2^{n+1}}} \sum_{i \in \{0,1\}^n/2} \left[|i\rangle |f(i)\rangle + |i \oplus a\rangle |f(i \oplus a)\rangle \right]$$

$$= \frac{1}{\sqrt{2^{n+1}}} \sum_{i \in \{0,1\}^n/2} \left[|i\rangle + |i \oplus a\rangle \right] |f(i)\rangle \tag{12.3}$$

since $f(i \oplus a) = f(i)$.

By measuring all the qubits i, we observe one of the qubits, say $i = i_0$, with some probability. The Born rule, discussed in Sect. 4.5, yields a *single* term from the sum given in Eq. 12.3, corresponding to some $i = i_0$, given by

$$|\psi_3\rangle \rightarrow |\psi_4(i_0)\rangle |f(i_0)\rangle \; ; \; \langle\psi_4|\psi_4\rangle = 1$$

The state vector of the first register, as a result of the measurement, is left in the state

$$|\psi_4(i_0)\rangle = \frac{1}{\sqrt{2}} \left[|i_0\rangle + |i_0 \oplus a\rangle \right]$$

Note that from Eq. 7.4, the Hadamard transformation for n qubits can be written as

$$H^{\otimes n}|x\rangle = \frac{1}{\sqrt{2^n}} \sum_{y \in \{0,1\}^n} (-1)^{xy} |y\rangle$$

Hence, applying the Hadamard transformation on the first register gives

$$|\psi_5\rangle = H^{\otimes n}|\psi_4(i_0)\rangle$$

$$= \frac{1}{\sqrt{2^{n+1}}} \sum_{j' \in \{0,1\}^n} \left[\left\{ (-1)^{i_0 \cdot j'} + (-1)^{i_0 \cdot j'} \right\} |j'\rangle \right] ; \; i_1 \equiv i_0 \oplus a \tag{12.4}$$

The Hadamard transformation introduces n-qubits labeled by j'. Let all the qubits denoted by j' be measured, and let the outcome of the measurement be given by j. The probability of obtaining this result is given by

$$P(i_0, j) = \text{Tr}\Big(\mathcal{M}(j)|\psi_5\rangle\langle\psi_5|\Big) = \frac{1}{2^{n+1}}\Big|(-1)^{i_0 \cdot j} + (-1)^{i_1 \cdot j}\Big|^2 \quad (12.5)$$

If the output is zero, the result is discarded. The output will be nonzero if

$$
\begin{aligned}
(-1)^{i_0 \cdot j} &= (-1)^{i_1 \cdot j} \\
\Rightarrow i_0 \cdot j &= i_1 \cdot j \\
\Rightarrow i_0 \cdot j &= (i_0 \oplus a) \cdot j \\
\Rightarrow i_0 \cdot j &= i_0 \cdot j \oplus a \cdot j \\
\Rightarrow a \cdot j &\equiv 0 \ (\text{mod} 2).
\end{aligned}
\quad (12.6)
$$

The string $j = j_1$ is recorded and the inner product with a is zero mod 2.

Repeating the algorithm roughly n times, we get n different values of j_i ($i = 1, \ldots, n$) satisfying

$$
\begin{aligned}
a \cdot j_1 &= 0 \\
a \cdot j_2 &= 0 \\
a \cdot j_3 &= 0 \\
&\vdots \\
a \cdot j_n &= 0.
\end{aligned}
\quad (12.7)
$$

From Eq. (12.7) a can be determined easily with classical methods (say Gaussian elimination).

This algorithm is a precursor to Fourier transform corresponding to a period of two. For the general case of the function having the periodic property of $f(x) = f(x + a)$ with addition not being binary but of two numbers, one needs the quantum Fourier transform for solving it and forms the core idea of the Shor algorithm.

12.2 An Illustrative Example

Let us now consider a particular example involving a function $f : \{0, 1\}^3 \to \{0, 1\}^3$. Let us also suppose that $a = 101$.

We first initialize two three-qubit registers to $|000\rangle_1 \otimes |000\rangle_2$.

Since $a = |101\rangle$, we have

$$
\begin{aligned}
f(|000\rangle) &= f(|101\rangle) = |000\rangle \\
f(|001\rangle) &= f(|100\rangle) = |101\rangle \\
f(|010\rangle) &= f(|111\rangle) = |010\rangle \\
f(|011\rangle) &= f(|110\rangle) = |111\rangle
\end{aligned}
$$

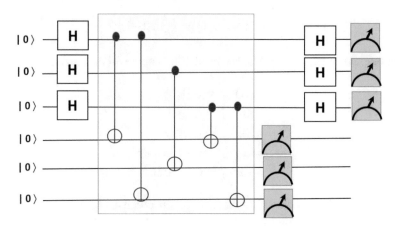

Fig. 12.1 Simon's algorithm for a function acting on three qubits. Published with permission of © Belal E. Baaquie and L. C. Kwek. All Rights Reserved

Figure 12.1 shows one such realization of the algorithm.

Finally, suppose the measurement output on the second register is $|010\rangle$, then the state of the first register becomes

$$\frac{1}{\sqrt{2}}(|010\rangle + |111\rangle) \xrightarrow{H\otimes H\otimes H} \frac{1}{2}(|000\rangle - |010\rangle + |101\rangle - |111\rangle) \qquad (12.8)$$

and a measurement of the first register yields any of the four states

$$\{|000\rangle, |010\rangle, |101\rangle, |111\rangle\}$$

with equal probability (in this case $1/4$). Therefore, after many measurements, we get

$$a \cdot 000 \equiv 0 \bmod 2$$
$$a \cdot 010 \equiv 0 \bmod 2$$
$$a \cdot 101 \equiv 0 \bmod 2$$
$$a \cdot 111 \equiv 0 \bmod 2.$$

This is a set of simultaneous equation. Assuming that $a = (x, y, z)$, then we see that except for the first equation above, the rest of the equations translate into:

$$y \equiv 0 \bmod 2$$
$$x + z \equiv 0 \bmod 2$$
$$x + y + z \equiv 0 \bmod 2,$$

from which it is not difficult to deduce that the solution is $a = 101$. Note that the first equation is not very useful. There is typically a minimum number of equations needed, but it is still better than existing methods with classical computers.

Reference

1. Simon DR (1997) On the power of quantum computation. SIAM J Comput 26(5):1474–1483

Chapter 13
Shor's Algorithm

13.1 Introduction

The Shor algorithm is widely regarded as the first non-trivial quantum algorithm that shows a potential of an 'exponential' speeding-up over its equivalent classical algorithms. Proposed by Peter Shor in 1994, the algorithm seriously challenges existing classical algorithms for integer factorization [1].

Why is Shor's algorithm so important?

For many years, the security of the many Internet transactions and the secure transfer of secret and confidential documents has relied on the fact that one could encrypt information using a publicly available key, but the reverse process of decoding the encrypted message relies on the difficulty of factoring a large (seriously large) integer. Such a scheme was first devised by a trio of computer scientists and mathematicians, Ron Rivest, Adi Shamir and Leonard Adleman in the seventies [2]. The scheme now known as the RSA public-key cryptosystem relies precisely on this difficulty of factoring a big number formed from the product of two large prime numbers. In short, although multiplying two large prime numbers is easy, factoring is difficult. The easy computation (multiplication) is used for encryption, and the difficult computation (factoring) is needed for decoding or decryption.

Peter Shor showed, in 1994, that if we have a workable quantum computing machine, then we should be able to factorize any product of two large primes more easily and faster than any classical algorithm. To factorize a number N into its prime factors, one needs to find cycle (period) r of a periodic function (defined below) that takes a classical computer about $\exp(N^{1/3})$ steps; in contrast, using Shor's algorithm, the algorithm takes only N^3 number of steps [3]. This exponential acceleration means that many encrypted codes based on factoring a large number N into its primes can be decoded by Shor's algorithm.

© The Author(s), under exclusive license to Springer Nature Singapore Pte Ltd. 2023
B. E. Baaquie and L.-C. Kwek, *Quantum Computers*,
https://doi.org/10.1007/978-981-19-7517-2_13

13.2   Understanding the Classical Algorithm

To understand the Shor algorithm, it is instructive to know how the classical algorithm works. Let us start with a simple example. Suppose we wish to find the prime factors of the number 15. The classical algorithm to factorize 15 goes as follows:

1. Choose **any** number that has no common divisor to 15, a say 2.
2. Consider the sequence of numbers $\{a^i\}$ for $i = 0, 1, 2, \cdots$:

$$\{\, 2^0 \, \text{Mod}(15) = 1, 2^1 \, \text{Mod}(15) = 2, 2^2 \, \text{Mod}(15) = 4,$$
$$2^3 \, \text{Mod}(15) = 8, 2^4 \, \text{Mod}(15) = 1, 2^5 \, \text{Mod}(15) = 2, \cdots \}$$

Note the crucial equation
$$2^4 \, \text{Mod}\,(15) = 1$$

We see that there is a repetition of the outcomes in the sequence; after every four numbers, the sequence returns to the number 1. We say that the sequence has a cycle length r of 4.

3. Determine $\text{GCD}(a^{r/2} \pm 1, 15) = \text{GCD}(2^2 \pm 1, 15)$ which obviously yield 3 or 5, where GCD is the greatest common divisor.

The difficult thing about this algorithm is getting the cycle length r. For the number 15, it was straightforward, in fact trivial. However, for a large prime number, this is not so. For instance, the 100th prime[1] is 541, and the 101st prime is 547. The product of these two primes is $N = 295827$. If we choose $a = 3$ and consider the sequence of numbers $3^r \, \text{Mod}\,(N)$, on a modestly fast laptop it takes at least 10 s before it yields the correct cycle length, which is $r = 1890$.

13.3   Quantum Algorithm

The classical algorithm relies on the determination of the cycle length. Let us now present Shor's elegant algorithm for finding this cycle length. The description follows closely the treatment in [4, 5]. The principal idea in the algorithm is the creation a state with periodicity r, and then apply Fourier transform over Z_Q, (the additive group of integers modulo Q), to reveal this periodicity. The quantum Fourier transform over the group Z_Q is defined as follows

$$|a\rangle \longrightarrow \frac{1}{\sqrt{Q}} \sum_{b=0}^{Q-1} e^{2\pi i ab/Q} |b\rangle = |\Psi_{Q,a}\rangle \tag{13.1}$$

---

[1] One can try to use the 2020th and 2021th prime. Their product is 308 915 767. However, a laptop, after five hours, fails to find the cycle length.

The algorithm to compute this quantum Fourier transform is described in Chap. 8.

Let N be the number we would like to factorize into its primes. Like the Grover search algorithm, we begin with two registers. Let us start with n qubits and let $Q = 2^n$; hence, the registers hold all the real integer from 0 to $Q - 1$; in particular, the first register can hold a number between 0 to $Q - 1$. It can be shown (discussed later) that we can choose $Q > N^2$: it is much larger than N, but still polynomial in N. The second register will also carry integers between 0 to $Q - 1$. Hence the two registers will consist of $O(\log(N))$ qubits. Once r has been determined, the prime N can be factorized using number theory.

13.3.1 How Then Does the Quantum Algorithm Work?

Shor's algorithm yields the value of cycle r.

Choose an integer a that is coprime with N. We know that if we consider the sequence:

$$\{a^0, a^1, a^2, a^3, a^4, \cdots\} \ \text{Mod} \ N$$

there exists an r such that $a^r \text{Mod} \ N = 1$. This is sometimes known as Euler's theorem, a generalization of Fermat's little theorem that states that for any integer N and a,

$$a^{\phi(N)} \equiv 1 \ \text{Mod} \ N$$

where $\phi(N)$ is Euler's quotient function.

Shor's algorithm relies on this result and starts with the function

$$f(x) = a^x \ \text{Mod} \ N$$

Note that

$$f(x + r) = a^{x+r} \ \text{Mod} \ N = a^x \ \text{Mod} \ N \times a^r \ \text{Mod} \ N = a^x \text{Mod} \ N = f(x)$$

In accordance with Euler's theorem, the function has the crucial property that it is a periodic function of x with some period r. Hence, for integer p

$$f(x + s) = f(x) \ ; \ \ s = pr$$

Note that due to the fact that $f(x) = a^x \ \text{Mod} \ N$, the function $f(x)$ never takes the same value within one period.

1. We first start with the state

$$|\psi_0\rangle = |0\rangle^{\otimes n} \otimes |0\rangle^{\otimes n}$$

Apply the Hadamard transformation $H \otimes H \otimes H \cdots H = H^{\otimes n}$ to the initial state to obtain

$$|\psi_1\rangle = \left(H^{\otimes n} |0^{\otimes n}\rangle \right) \otimes |0\rangle^{\otimes n} = \frac{1}{\sqrt{Q}} \sum_{x=0}^{Q-1} |x\rangle \otimes |0\rangle^{\otimes n} \qquad (13.2)$$

Note the quantum algorithm is based on the input state $|\psi_1\rangle$ that can be created by the hardware.

2. Use unitary transformation U_f to create the state

$$|\psi_f\rangle = U_f |\psi_1\rangle = \frac{1}{\sqrt{Q}} \sum_{x=0}^{Q-1} |x\rangle \otimes |f(x)\rangle \qquad (13.3)$$

3. Let the total number of qubit states be divided into periods of length r. The number of states in each cycle is given by A where

$$A = [\frac{Q}{r}] \quad \text{or} \quad A = [\frac{Q}{r}] + 1 \; ; \quad Q = 2^n$$

4. Using $f(x + kr) = f(x)$, Eq. 13.3 yields

$$|\psi_f\rangle = U_f |\psi_1\rangle = \frac{1}{\sqrt{Q}} \sum_{i=0}^{r-1} \sum_{k=0}^{A-1} |x_i + kr\rangle \otimes |f(x_i + kr)\rangle$$

$$= \frac{1}{\sqrt{Q}} \sum_{i=0}^{r-1} \left[\sum_{k=0}^{A-1} |x_i + kr\rangle \right] \otimes |f(x_i)\rangle \qquad (13.4)$$

Note the important fact that the quantum algorithm is based on the explicit representation of the input state given in Eq. 13.2, for which we do not need to know the numerical value of r. To rewrite the summation as in Eq. 13.4 is a mathematical transformation of the input state, and for doing so all that we need to know is that $Q > A > N$, which is guaranteed since $Q \gg N^2$.

5. The auxiliary state vector $|f(x_i)\rangle$ is available to the device, although the prefactor that contains a sum over r is not known. A measurement is carried out on the $|x\rangle$ qubit states; the measurement yields, with some likelihood, the state vector to which $|\psi_f\rangle$ has collapsed to and yields a particular value of the state vector which is denoted by $|x_i\rangle \rightarrow |x_0\rangle$. Note the important fact that to perform this measurement, no knowledge of r is required.

The state vector given in Eq. 13.4 is **entangled**: the state vector's dependence on the degrees of freedom x_i and k is not factorizable into product state vectors. The generalized Born rule for measurement, discussed in Sect. 4.10, states that if the measurement yields the state vector to be, with some likelihood, in the state $|x_0\rangle$, then the resultant post-measurement state vector is given by

$$|\psi_f(x)\rangle \rightarrow |\psi_3(x_0)\rangle |f(x_0)\rangle \ ; \ \langle\psi_3(x_0)|\psi_3(x_0)\rangle = 1 \tag{13.5}$$

where

$$|\psi_3\rangle \equiv |\psi_3(x_0)\rangle = \frac{1}{\sqrt{A}} \sum_{k=0}^{A-1} |x_0 + kr\rangle \tag{13.6}$$

The post-measurement state given in Eq. 13.5, for different values of x_0 occurs with different likelihoods, as discussed in Sect. 4.10. Since the analysis of the state $|\psi_3\rangle$ is valid for all values of x_0, we do not need to know the likelihood of the various post-measurement states for the different values of x_0.

The generalized Born rule does the magic of picking out the state vector $|\psi_3\rangle$, which contains the number A that has information about r. The auxiliary state has factored out and will be ignored henceforth. It is now a matter of another measurement that allows us to isolate r and hence factorize N into a product of two primes.

6. Apply the quantum Fourier Transform on the first register

$$|\psi_3\rangle = \frac{1}{\sqrt{Q}} \sum_{k=0}^{Q-1} \left(\frac{1}{\sqrt{A}} \sum_{j=0}^{A-1} e^{2\pi i (jr + x_0)k/Q} \right) |k\rangle$$

7. Measure only the first register. Let $|k_1\rangle$ be the outcome of the measurement; the probability is given by

$$\mathrm{Tr}\left(\mathcal{M}(k_1) |\psi_3\rangle\langle\psi_3| \right)$$

$$= \left| \left(\frac{1}{\sqrt{QA}} \sum_{j=0}^{A-1} \exp\left[2\pi i (jr + x_0)k_1/Q\right] \right) \right|^2$$

$$= \left| \frac{1}{\sqrt{QA}} \exp(2\pi i \frac{x_0 k_1}{Q}) \sum_{j=0}^{A-1} \exp(2\pi i \frac{jrk_1}{Q}) \right|^2$$

$$= \left| \exp(2\pi i \frac{x_0 k_1}{Q}) \right|^2 \left| \frac{1}{\sqrt{QA}} \sum_{j=0}^{A-1} \exp(2\pi i \frac{jrk_1}{Q}) \right|^2$$

$$= \left| \frac{1}{\sqrt{QA}} A \right|^2 = \frac{A}{Q} \simeq \frac{1}{r}, \tag{13.7}$$

whenever (since j is an integer)

$$\frac{rk_1}{Q} \text{ is some integer } m_1$$

Note that $\left|\exp(2\pi i \frac{x_0 k_1}{Q})\right| = 1$ and $\sum_{j=0}^{A-1} \exp(2\pi i \frac{jrk_1}{Q}) = \sum_{j=0}^{A-1} 1 = A$.

The fraction k_1/Q is approximated by a fraction with denominator smaller than N, using the (classical) method of continued fractions. Thus, we see that

$$\frac{k_1}{Q} \simeq \frac{m_1}{r} \quad \Longleftrightarrow \quad rk_1 \equiv 0 \, \text{Mod} \, Q$$

8. Repeat all previous steps poly$(\log(N))$ number of times to get

$$k_1 = \frac{m_1}{r_1} Q, \quad k_2 = \frac{m_2}{r_2} Q, \cdots$$

This gives possible rs as r_1, r_2 which we can check to see if it is the correct r (see analysis with continued fractions later).

It is interesting to note that in the second step of the algorithm, all numbers between 0 and $Q - 1$ are present in the superposition, with equal weights. Subjecting this linearly superposed state to a unitary transformation associated with the function, generating the sequences of numbers, $\{a^0, a^1, \ldots a^{r-1}\}$ is analogous to a parallel processing of all numbers from 0 to $Q - 1$. The third step of the algorithm groups the numbers in the sequence into sets, each with periodicity r. This is done as follows: there are r possible values written on the second register: $a \in \{a^0, a^1, \ldots a^{r-1}\}$. The state in the third step can thus be written as:

$$\frac{1}{\sqrt{Q}} \left(\left(\sum_{x=0|a^l=a}^{Q-1} |l\rangle \otimes |a\rangle \right) + \left(\sum_{x=0|a^l=a^2}^{Q-1} |l\rangle \otimes |a^2\rangle \right) + \cdots \right.$$
$$\left. + \left(\sum_{x=0|a^l=a^r}^{Q-1} |l\rangle \otimes |a^r\rangle \right) \right) = 1 \quad (13.8)$$

Note that the values l that give $a^l = a$ have periodicity r: If the smallest such l is x_0, then $x = x_0 + r, l_0 + 2r, \ldots$ will also give $a^x = a$. Hence, each term in the brackets has periodicity r. Each set of x's, with periodicity r, is attached to a different state of the second register. Before the computation of a^x, all x's appeared equally in the superposition.

Writing down the a^x on the second register can be thought of as giving a different 'color' to each periodic set in $[0, Q - 1]$.

The measurement of the second register picks randomly one of these sets. The state then collapses to a superposition of l's with periodicity r and an arbitrary shift l_0. To determine the the periodicity, we apply Fourier transform.

We then measure the first register, to obtain k. To find the probability to measure each k, we need to sum up the weights coming from all the j's in the periodic set.

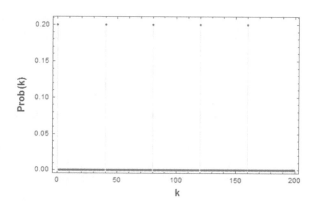

$$\text{Prob}(k) = \left| \frac{1}{\sqrt{QA}} \sum_{j=0}^{A-1} e^{2\pi ik(jr+x_0)/Q} \right|^2 \tag{13.9}$$

$$= \left| \frac{1}{\sqrt{QA}} \sum_{j=0}^{A-1} (e^{2\pi ikr/Q})^j \right|^2 \tag{13.10}$$

This is a geometrical series.

There are two cases:

- exact periodicity and
- imperfect periodicity.

In the case of exact periodicity, i.e., that r divides Q exactly, we have $A = Q/r$. In this case, the above geometrical series is equal to zero, unless $e^{2\pi ikr/Q} = 1$. Thus, we measure with probability 1 only k's such that $kr = 0 \mod Q$. We can write $kr = mQ$, for some integer m, or $k/Q = m/r$. We know Q, and we know k since we have measured it. Therefore, we can reduce the fraction k/Q. If m and r are coprime. the denominator will be exactly r which we are looking for! By the prime number theorem, there are approximately $n/\log(n)$ numbers smaller than n and coprime with n, so since m is chosen randomly, repeating the experiment a large enough number of times we will with very high probability eventually get m coprime to r. Sometimes, one says that this is the case of constructive interference: and the ones that we see are all integer multiples of Q/r.

For concreteness, if we consider an example of $Q = 200$ with $r = 5$ and plot the values of Prob(k) as a function of k, we get Fig. 13.1 which give nonzero probabilities for $k = \{0, 40, 80, 120, 160\}$. Dividing by Q gives the fractions

$$\left\{ 0, \frac{1}{5}, \frac{2}{5}, \frac{3}{5}, \frac{4}{5} \right\}$$

Clearly, $r = 5$ in this case.

In the case of imperfect periodicity, r does not divide Q, it turns out that with high probability, we will measure only k's which satisfy an *approximate* criterion $kr \approx 0 \bmod Q$. In particular, consider k's which satisfy:

$$-r/2 \le kr \ \mathrm{Mod} \ Q \le r/2 \tag{13.11}$$

There are exactly r values of k satisfying this requirement, because k runs from 0 to $Q - 1$, therefore kr runs from 0 to $(Q - 1)r$, and this set of integers contains exactly r multiples of Q. Now the probability to measure such a k is bounded below, by choosing the largest exponent possible. It is interesting to note that a slightly different way of arriving at the same conclusion is:

$$
\begin{aligned}
\mathrm{Prob}(k) &= |\frac{1}{\sqrt{QA}} \sum_{j=0}^{A-1} (e^{2\pi i k r/Q})^j|^2 \\
&= \frac{1}{QA} |\frac{1 - e^{i 2 k \pi r A/Q}}{1 - e^{i 2 k \pi r/Q}}|^2 \\
&= \frac{1}{QA} |\frac{e^{i k \pi r A/Q}(e^{-i k \pi r A/Q} - e^{i k \pi r A/Q})}{e^{i k \pi r/Q}(e^{-i k \pi r/Q} - e^{i k \pi r/Q})}|^2 \\
&= \frac{1}{QA} |\frac{\sin(\frac{k \pi r A}{Q})}{\sin(\frac{k \pi r}{Q})}|^2 \\
&\approx \frac{1}{QA} |\frac{\sin(k\pi)}{\frac{k \pi r}{Q}}|^2 \ \text{where we have assumed} \ A \approx \frac{Q}{r} \ \text{and} \ Q \gg r \\
&\approx \frac{1}{QA} |\frac{1}{\frac{r}{Q}}|^2 \ \text{choosing} \ k \ \text{so that} \ \frac{\sin(k\pi)}{k\pi} \approx 1 \\
&= \frac{1}{QA} \frac{Q^2}{r^2} \\
&\approx \frac{1}{r} \ \text{using} \ r \approx \frac{Q}{A}
\end{aligned}
$$

The approximation arises due to the fact that Q is chosen to be much larger than $N > r$, in which case the sine term in the numerator is close to 1 with negligible correction of the order of r/Q. In the denominator, we use the approximation: $\sin(x) \approx x$ for small x, and the correction is again of the order of r/Q. The probability to measure any k which satisfies 13.11 is approximately $1/r$.

At this juncture, let us consider an example of a specific case where $Q = 200$ and $r = 13$. Let $A = [Q/r] = 15$. We can now plot the probabilities $\mathrm{Prob}(k)$ for $1 \le k \le 199$. From Fig. 13.2, we see that there are some k values that give higher probabilities than others. Here $k = 15, 31, 46, 77$ and 92 gives $\mathrm{Prob}(k) > 0.04$. Suppose a measurement of the probabilities yield $k = 77$. In this case, a good approximation is to consider the continued fraction of $\frac{k}{Q} = \frac{77}{200} = \{0, 2, 1, 1, 2, 15\}$ which

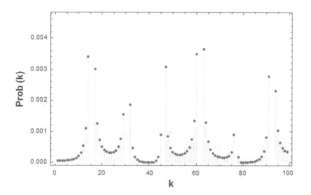

Fig. 13.2 Plots of the probabilities Prob(k) as a function of k. Published with permission of © Belal E. Baaquie and L. C. Kwek. All Rights Reserved

occurs with a probability approximately $1/13 \approx 0.07$. From the continued fraction, $\{0, 2, 1, 1, 2, 15\}$, we see that the fractions below all approximate $\dfrac{77}{200}$:

$$0 + 1/2 = \frac{1}{2},$$
$$0 + 1/(2 + 1/1) = \frac{1}{3},$$
$$0 + 1/(2 + 1/(1 + 1/1)) = \frac{2}{5},$$
$$0 + 1/(2 + 1/(1 + 1/(1 + 1/2))) = \frac{5}{13}$$

We clearly see that $r = 13$ is one of the possibilities. It is then possible to figure out the correct r from several measurements.

Why are such k's 'good'? Given an integer k which satisfies the criterion 13.11, we can find r with reasonably high probability. Note that for 'good' k's, there exists an integer m such that:

$$\left| \frac{k}{Q} - \frac{m}{r} \right| \leq \frac{1}{2Q}.$$

In general, recall that Q is chosen to be much larger than N, say $Q \geq N^2$. This means that $\frac{k}{Q}$, a fraction with denominator $\geq N^2$, can be approximated by $\frac{m}{r}$, a fraction with denominator smaller than N, to within $\frac{1}{N^2}$. Umesh Vazirani presents another simple argument. In this argument, one notes that the number of copies of the period is $\dfrac{Q}{r}$ and as long as this ratio is greater than the period, r, i.e., $\dfrac{Q}{r} > r$ or $Q > r^2$. But then $r < N$, so if we ensure that $Q > N^2 > r^2$, we are done.

There is only one fraction with such a small denominator that approximates a fraction so well with such large denominator. Given k/Q, the approximating fraction, $\frac{m}{r}$, can be found efficiently, using the method of continued fractions:

$$a = a_0 + \cfrac{1}{a_1 + \cfrac{1}{a_2 + \cdots}},$$

where a_i are all integers. Finding this fraction, the denominator will be r! There is high probability that m is coprime to r. The probability is greater than $1/\log(r)$. If not, we repeat the iteration.

This concludes Shor's algorithm.

References

1. Shor PW (1994) Algorithms for quantum computation: discrete logarithms and factoring. In: Proceedings 35th annual symposium on foundations of computer science. IEEE, pp 124–134
2. Calderbank M (2007) The rsa cryptosystem: history, algorithm, primes. Chicago: Math. Uchicago. Edu
3. Mermin ND (2007) Quantum computer science. Cambridge University Press, UK
4. Michel LB (2006) A short introduction to quantum information and computation. Cambridge University Press, UK
5. Artur E, Richard J (1996) Quantum computation and shor's factoring algorithm. Rev Mod Phys 68(3):733

Part III
Applications

Chapter 14
Quantum Algorithm for Option Pricing

One of the main areas for the application of quantum algorithms is in the pricing of options. An introductory text cannot do justice to this fast-evolving subject, and hence the discussion will focus on only the pricing of a European call or a put option. Other more complex financial instruments are left for further reading.

The approach of this chapter is the one taken in [1–3]. The principles of finance are taken from the existing theories of finance, and financial instruments are not assumed to have any quantum indeterminacy—but rather, the uncertainty in the future values of financial instruments is determined by classical probability theory. Quantum algorithms are focused on more efficient procedures for evaluating option prices.

The price of all options is the discounted value of the expectation value of a payoff function in the future, with the future uncertain price of financial instruments, like stocks or bonds, being represented by classical random variables. Most numerical studies of option pricing are based either on solving the relevant partial differential equations or on Monte Carlo simulations of the stochastic equations [4].

14.1 Review of Option Pricing

Consider the price of a path independent option $C(x, t, T)$ at present time t, and which matures at a future time T. The payoff function of the option $v(x)$ is defined to be the price of the option when it matures at a future time T and is given by

$$C(x, T, T) = e^{-r\tau} v(x)$$

The evolution of the stock price $S = e^{x(t)}$ is given by a stochastic differential equation. The payoff function of a call option is given by

© The Author(s), under exclusive license to Springer Nature Singapore Pte Ltd. 2023
B. E. Baaquie and L.-C. Kwek, *Quantum Computers*,
https://doi.org/10.1007/978-981-19-7517-2_14

$$v(x) = [e^x - K]_+$$

Let the initial stock price be $S = e^{x_0}$, with spot interest rate given by r and remaining time for maturity of option given by $\tau = T - t$; let the price of the option for remaining time be given by $C(x, \tau)$. Let $P(x_0, x; \tau)$ be the conditional probability that given the stock price is $S = e^{x_0}$ today, it has a value of e^x after time τ; the discounted price of the payoff yields the option price is given by

$$C(x_0, \tau) = e^{-r\tau} \int dx\, P(x_0, x; \tau) v(x) = e^{-r\tau} \int dx p(x) v(x)$$

$$= e^{-r\tau} E[v] \qquad (14.1)$$

where

$$p(x) = P(x_0, x; \tau) \qquad (14.2)$$

Ignoring for the rest of the discussion the discounting by the spot interest rate. The price of the call option is given by evaluating the expectation value $E[v]$. Equation 14.1 yields the following expectation value

$$E[v] = \int dx p(x) v(x) \qquad (14.3)$$

Note that an European and American call option's price is always less than the security $S = e^{x_0}$ [4]; hence by dividing out by $S = e^{x_0}$, one can always redefine the payoff function so that

$$\mu = E[e^{-x_0} v] < 1 \qquad (14.4)$$

For ease of notation, we use v and $e^{-x_0} v$ interchangeably, and the difference will be known from the context of the equation.

If one does a Monte Carlo simulation for the evolution of the stock price using its stochastic differential equation, then after N trials of evolving the initial value of the log of stock x_0 to its final random values $x^{(i)}$, the Monte Carlo estimate of the expectation value, from Eq. 14.3, is given by

$$E_{MC}[v] = \frac{1}{N} \sum_{i=1}^{N} v(x^{(i)}) \pm \frac{\sigma_v}{\sqrt{N}} \quad \text{with } 66\% \text{ likelihood}$$

The quantum algorithm, after amplitude amplification, gives a *quadratic improvement* and yields

$$E[v] = \sum_{i=0}^{N-1} p(x^{(i)}) v(x^{(i)}) \pm \frac{\sigma_v}{N} \quad \text{with } 66\% \text{ likelihood}$$

In simplified notation, the summation is rewritten as

$$E[v] = \sum_{x=0}^{N-1} p(x)v(x) \pm \frac{\sigma_v}{N} \text{ with } 66\% \text{ likelihood} \qquad (14.5)$$

This improvement by a factor of a square root is similar to the improvement of the Grover algorithm, and the quantum algorithm discussed in this chapter is closely related to the Grover algorithm.

14.2 Quantum Algorithm

The basic strategy for applying quantum algorithms to option pricing is to map the computation required for evaluating the option's price to an equivalent quantum algorithm. One is then free to use the resources of both quantum entanglement and superposition in improving the efficiency of the quantum algorithm, which for option pricing achieves a quadratic improvement when compared to the classical Monte Carlo simulation. It is worth noting that one is evaluating a classical quantity using the computational tools of a quantum algorithm.

A quantum algorithm for pricing options has the following distinct steps.

1. The classical random variable x with the probability distribution function $p(x)$ given by Eq. 14.2 is discretized and replaced by n binary quantum degrees of freedom, which yields the basis states $|x\rangle$ with $x = 0, 1, \ldots, 2^n - 1$, with $N = 2^n$.
2. The basis states are combined into a superposed state to encode the option's data.
3. The probability distribution function $p(x)$ defines the function $\alpha(x) = \sqrt{p(x)}$, and the initial data of the option is encoded into the state vector $|\psi\rangle$.
4. The superposed basis states $|x\rangle$ yield the n qubits state vector

$$|\psi\rangle = \sum_{x=0}^{2^n-1} a(x)|x\rangle \; ; \; \sum_{x=0}^{2^n-1} |a(x)|^2 = 1$$

5. A state that entangles the state vector $|\psi\rangle$ with the payoff function v is created using an auxiliary qubit and is given by

$$|\psi\rangle \rightarrow |\eta\rangle = \sum_{x=0}^{2^n-1} a(x)|x\rangle \left(\sqrt{1 - v(x)}|0\rangle + \sqrt{v(x)}|1\rangle \right)$$

6. In Sect. 14.4 onwards, amplitude amplification is applied to the quantum state $|\eta\rangle$. A quantum gate (oracle) is applied simultaneously to all the basis states and the auxiliary qubit. At the end of the application of the gate, one of the states that hold the correct answer is marked, as in the case of Grover's algorithm.

7. The probability of measuring the marked state is amplified using quantum gates that achieve amplification using quantum interference.
8. A generalized Born measurement is carried out on the final qubits.

14.3 Quantum Algorithm for Expectation Value

We discuss a quantum algorithm that is valid for any expectation value. The algorithm for amplitude amplification is later built on this algorithm.

The required quantum algorithm yields the following expectation value

$$E[v] = \sum_{x=0}^{2^n-1} |a(x)|^2 v(x)$$

The algorithm is defined by the following [5]. Consider an ancillary qubit state $|0\rangle$ and a conditional gate R such that

$$R\Big(|x\rangle|0\rangle\Big) = |x\rangle\Big(\sqrt{1 - v(x)}|0\rangle + \sqrt{v(x)}|1\rangle\Big)$$

The algorithm $\mathcal{A}$ acts on the input qubits state and yields

$$\mathcal{A}|0\rangle^{\otimes n} = \sum_{x=0}^{2^n-1} a(x)|x\rangle \tag{14.6}$$

Hence

$$|\eta\rangle = R\Big(\mathcal{A}|0\rangle^{\otimes n}|0\rangle\Big) = R\Big(\sum_{x=0}^{2^n-1} a(x)|x\rangle|0\rangle\Big) \;\; ; \;\; \langle\eta|\eta\rangle = 1$$

$$\Rightarrow |\eta\rangle = \sum_{x=0}^{2^n-1} a(x)|x\rangle\Big(\sqrt{1 - v(x)}|0\rangle + \sqrt{v(x)}|1\rangle\Big) \tag{14.7}$$

The state vector $|\eta\rangle$ is a rotation of state vector $|\psi\rangle$ through an angle γ and shown in Fig. 14.1. The unitary transformation $\mathcal{F}$ is given by

$$\mathcal{F} = R(\mathcal{A} \otimes \mathbb{I}_2) \;\; : \;\; \mathcal{F}\Big(|0\rangle^{\otimes n}|0\rangle\Big) = \Big(R(\mathcal{A} \otimes \mathbb{I}_2)\Big)\Big(|0\rangle^{\otimes n}|0\rangle\Big) = |\eta\rangle \quad (14.8)$$

Note no amplification of any states is carried out. In preparation for a generalized Born measurement on only the ancillary qubits, we rewrite the state $|\eta\rangle$ as follows

$$|\eta\rangle = \sqrt{p(0)}|0\rangle|\Phi_0\rangle + \sqrt{p(1)}|0\rangle|\Phi_1\rangle$$

where

$$|\Phi_0\rangle = \frac{1}{\sqrt{p(0)}} \sum_{x=0}^{2^n-1} a(x)|x\rangle\sqrt{1-v(x)} \; ; \; |\Phi_1\rangle = \frac{1}{\sqrt{p(1)}} \sum_{x=0}^{2^n-1} a(x)|x\rangle\sqrt{v(x)}$$

and

$$p(0) = \sum_{x=0}^{2^n-1} |a(x)|^2(1-v(x)) \; ; \; p(1) = \sum_{x=0}^{2^n-1} |a(x)|^2 v(x)$$

Performing a generalized Born measurement on only the ancillary qubit state, discussed in Sect. 4.10, yields

$$|\eta\rangle \; \rightarrow \; \text{Measurement} \; \rightarrow \; |a\rangle|\Phi_a\rangle \; : \; a = 0, 1$$

The probability of observing the ancillary qubit $|1\rangle$ is given by the following expectation value

$$\mu = p(1) = \langle\eta|\big(I_n \otimes |1\rangle\langle1|\big)|\eta\rangle = \langle\eta|1\rangle\langle1|\eta\rangle = \sum_{x=0}^{2^n-1} |a(x)|^2 v(x)$$

$$\Rightarrow \mu = E[v] \approx \int dx\, p(x)v(x) \tag{14.9}$$

Hence, we see from Eq. 14.9 that any expectation value can be re-expressed as a quantum algorithm. The only constraint in finding the expectation value is that, since $0 \le \mu \le 1$, we must have that $v(x) \le 1$ for all x.

Equation 14.9 is based on sampling a random outcome many times and hence reproduces the classical Monte Carlo result. In observing the ancillary qubit, we have two outcomes: $|1\rangle$ with probability μ and $|0\rangle$ with probability $1 - \mu$; this is a Bernoulli random variable with $\sigma_\mu^2 = \mu(1 - \mu)$. From the central limit theorem, performing the measurement N, the accuracy of the estimate is given by

$$\tilde{\mu} = \mu \pm \frac{\sigma_\mu}{\sqrt{N}} \text{ with } 66\% \text{ likelihood}$$

The accuracy and speed are the same as the classical Monte Carlo calculation.

14.4 Algorithm for Quadratic Improvement

Application of quantum algorithms to option price has been discussed in [5–8]. To get a quadratic improvement in the accuracy of the computed value of μ, we discuss the quantum algorithm for option price developed by Patrick et al. [5]—based on an oscillating phase.

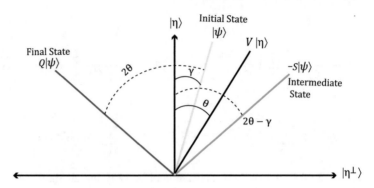

Fig. 14.1 Rotations of the initial qubit state $|\psi\rangle$ to the final state $Q|\psi\rangle$. Published with permission of © Belal E. Baaquie and L. C. Kwek. All Rights Reserved

The implementation of the oscillating phase approach is similar in spirit to the Grover algorithm, but the details are quite different. Note that the angle θ (defined below) is not known for option pricing, whereas in the Grover algorithm the angle θ, as in Eq. 11.16, is fixed by the total number of configurations of the degrees of freedom. What is similar to both approaches is that there is a rotation of an angle 2θ. For the option price, the rotation yields the final output qubits for which the qubit that is most likely to be observed is the qubit yields the best value for θ.

To perform the rotation, define the following unitary operator

$$V = I_{2^{n+1}} - 2I_{2^n} \otimes |1\rangle\langle 1| = V^\dagger \; ; \quad V^2 = I_{2^{n+1}}$$

The state vector $V|\eta\rangle$ is a reflection of $|\eta\rangle$ about the axis defined by the state vector $|\psi\rangle$, and is shown in Fig. 14.1.

The angle between the state $|\eta\rangle$, defined in Eq. 14.7, and $V|\eta\rangle$ yields the angle θ, as shown in Fig. 14.1, and given by the following

$$\cos\left(\frac{2\pi\theta}{2}\right) \equiv \left(\langle\eta|\right)\left(V|\eta\rangle\right) = \langle\eta|V|\eta\rangle$$
$$= \langle\eta|\eta\rangle - 2\langle\eta|1\rangle\langle 1|\eta\rangle = 1 - 2\langle\eta|1\rangle\langle 1|\eta\rangle$$
$$= 1 - 2\sum_{x=0}^{2^n-1}|a(x)|^2 v(x) = 1 - 2\mu$$
$$\Rightarrow \mu = \frac{1}{2}(1 - \cos(\pi\theta)) \tag{14.10}$$

Once θ is evaluated, the price of the option μ can be obtained from it.

The rotation operator $\mathcal{Q}$ needs to be defined that will act on the input qubit $|\psi\rangle$ and rotate it in the plane spanned by $|\eta\rangle$ and $V|\eta\rangle$ through an angle 2θ to yield state vector $\mathcal{Q}|\psi\rangle$. Following the derivation given in Patrick et al. [5], define another operator

$$U = I_{2^{n+1}} - 2|\eta\rangle\langle\eta|$$

Note that

$$U|\eta\rangle = -|\eta\rangle \; ; \; U|\eta^\perp\rangle = |\eta^\perp\rangle \; : \; \langle\eta|\eta^\perp\rangle = 0$$

and hence $-U$ reflects about the axis defined by $|\eta\rangle$, leaving the state vector $|\eta\rangle$ unchanged. Define the operator

$$\mathcal{Z} = I_{2^{n+1}} - 2|0\rangle^{\otimes(n+1)}\langle 0|^{\otimes(n+1)}$$

In terms of operator $\mathcal{F}$ introduced in Eq. 14.8, we have

$$U = \mathcal{F}\mathcal{Z}\mathcal{F}^\dagger$$

since

$$
\begin{aligned}
U &= \left[R(\mathcal{A} \otimes \mathbb{I}_2)\right]\left[I_{2^{n+1}} - 2|0\rangle^{\otimes(n+1)}\langle 0|^{\otimes(n+1)}\right]\left[(\mathbb{I}_2 \otimes \mathcal{A}^\dagger)R^\dagger\right] \\
&= I_{2^{n+1}} - 2|\eta\rangle\langle\eta|
\end{aligned}
\tag{14.11}
$$

Define the diffusion operator Q, similar to the case of Grover's rotation operator, that rotates the initial state vector $|\psi\rangle$ and is given by Patrick et al. [5]

$$
\begin{aligned}
\mathcal{Q} &= U(VUV) \equiv US \\
&= \left(I_{2^{n+1}} - 2|\eta\rangle\langle\eta|\right)V\left(I_{2^{n+1}} - 2|\eta\rangle\langle\eta|\right)V \\
&= \left(I_{2^{n+1}} - 2|\eta\rangle\langle\eta|\right)\left(I_{2^{n+1}} - 2V|\eta\rangle\langle\eta|V\right)
\end{aligned}
\tag{14.12}
$$

since $V^2 = I_{2^{n+1}}$.

The action of $Q = US$, with $S = VUV$ and $U = \mathcal{F}\mathcal{Z}\mathcal{F}^\dagger$, on an arbitrary state $|\psi\rangle$ in the span of $|\eta\rangle$ and $V|\eta\rangle$ is shown in Fig. 14.1. First, the action of $-S$ on $|\psi\rangle$ is to reflect along $V|\eta\rangle$, resulting in the intermediate $-S|\psi\rangle$. Then, $-U$ acts on $-S|\psi\rangle$ by reflecting along $|\eta\rangle$. The resultant state $\mathcal{Q}|\psi\rangle$ has been rotated anticlockwise by an angle 2θ in the hyperplane of $|\eta\rangle$ and $V|\eta\rangle$.

From the expression above, it can be seen that the operator $\mathcal{Q}$ rotates vectors in the two-dimensional plane defined by the vectors $|\eta\rangle$ and $V|\eta\rangle$, as shown in Fig. 14.1.

Note that

$$\mathcal{Q} = UVUV = \mathcal{F}\mathcal{Z}\mathcal{F}^\dagger \cdot V \cdot \mathcal{F}\mathcal{Z}\mathcal{F}^\dagger \cdot V$$

The gates for operator $\mathcal{Q}$ are shown in Fig. 14.2.

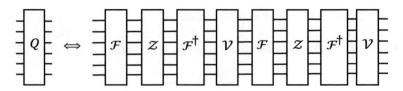

14.5 Eigenvalues of Diffusion Operator Q

The repeated application of the Grover diffusion operator G was evaluated by a recursion equation, given in Sect. 11.6. For the option pricing diffusion operator Q, the eigenvalues are evaluated, as this will allow us to isolate the angle θ that, via the relation $\cos \pi\theta = 1 - 2\mu$, will yield the option price.

Define the vector $|\eta^{\perp}\rangle$ by

$$\langle \eta | \eta^{\perp} \rangle = 0$$

Using the identities

$$\langle \eta | V | \eta \rangle = 1 - 2\mu = \cos \pi\theta \;\; ; \;\; \langle \eta | \eta \rangle = \langle \eta | V^2 | \eta \rangle = 1$$

yields

$$V|\eta\rangle = \cos(\pi\theta)|\eta\rangle + e^{i\chi}(\sin(\pi\theta))|\eta^{\perp}\rangle$$

From above equation it follows that, even though $|\eta\rangle$ and $V|\eta\rangle$ are not orthogonal, they are linearly independent.

Consider an arbitrary state vector $|\psi\rangle$ in the span of state vectors $|\eta\rangle$ and $V|\eta\rangle$; since these two vectors are linearly independent we have

$$|\psi\rangle = a|\eta\rangle + bV|\eta\rangle$$

The action of Q on this state vector is given by

$$Q|\psi\rangle = aQ|\eta\rangle + bQV|\eta\rangle \tag{14.13}$$

Recall from Eq. 14.12

$$\mathcal{Q} = \left(I_{2^{n+1}} - 2|\eta\rangle\langle\eta| \right) \left(I_{2^{n+1}} - 2V|\eta\rangle\langle\eta|V \right) \tag{14.14}$$

Hence

$$Q|\eta\rangle = \left(-1 + 4\cos^2(\pi\theta)\right)|\eta\rangle - 2\cos(\pi\theta)V|\eta\rangle$$
$$QV|\eta\rangle = 2\cos(\pi\theta)|\eta\rangle - V|\eta\rangle \tag{14.15}$$

and from Eqs. 14.13 and 14.15 we obtain

$$Q|\psi\rangle = \left(a(-1 + \cos^2(\pi\theta)) + 2b\cos(\pi\theta)\right)|\eta\rangle$$
$$+ \left(-2a\cos(\pi\theta) - b\right)V|\eta\rangle \tag{14.16}$$

The action of the diffusion operator Q on the space spanned by $|\eta\rangle$ and $V|\eta\rangle$ is closed and yields

$$Q|\psi\rangle = aQ|\eta\rangle + bQV|\eta\rangle$$
$$= \tilde{a}|\eta\rangle + \tilde{b}V|\eta\rangle \tag{14.17}$$

On the subspace spanned by $|\eta\rangle$ and $V|\eta\rangle$, Q is equal to the matrix M and yields the following linear transformation on the coefficients a, b

$$\begin{bmatrix} \tilde{a} \\ \tilde{b} \end{bmatrix} = M \begin{bmatrix} a \\ b \end{bmatrix} \tag{14.18}$$

The matrix M, from Eqs. 14.16 and 14.17, is given by

$$Q \equiv M = \begin{bmatrix} 4\cos^2(\pi\theta) - 1 & 2\cos(\pi\theta) \\ -2\cos(\pi\theta) & -1 \end{bmatrix} \; ; \quad \det M = 1 \tag{14.19}$$

The eigenvalues of M, given by λ, satisfy the equation

$$0 = \det(M - \lambda\mathbb{I}_2) = \det \begin{vmatrix} 4\cos^2(\pi\theta) - 1 - \lambda & 2\cos(\pi\theta) \\ -2\cos(\pi\theta) & -1 - \lambda \end{vmatrix} \tag{14.20}$$

and yields

$$(1 + \lambda)^2 - (1 + \lambda)\cos^2(\pi\theta) + \cos^2(\pi\theta) = 0$$

Hence the eigenvalues of Q are given by

$$\lambda_\pm = \frac{1}{2}\left(\cos^2(\pi\theta) - 2 \pm i\cos(\pi\theta)\sqrt{4 - \cos^2(\pi\theta)}\right) \tag{14.21}$$

Note that

$$\cos^2(\pi\theta) - 2 = 2\cos(2\pi\theta) \; ; \quad \cos(\pi\theta)\sqrt{4 - \cos^2(\pi\theta)} = 2\sin(2\pi\theta)$$

Hence, simplifying the eigenvalues given in Eq. 14.21 yields

$$\lambda_\pm = \cos(2\pi\theta) \pm i\sin(2\pi\theta) = \exp\{\pm i2\pi\theta\}$$

For completeness, note that the matrix $\mathcal{Q}$ in the $\{|\eta\rangle, |\eta^\perp\rangle\}$ basis is given by

$$\mathcal{Q} = \begin{bmatrix} \cos(2\pi\theta) & e^{i\chi}\sin(2\pi\theta) \\ -e^{-i\chi}\sin(2\pi\theta) & \cos(2\pi\theta) \end{bmatrix}$$

The matrix elements of $\mathcal{Q}$ can be obtained easily from Eq. (14.14).

To get the matrix elements of $\mathcal{Q}$, we can also evaluate the matrix elements $\langle\eta|\mathcal{Q}|\eta\rangle$, $\langle\eta|\mathcal{Q}|\eta^\perp\rangle$ (and also the other off-diagonal term, $\langle\eta^\perp|\mathcal{Q}|\eta\rangle$) and $\langle\eta^\perp|\mathcal{Q}|\eta^\perp\rangle$: For instance,

$$\begin{aligned} \langle\eta|\mathcal{Q}|\eta\rangle &= (4\cos^2(\pi\theta) - 1) - 2\cos(\pi\theta)\langle\eta|V|\eta\rangle \\ &= (4\cos^2(\pi\theta) - 1) - 2\cos^2(\pi\theta) \\ &= 2\cos^2(\pi\theta) - 1 = \cos(2\pi\theta) \end{aligned} \tag{14.22}$$

The matrix representation of $\mathcal{Q}$ is given by

$$\begin{aligned} \mathcal{Q} &= \left\{ \begin{bmatrix} 1 & 0 \\ 0 & 1 \end{bmatrix} - 2\begin{bmatrix} 1 & 0 \\ 0 & 0 \end{bmatrix} \right\} \\ &\quad \times \left\{ \begin{bmatrix} 1 & 0 \\ 0 & 1 \end{bmatrix} - 2\begin{bmatrix} \cos(\pi\theta) \\ e^{i\chi}\sin(\pi\theta) \end{bmatrix} \begin{bmatrix} \cos(\pi\theta) & e^{-i\chi}\sin(\pi\theta) \end{bmatrix} \right\} \\ &= \begin{bmatrix} -1 & 0 \\ 0 & 1 \end{bmatrix} \begin{bmatrix} 1 - 2\cos^2(\pi\theta) & -2e^{-i\chi}\cos(\pi\theta)\sin(\pi\theta) \\ -2e^{-i\chi}\cos(\pi\theta)\sin(\pi\theta) & 1 - 2\sin^2(\pi\theta) \end{bmatrix} \\ &= \begin{bmatrix} -1 & 0 \\ 0 & 1 \end{bmatrix} \begin{bmatrix} -\cos(2\pi\theta) & -e^{-i\chi}\sin(2\pi\theta) \\ -e^{i\chi}\sin(2\pi\theta) & \cos(2\pi\theta) \end{bmatrix} \\ &= \begin{bmatrix} \cos(2\pi\theta) & e^{-i\chi}\sin(2\pi\theta) \\ -e^{i\chi}\sin(2\pi\theta) & \cos(2\pi\theta) \end{bmatrix} \end{aligned}$$

where we have represented $\{|\eta\rangle, |\eta^\perp\rangle\}$ as $\left\{ \begin{bmatrix} 1 \\ 0 \end{bmatrix}, \begin{bmatrix} 0 \\ 1 \end{bmatrix} \right\}$, respectively, so that

$$V|\eta\rangle = \begin{bmatrix} \cos(\pi\theta) \\ e^{i\chi}\sin(\pi\theta) \end{bmatrix}$$

The eigenvectors of $\mathcal{Q}$ are $|\psi_\pm\rangle$ with eigenvalues $\exp\{\pm i2\pi\theta\}$. The explicit representation of the eigenvectors is not required, but they are given for completeness. The eigenvectors of $\mathcal{Q}$ are given by

$$|\psi_\pm\rangle = a_\pm|\eta\rangle + b_\pm V|\eta\rangle$$

with

$$M \begin{bmatrix} a_\pm \\ b_\pm \end{bmatrix} = \begin{bmatrix} 4\alpha^2 - 1 & 2\alpha \\ -2\alpha & -1 \end{bmatrix} \begin{bmatrix} a_\pm \\ b_\pm \end{bmatrix} = e^{\pm i 2\pi\theta} \begin{bmatrix} a_\pm \\ b_\pm \end{bmatrix} \tag{14.23}$$

Using Eq. 14.23, the eigenvectors are the following

$$|\psi_+\rangle = \frac{1}{\sqrt{2}} \left[|\eta\rangle - e^{-i\pi\theta} V |\eta\rangle \right] \; ; \; |\psi_-\rangle = \frac{1}{\sqrt{2}} \left[e^{i\pi\theta} |\eta\rangle + V |\eta\rangle \right] \tag{14.24}$$

Note $|\psi_\pm\rangle$ are not normal or orthogonal since $|\eta\rangle$ and $V|\eta\rangle$ are not orthogonal. The amplitude amplification does not require that the eigenfunctions $|\psi_\pm\rangle$ be orthogonal, and hence, the state vector $|\psi\rangle$ needs not be expressed in an orthogonal basis.

14.6 Amplitude Amplification

The n-qubit $|\eta\rangle$ has the following eigenfunction expansion

$$|\eta\rangle = \beta_+ |\psi_+\rangle + \beta_- |\psi_-\rangle$$

Consider a m-qubit state $|y\rangle$ and the $n + 1$-qubit $|\eta\rangle$; the input tensor m-qubit $\otimes$ $n + 1$-qubit state is given by

$$|y\rangle |\eta\rangle$$

where recall from Eq. 14.7 that

$$|\eta\rangle = \sum_{x=0}^{2^n - 1} a(x) |x\rangle \left(\sqrt{1 - v(x)} |0\rangle + \sqrt{v(x)} |1\rangle \right) \; : \; \langle \eta | \eta \rangle = 1$$

Extend the definition of $\mathcal{Q}$ to $\mathcal{Q}_c$ such that it has a *conditional application*, which depends on the m-qubit state $|y\rangle$, and is given by

$$\mathcal{Q}_c = I_{2^m} \otimes \mathcal{Q} \; \Rightarrow \; \mathcal{Q}_c \left(|y\rangle |\psi\rangle \right) = |y\rangle \mathcal{Q}^y |\psi\rangle$$

where

$$\mathcal{Q}^y \equiv (\mathcal{Q})^y$$

The phase angle θ is evaluated in the following manner. Take a copy of input $|0\rangle^{\otimes(n+1)}$ and applying $\mathcal{F}$ yields

$$\mathcal{F} |0\rangle^{\otimes(n+1)} = |\eta\rangle$$

The m-qubit register is prepared in the uniform superposition state using the Hadamard gate and yields

$$H|0\rangle^{\otimes m}|\eta\rangle = \frac{1}{\sqrt{2^m}} \sum_{y=0}^{2^m-1} |y\rangle|\eta\rangle$$

Applying the controlled $\mathcal{Q}_c$ to the input state yields the following entangled state

$$|\Psi\rangle = \mathcal{Q}_c \left[\frac{1}{\sqrt{2^m}} \sum_{y=0}^{2^m-1} |y\rangle|\eta\rangle \right] = \frac{1}{\sqrt{2^m}} \sum_{y=0}^{2^m-1} |y\rangle \mathcal{Q}^y|\eta\rangle$$

$$= \frac{1}{\sqrt{2^m}} \sum_{y=0}^{2^m-1} |y\rangle \mathcal{Q}^y \Big(\beta_+|\psi_+\rangle + \beta_-|\psi_-\rangle \Big)$$

$$\Rightarrow |\Psi\rangle = \frac{1}{\sqrt{2^m}} \sum_{y=0}^{2^m-1} |y\rangle \Big(e^{2\pi i y\theta} \beta_+|\psi_+\rangle + e^{-2\pi i y\theta} \beta_-|\psi_-\rangle \Big) \qquad (14.25)$$

The inverse Fourier transform on the $|y\rangle$ qubits is given by

$$|y\rangle = \frac{1}{\sqrt{2^m}} \sum_{x=0}^{2^m-1} e^{-2\pi i x y/2^m} |x\rangle$$

The full quantum circuit for the phase estimation is shown in Fig. 14.3

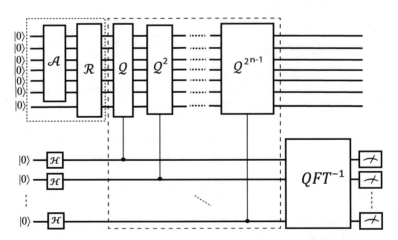

Fig. 14.3 Quantum circuit for estimation of the phase θ. Published with permission of © Belal E. Baaquie and L. C. Kwek. All Rights Reserved

Hence, from Eq. 14.25

$$|\Psi\rangle$$
$$= \frac{1}{\sqrt{2^m}} \sum_{y=0}^{2^m-1} \frac{1}{\sqrt{2^m}} \sum_{x=0}^{2^m-1} |x\rangle \left(e^{-2\pi i(x/2^m-\theta)y} \beta_+ |\psi_+\rangle + e^{-2\pi i(x/2^m+\theta)y} \beta_- |\psi_-\rangle \right)$$
$$\Rightarrow |\Psi\rangle = \sum_{x=0}^{2^m-1} |x\rangle \left(\beta_+(x) |\psi_+\rangle + \beta_-(x) |\psi_-\rangle \right) \tag{14.26}$$

where

$$\beta_\pm(x) = \frac{1}{2^m} \sum_{y=0}^{2^m-1} \exp\{-2\pi i(x/2^m \mp \theta)y\} \beta_\pm$$

Note that from Eq. 2.2 x is given by

$$x = x_m 2^0 + x_{m-1} 2^1 + x_{m-2} 2^2 \cdots + x_i 2^i + \cdots + x_1 2^{m-1} \ : \ 0 \le x \le 2^m - 1$$

Recall from Eq. 14.10 that the angle $\theta \in [0, 1]$ is defined by the payoff function and is not an integer. Consider $\theta = \theta_m + \delta$, where θ_m is a binary decimal closest to θ. Then, for example

$$\beta_+(x) = \frac{1}{2^m} \sum_{y=0}^{2^m-1} \exp\{-2\pi i(x/2^m - \theta_m - \delta)y\} \beta_+$$

Note

$$\frac{1}{2^m} \sum_{y=0}^{2^m-1} \exp\{-2\pi i(x/2^m - \theta_m)y\} \simeq \delta_{x-2^m\theta_m} + O(\delta/2^m)$$

For $x = 2^m \theta_m$, we have

$$\beta_+(2^m \theta_m) = \frac{1}{2^m} \sum_{y=0}^{2^m-1} \exp\{2\pi i\delta y\} \beta_+ = \frac{1}{2^m} \frac{1 - e^{2\pi i\delta 2^m}}{1 - e^{2\pi i\delta}} \beta_+ \simeq \beta_+ + O(\delta/2^m)$$

Note that, from Eq. 14.26

$$|\Psi\rangle = \sum_{x=0}^{2^m-1} \beta(x)|x\rangle \cdot \frac{1}{\beta(x)} \left(\beta_+(x)|\psi_+\rangle + \beta_-(x)|\psi_-\rangle \right)$$
$$\Rightarrow |\Psi\rangle \equiv \sum_{x=0}^{2^m-1} \beta(x)|x\rangle |\Phi(x)\rangle \ ; \ \beta(x) = \sqrt{\beta_+^2(x) + \beta_-^2(x)} \tag{14.27}$$

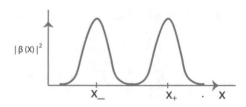

On performing a generalized Born measurement of only the x qubits in Eq. 14.27,
we have

$$|\Psi\rangle \;\rightarrow\; \text{Measurement} \;\rightarrow\; |x\rangle|\Phi(x)\rangle \;:\; x = 0, 1, \ldots, 2^m - 1$$

The probability $\beta^2(x)$ to find the system in qubit basis state $|x\rangle$ is given by

$$|\beta(x)|^2 = \text{Tr}\Big(|x\rangle\langle x| \otimes I_{2^n}\Big)\Big(|\Psi\rangle\langle\Psi|\Big) = |\beta_+(x)|^2 + |\beta_-(x)|^2$$

The maximum likelihood of occurrence for a qubit x, as shown in Fig. 14.4, is given
by

$$x_\pm/2^m = \pm\theta_m$$

Since the option price is the function $\cos(\pm 2\pi\theta_m)$, the value of either of the qubits
$x_\pm$ that occur with maximum likelihood gives the option price.

It is shown in Patrick et al. [5] that the estimate of the option price has the expected
quadratic speeding up given in Eq. 14.5. An intuitive discussion of this result is given
in Sect. 14.8.

14.7   Call Option

The price of a European call option is given by [9]

$$C(x_0, \tau) = \frac{e^{-r\tau}}{\sqrt{2\pi\tau\sigma^2}} \int_{-\infty}^{+\infty} dx\, e^{-\frac{1}{2\tau\sigma^2}(x - x_0 - \tau(r - \frac{1}{2}\sigma^2))^2}[e^x - K]_+$$

$$= e^{-r\tau} \int_{-\infty}^{+\infty} dx\, \frac{e^{-\frac{1}{2}x^2}}{\sqrt{2\pi}}\Big[e^{\sqrt{\tau}\sigma x + x_0 + \tau(r - \frac{1}{2}\sigma^2)} - K\Big]_+$$

$$\simeq e^{-r\tau} \int_{-x_{\max}}^{+x_{\max}} dx\, \frac{e^{-\frac{1}{2}x^2}}{\sqrt{2\pi}}\Big[e^{\sqrt{\tau}\sigma x + x_0 + \tau(r - \frac{1}{2}\sigma^2)} - K\Big]_+ \tag{14.28}$$

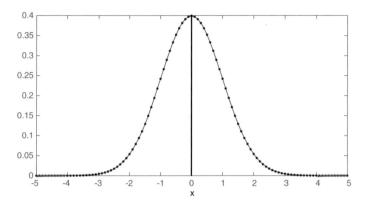

Fig. 14.5 Discretized Gaussian probability distribution function. Published with permission of © Belal E. Baaquie and L. C. Kwek. All Rights Reserved

In the approximation given in Eq. 14.28, one replaces $[-\infty, +\infty]$ by $[-x_{max}, +x_{max}]$; it is sufficient to take x_{max} to be 1. The probability density for the Gaussian random variable is given by

$$p(x) = \frac{1}{\sqrt{2\pi}} e^{-\frac{1}{2}x^2}$$

The random variable x is discretized into 2^n points with

$$x_j = -x_0 + j\Delta x \ ; \ \ \Delta x = 2x_{max}/(2^n - 1) \ ; \ \ j = 0, 1, \ldots, 2^n - 1$$

The discrete probabilities are given by the following

$$p(x) \rightarrow p_j = \frac{1}{c} p(x_j) \ ; \ \ c = \sum_{j=0}^{2^n-1} p(x_j)$$

The discretized probability distribution is shown in Fig. 14.5.

The payoff function

$$v(x) = \left[e^{\sqrt{\tau}\sigma x + x_0 + \tau(r - \frac{1}{2}\sigma^2)} - K \right]_+$$

is discretized to

$$v_j = v(x_j)$$

A binary approximation $\tilde{v}_j$ is made of the discretized payoff function such that the error is given by

$$|v_j - \tilde{v}_j| < \frac{1}{2^n}$$

Hence, from Eq. 14.28, the approximate call option price is given by

$$C(x_0, \tau) \simeq e^{-r\tau} \sum_{j=0}^{2^n-1} p_j \tilde{v}_j \tag{14.29}$$

Equation 14.6 is specialized to

$$\mathcal{A}|0\rangle^{\otimes n} = \sum_{j=0}^{2^n-1} \sqrt{p_j}|j\rangle \tag{14.30}$$

The input qubit state, as in the general case, is prepared to yield

$$|\eta\rangle = \sum_{j=0}^{2^n-1} \sqrt{p_j}|j\rangle \left(\sqrt{1 - \tilde{v}_j}|0\rangle + \sqrt{\tilde{v}_j}|1\rangle \right) \ : \ \langle\eta|\eta\rangle = 1 \tag{14.31}$$

The option price, as in the general case, is given by

$$\mu = \langle\eta|\left(I_{2^n} \otimes |1\rangle\langle 1| \right)|\eta\rangle = \sum_{j=0}^{2^n-1} \tilde{v}_j p_j$$

and we have recovered the approximate call option price given in Eq. 14.29. The quadratic improvement discussed in Sect. 14.4 can be applied for pricing the call option.

14.8   Discussion

The Grover and option price algorithms are both based on amplitude amplification. There is however an interesting difference between the two. For Grover's case, one is seeking a marked state vector, $|\xi\rangle$, as in Eq. 11.32. Hence, there is no need for a Fourier transform for isolating the marked state vector, since doing the recursion enough times makes the coefficient of $|\xi\rangle$ in the superposed state vector almost equal to 1.

In contrast, the algorithm for finding the option price reduces the problem to finding an unknown angle θ. The Grover approach does not work, since for Grover's algorithm, the angle θ is known: it is determined by the number of qubits being used for the algorithm. The number of recursions required to find the marked state $|\xi\rangle$ is determined by θ.

For finding the option price, the Fourier transform is required to pick out the most likely state, which after sufficient iterations of the diffusion operator Q, is seen to yield the sought for angle.

Intuitively, for classical probabilistic models, the probability of success increases, roughly, by a constant on each iteration; by contrast, amplitude amplification roughly increases the amplitude of success by a constant on each iteration. Because amplitudes correspond to square roots of probabilities, it suffices to repeat the amplitude amplification process approximately $\sqrt{N}$ times to achieve the same accuracy, with high probability, as the classical case with N iterations.

In summary, the quantum algorithm discussed in this chapter computes the option price more efficiently than the Monte Carlo simulations. For a simulation consisting of N trials, classical Monte Carlo yields an error that falls like $1/\sqrt{N}$, whereas in the case of a quantum algorithm there is a quadratic improvement and the error falls like $1/N$.

References

1. Baaquie BE (2004) Quantum finance: path integrals and Hamiltonians for options and interest rates. Cambridge University Press, UK
2. Baaquie BE (2010) Interest rates and coupon bonds in quantum finance, 1st edn. Cambridge University Press, UK
3. Baaquie BE (2018) Quantum field theory for economics and finance. Cambridge University Press, UK
4. Hull JC (2000) Options, futures, and other derivatives, 4th edn. Prentice Hall, New Jersey
5. Rebentrost P, Gupt B, Bromley TR (2018) Quantum computational finance: Monte carlo pricing of financial derivatives. Phys Rev A 98:022321
6. Orús SMR, Lizaso E (2019) Quantum computing for finance: overview and prospects. Rev Phys (4):100028
7. Brassard MMG, Høyer P, Tapp A (2000) Quantum amplitude amplification and estimation. arXiv:quant-ph/0005055v1
8. Kaneko NTK, Miyamoto K, Yoshinox K (2020) Quantum computing for finance: overview and prospects. arXiv:1905.02666v4
9. Baaquie BE (2020) Mathematical methods and quantum mathematics for economics and finance. Springer, Singapore

Chapter 15
Solving Linear Equations

15.1 Introduction

Historically, some of the most ancient mathematical problems are directly related to the solutions of linear equations. As a matter of fact, a long time ago, the Chinese mentioned their techniques for solving linear equations with remarkable resemblance to our modern Gaussian elimination method [1]. Published nearly two thousand years ago, the method was described in a book entitled 'The Nine Chapters on the Mathematical Art' (Jiuzhang Suanshu) [1].[1] Like many ancient publication, nobody really knows the true authors of the book. Yet, it constitutes an important mathematical treatise that well precedes the magnificent works of Carl Friedrich Gauss (1777–1855). It is also interesting to note that one of the first digital computers, the Atanasoff–Berry Computer (ABC), is designed specifically for solving linear equations.

Nowadays, the solution of linear equations features prominently in many science, engineering or even finance problems. A system of linear equations is typically cast into matrix form:

$$A\mathbf{x} = \mathbf{b}, \tag{15.1}$$

where $\mathbf{x}$ is the vector of unknown variables that need to be determined. Essentially, if we can invert the matrix A, we want to find $\mathbf{x} = A^{-1}\mathbf{b}$. Solving a system of three or four equations can be done quite easily by hand, for example with Gaussian elimination. However, when the number of linear equations is large, efficient computational algorithms are needed, and typically such algorithms are no longer based on Gaussian elimination. Most computers use algorithms based on what is technically called Lower-Upper (LU) decomposition of the matrix A. For symmetric positive definite matrix A, one can resort to Cholesky decomposition. For Toeplitz matrices, Levinson recursion is used. Moreover, it often pays if we know if the matrix A is a sparse or not.

---

[1] Jiuzhang is also the name given to one of the latest quantum computers in China.

© The Author(s), under exclusive license to Springer Nature Singapore Pte Ltd. 2023   239
B. E. Baaquie and L.-C. Kwek, *Quantum Computers*,
https://doi.org/10.1007/978-981-19-7517-2_15

In 2009, Aram Harrow, Avinatan Hassidim and Seth Lloyd proposed a quantum algorithm for solving a linear set of equations [2]. This algorithm is nowadays widely known by its acronym, HHL. The classical computer takes polynomial time to solve a system of n equations. For a set of linear equations of size n with condition number κ, sparsity s and desired precision ϵ, the time complexity is known to be $\mathcal{O}(\log(n)s^2\kappa^2/\epsilon)$. This is not exponential speedup in the computational time over the classical algorithms, for instance, for the Gaussian elimination, the time complexity is roughly $\mathcal{O}(n^3)$, and if it is sparse, that complexity reduces to $\mathcal{O}(\kappa sn \log n/\log \epsilon)$ by gradient descent. Yet like Grover's algorithm, it can still provide an advantage over classical algorithm if used appropriately.

15.2 Harrow–Hassidim–Lloyd Algorithm

We first start with the quantum mechanical version of Eq. (15.1),

$$A|x\rangle = |b\rangle. \tag{15.2}$$

or $|x\rangle = A^{-1}|b\rangle$. So if A is Hermitian so that $A = A^\dagger$, then we are looking for the solution of $e^{-iAt}|b\rangle$. Even if A is not Hermitian (or even a square matrix), we can modify the problem into

$$\begin{pmatrix} 0 & A \\ A^\dagger & 0 \end{pmatrix} \begin{pmatrix} 0 \\ |x\rangle \end{pmatrix} = \begin{pmatrix} |b\rangle \\ 0 \end{pmatrix}. \tag{15.3}$$

Suppose we know how to diagonalize A, i.e., $UAU^\dagger = D$ where D is a diagonal matrix given by

$$D = \begin{pmatrix} \lambda_1 & 0 & 0 & 0 \\ 0 & \lambda_2 & 0 & 0 \\ & & \cdots & \\ 0 & 0 & 0 & \lambda_n \end{pmatrix},$$

if A is an $n \times n$ square matrix and $\{\lambda_i\}$ is the set of eigenvalues of A. We next apply the quantum phase estimation algorithm, which is simply a Von Neumann projective measurement in quantum mechanics. In particular, one could think of the quantum phase estimation part as follows: for the matrix A, if the eigenvalues of A are $\{\lambda_i\}$ with corresponding eigenvectors $\{u_i\}$, a projective measurement maps the eigenvalues λ_i into the register into the state This is somewhat analogous to the Von Neumann measurement process where

$$e^{iAt\otimes\hat{p}}|u_i\rangle \otimes |0\rangle_r$$

gives $|u_i\rangle \otimes |\lambda_i\rangle_r$ where the little subscript r denotes the register and $\hat{p}$ is some momentum operator.

We are now in a position to describe the entire process involved in HHL. We start with the state $|0\rangle_a |0\rangle_r^{\otimes p} |b\rangle_m$, where the subscript m denotes the state stored in the memory, r denotes the register of p qubits (for storing the binary fraction form of the phase associated with the eigenvalue of A, and a denotes the ancillary qubit needed for the operation, i.e., $|0\rangle_a |0\rangle_r^{\otimes p} |b\rangle_m$.

We then perform the quantum phase estimation using the unitary transformation e^{iA}:

$$|0\rangle_a |0\rangle_r^{\otimes p} |b\rangle_b \rightarrow e^{i2\pi A} |0\rangle_a |0\rangle_r^{\otimes p} |b\rangle_m = \sum_{j=1}^{n} \beta_j e^{2\pi i \tilde{\lambda}_j} |0\rangle_a |\tilde{\lambda}_j\rangle_r |u_j\rangle_m \qquad (15.4)$$

for some coefficients β_j. Since A is a known Hermitian matrix, the matrix $e^{i2\pi A} \equiv U$ is unitary so the eigenvalues $|e^{i2\pi \tilde{\lambda}_j}| = 1$ where $\tilde{\lambda}_j$ is the j-th eigenvalue of A with eigenvector $|u_j\rangle$. Note that any unitary transformation U can be easily realized in the circuit model since any unitary matrix can be decomposed into three Pauli rotation operators,

$$\{\exp(i\theta_k \sigma_k / 2), \ k = (x, y, z)\}$$

along any two axes, i.e., for example,

$$U = \exp(i\theta \sigma_x / 2) \exp(i\eta \sigma_z / 2) \exp(i\phi \sigma_x / 2).$$

Note that we have also expanded $|b\rangle$ in the eigenvector basis.

As noted earlier, the eigenvalues of U are $e^{i2\pi \tilde{\lambda}_j}$ with $0 \leq \tilde{\lambda}_j \leq 1$. By applying an inverse Fourier transform, we encode $\tilde{\lambda}_j$ as a binary fraction in the p qubits in the register, i.e., assuming $2^{-p} \leq \tilde{\lambda}_j \leq 2^{-1}$ so that the state in Eq. (15.4) becomes

$$\sum_{j=1}^{n} \beta_j |0\rangle_a |\tilde{\lambda}_j\rangle_r |u_j\rangle_m$$

using inverse quantum Fourier transform (see Chap. 8).

We then apply a conditional rotation depending on the state $|\tilde{\lambda}_j\rangle$ and rotate the ancillary qubit to

$$\sqrt{1 - \frac{C^2}{\tilde{\lambda}_j^2}} |0\rangle_a + \frac{C}{\tilde{\lambda}_j} |1\rangle_a \text{ for each } \tilde{\lambda}_j :$$

$$\sum_{j=1}^{n} \beta_j |0\rangle_a |\tilde{\lambda}_j\rangle_r |u_j\rangle_m \rightarrow \sum_{j=1}^{n} \beta_j \left(\sqrt{1 - \frac{C^2}{\tilde{\lambda}_j^2}} |0\rangle_a + \frac{C}{\tilde{\lambda}_j} |1\rangle_a \right) |\tilde{\lambda}_j\rangle_r |u_j\rangle_m \qquad (15.5)$$

followed by a reverse of the quantum phase transformation resetting the register to $|0\rangle_r^{\otimes p}$:

$$\sum_{j=1}^{n} \beta_j \left(\sqrt{1 - \frac{C^2}{\tilde{\lambda}_j^2}} |0\rangle_a + \frac{C}{\tilde{\lambda}_j} |1\rangle_a \right) |0\rangle_r^{\otimes p} |u_j\rangle_m. \tag{15.6}$$

In terms of gates, the reverse of the quantum phase transformation is just the symmetrical reversal of all the gates in the quantum phase estimation and quantum Fourier transform stages.

Measuring the ancillary qubit in the computational basis, i.e., $\{|0\rangle, |1\rangle\}$, yields

$$\sum_{j=1}^{n} C \frac{\beta_j}{\tilde{\lambda}_j} |u_j\rangle_m \approx A^{-1} |b\rangle_m \sim |x\rangle_a \tag{15.7}$$

if the measurement outcome is $|1\rangle_a$.

15.3   Specific Example

As a specific example, we consider the case of a simple set of linear equations

$$A|x\rangle = \begin{pmatrix} 3/8 & 1/8 \\ 1/8 & 3/8 \end{pmatrix} |x\rangle = \begin{pmatrix} 0 \\ 1 \end{pmatrix} = |b\rangle \tag{15.8}$$

It is **not** necessary in the HHL algorithm to compute the eigenvectors and eigenvalues of the matrix A, but for the purpose of illustrating the algorithm, we will do it: Matrix A has eigenvalues $\tilde{\lambda}_1 = 1/2$ and $\tilde{\lambda}_2 = 1/4$ with eigenvectors $|u_1\rangle = \frac{1}{\sqrt{2}} \begin{pmatrix} 1 \\ 1 \end{pmatrix} = 1/\sqrt{2}(|0\rangle + |1\rangle)$ and $|u_2\rangle = \frac{1}{\sqrt{2}} \begin{pmatrix} 1 \\ -1 \end{pmatrix} = 1/\sqrt{2}(|0\rangle - |1\rangle)$, respectively. In the new basis $\{|u_i\rangle\}$ $(i = 1, 2)$,

$$|b\rangle = |1\rangle = 1/\sqrt{2}(|u_1\rangle - |u_2\rangle). \tag{15.9}$$

Note that in binary notation, the eigenvalues $\lambda_1 \equiv 0.1$ (or in state notation, $|10\rangle$) and $\lambda_2 \equiv 0.01$ (or in state notation, $|01\rangle$).

We then apply the quantum phase estimation algorithm using the unitary evolution matrix

$$U = \exp(i2\pi A) = \sum_{j=1}^{2} \exp(2\pi i \tilde{\lambda}_j) |u_j\rangle \langle u_j|. \tag{15.10}$$

Note that like quantum phase estimation where we wish to estimate the unknown phase of a unitary matrix, we do not really need to know these eigenvalues and eigenvectors (see Sect. 15.4). It is useful to decompose the vector $|b\rangle$ in terms of the eigenvectors of A since the circuit in Fig. 15.1 acts only on $|b\rangle$ and not on the eigenvectors $|u_1\rangle$ or $|u_2\rangle$.

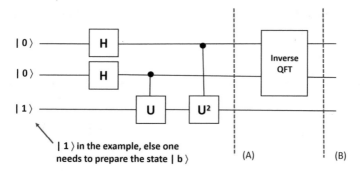

Fig. 15.1 Circuit diagram for the first part of HHL algorithm. Published with permission of © Belal E. Baaquie and L. C. Kwek. All Rights Reserved

We have chosen $|b\rangle$ to be a computational state $|1\rangle$. In principle, it could be more complicating and one needs to prepare the state. In Fig. 15.1, it is not difficult to see that at step (A), the state of the system is

$$|\text{state at (A)}\rangle = \frac{1}{\sqrt{2}}\left(\frac{1}{\sqrt{2}}\right)^2 \sum_{j=1}^{2}] \left\{ |00\rangle + \exp(i2\pi\tilde{\lambda}_j)|01\rangle \right.$$

$$\left. + \exp(i4\pi\tilde{\lambda}_j)|10\rangle + \exp(i6\pi\tilde{\lambda}_j)|11\rangle \right\} \otimes |u_j\rangle,$$

$$= \left(\frac{1}{\sqrt{2}}\right)^3 \sum_{j=1}^{2} \sum_{k=0}^{3} \exp(i2\pi\tilde{\lambda}_j k)|k\rangle|u_j\rangle. \tag{15.11}$$

If we now perform an inverse Fourier transform on the first two qubits, we get

$$|\text{state at (B)}\rangle = \left(\frac{1}{\sqrt{2}}\right)^3 \sum_{j=1}^{2} \sum_{k=0}^{3} \exp(i2\pi\tilde{\lambda}_j k)\left(\frac{1}{\sqrt{2}}\right)^2 \sum_{\ell=0}^{3} \exp\left(-\frac{2\pi i k\ell}{4}\right)|\ell\rangle|u_j\rangle$$

$$= \left(\frac{1}{\sqrt{2}}\right)^3 \left(\frac{1}{\sqrt{2}}\right)^2 \sum_{j=1}^{2} \sum_{k,\ell=0}^{3} \exp\left(i2\pi\tilde{\lambda}_j k - \frac{2\pi i k\ell}{4}\right)|\ell\rangle|u_j\rangle,$$

$$= \left(\frac{1}{\sqrt{2}}\right)^5 \sum_{j=1}^{2} \sum_{k,\ell=0}^{3} \exp\left(i\frac{2\pi}{4}(4\tilde{\lambda}_j - \ell)k\right)|\ell\rangle|u_j\rangle \tag{15.12}$$

Noting that $\tilde{\lambda}_1 = 1/2$ and $\tilde{\lambda}_2 = 1/4$, we see that whenever

$$\ell_1 = 4\tilde{\lambda}_1 = 2,$$
$$\ell_2 = 4\tilde{\lambda}_2 = 1,$$

the expression in Eq. (15.12) peaks. To see this, we tabulate the coefficients in the expression in Eq. (15.12) for each j:

k ℓ			
$|00\rangle$	$|01\rangle$	$|10\rangle$	$|11\rangle$
0 1	1	1	1
1 $\exp\left(i\frac{2\pi}{4}(4\tilde{\lambda}_j)\right)$	$\exp\left(i\frac{2\pi}{4}(4\tilde{\lambda}_j-1)\right)$	$\exp\left(i\frac{2\pi}{4}(4\tilde{\lambda}_j-2)\right)$	$\exp\left(i\frac{2\pi}{4}(4\tilde{\lambda}_j-3)\right)$
2 $\exp\left(i\frac{2\pi(2)}{4}(4\tilde{\lambda}_j)\right)$	$\exp\left(i\frac{2\pi(2)}{4}(4\tilde{\lambda}_j-1)\right)$	$\exp\left(i\frac{2\pi(2)}{4}(4\tilde{\lambda}_j-2)\right)$	$\exp\left(i\frac{2\pi(2)}{4}(4\tilde{\lambda}_j-3)\right)$
3 $\exp\left(i\frac{2\pi(3)}{4}(4\tilde{\lambda}_j)\right)$	$\exp\left(i\frac{2\pi(3)}{4}(4\tilde{\lambda}_j-1)\right)$	$\exp\left(i\frac{2\pi(3)}{4}(4\tilde{\lambda}_j-2)\right)$	$\exp\left(i\frac{2\pi(3)}{4}(4\tilde{\lambda}_j-3)\right)$

In particular for $j = 1$, $\tilde{\lambda}_1 = 1/2$, this table becomes

k	ℓ			
	$|00\rangle$	$|01\rangle$	$|10\rangle$	$|11\rangle$
0	1	1	1	1
1	$\exp(i\pi)$	$\exp(i\frac{\pi}{2})$	1	$\exp(-i\frac{\pi}{2})$
2	$\exp(2i\pi)$	$\exp(i\pi)$	1	$\exp(-i\pi)$
3	$\exp(i3\pi)$	$\exp(i\frac{3\pi}{4})$	1	$\exp(-i\frac{3\pi}{2})$
Sum	0	0	4	0

Finally, we see that

$$|\text{state at (B)}\rangle = \left(\frac{1}{\sqrt{2}}\right)(|10\rangle|u_1\rangle + |01\rangle|u_2\rangle)$$

Note that the state $|10\rangle$ and $|01\rangle$ are precisely the binary fractions $1/2$ and $1/4$, respectively. In this case, $p = 2$ suffices to nail the precision needed in the binary fractions.

We are now ready for one last step. We attached another ancillary qubit and apply a rotation to the ancillary qubit conditioned on the first two qubits above. Essentially we apply a rotation $\exp(i\theta_i/2\sigma_y)$ on the ancillary qubit where the angle $\theta_i = 2\sin^{-1}(C/\lambda_i)$ for some suitably chosen parameter C (Fig. 15.2).

So our final state is

$$|\text{state at (B)}\rangle|0\rangle$$
$$\rightarrow |\text{final state}\rangle$$
$$= \frac{1}{\sqrt{2}}\left(|10\rangle|u_1\rangle(\sqrt{1-\frac{C^2}{\tilde{\lambda}_1}}|0\rangle + \frac{C}{\tilde{\lambda}_1}|1\rangle)\right.$$
$$\left. -|01\rangle|u_2\rangle(\sqrt{1-\frac{C^2}{\tilde{\lambda}_2}}|0\rangle + \frac{C}{\tilde{\lambda}_2}|1\rangle)\right) \qquad (15.13)$$

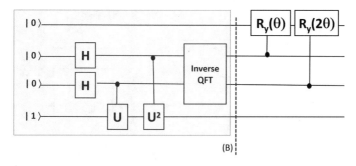

Fig. 15.2 Conditional rotation depending on the eigenvalues. Published with permission of © Belal E. Baaquie and L. C. Kwek. All Rights Reserved

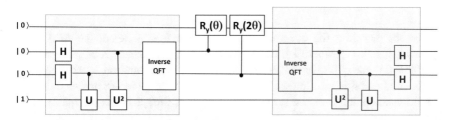

Fig. 15.3 Circuit diagram for the full HHL algorithm. Published with permission of © Belal E. Baaquie and L. C. Kwek. All Rights Reserved

We then reset the qubits used for the phase estimation $|01\rangle$ and $|10\rangle$ to $|00\rangle$ (using a circuit that is exactly the mirror image of the one used earlier to generate the eigenvalues, see Fig. 15.3) so that a measurement on the top ancillary qubit yielding the state $|1\rangle$ (success) gives the final result as

$$|x\rangle \propto \frac{1}{\sqrt{2}} \left(\frac{1}{\tilde{\lambda}_1} |u_1\rangle - \frac{1}{\tilde{\lambda}_2} |u_2\rangle \right). \qquad (15.14)$$

In hindsight, this is just a way to implement $A^{-1}|b\rangle$ in a somewhat roundabout way. Yet, we do not wish to compute A^{-1} directly. We reiterate once again that there is no need to know the exact eigenstates of A even though we use it implicitly in the description of the algorithm.

15.4 Why Do We Not Need the Eigenvalues?

An arbitrary input state $|b\rangle$ can always be regarded as a superposition of eigenstates $|u_i\rangle$ of the matrix A. We next apply HHL algorithm to solve trivially the equation $A|x\rangle = |u\rangle$ for some $|u\rangle \in \{|u_i\rangle\}$. Figure 15.4 shows a schematic diagram of the protocol acting on an eigenstate.

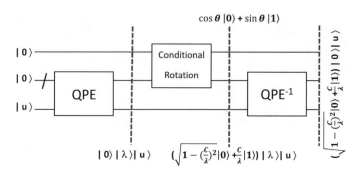

Fig. 15.4 Schematic perspective of the HHL algorithm acting on an eigenstate of A. Published with permission of © Belal E. Baaquie and L. C. Kwek. All Rights Reserved

We see that after the quantum phase estimator is completed, the state of the system is $|0\rangle_a|\lambda\rangle|u\rangle$, where λ is the corresponding eigenvalue of A with eigenvector $|u\rangle$. Note that the qubits in the register are a binary representation of the eigenvalue λ, i.e., the first bit is $1/2$, the second bit is $1/4$, the third bit is $1/8$ and so forth. The conditional rotation uses the same state λ but this time it labels the qubits in the register differently as $2, 2^2, 2^3$, etc., effectively rotating the ancillary qubit set initially at $|0\rangle$ to $\cos\theta|0\rangle + \sin\theta|1\rangle$ where $\sin\theta \propto \lambda^{-1}$. This is equivalent to extracting the eigenvalue of A and multiplying the input state $|u\rangle$ by λ^{-1}.

For an arbitrary state $|b\rangle$, we can regard this state as a superposition of $\{|u_i\rangle\}$; i.e., we can say that such a state is in state $|u_i\rangle$ with probability $|\langle u_i|b\rangle|^2$ since[2]

$$|b\rangle = \sum_i |u_i\rangle\langle u_i|b\rangle.$$

In this way, the output state takes on $\lambda_i^{-1}|u_i\rangle$ with the same probability $|\langle u_i|b\rangle|^2$. Thus, we never need to know which eigenstate of A is being operated, but our final output state simply yields

$$\sum_i \lambda_i^{-1}|u_i\rangle\langle u_i|b\rangle,$$

which is the desired output state.

---

[2] Note that $\sum_i |u_i\rangle\langle u_i|$ is just the identity matrix due to the completeness relation of the orthonormal eigenstates of $\exp(iA)$.

15.5 Other Applications

The HHL algorithm is sufficiently flexible for other applications. In particular, it can be exploited for applications in machine learning [3] where the algorithm is used as a subroutine. Examples of such applications are:

- quantum support vector machine (SVM) [4]—are supervised learning models with associated learning algorithms that analyze data for classification and regression analysis;
- quantum linear regression [5]—somewhat similar to SVM, linear regression seeks to predict the relationship between the input variables and the target variables;
- quantum recommendation systems—primarily used in commercial applications to predict products or items that the users have shown inherent interests [6, 7];
- quantum singular value thresholding [8]—an algorithm often used in solving the image classification problem in machine learning.

References

1. Dauben JW (1998) Ancient Chinese mathematics: the (jiu zhang suan shu) vs euclid's elements. Aspects of proof and the linguistic limits of knowledge. Int J Eng Sci 36(12–14):1339–1359
2. Harrow AW, Hassidim A, Lloyd S (2009) Quantum algorithm for linear systems of equations. Phys Rev Lett 103(15):150502 (2009)
3. Duan B, Yuan J, Yu C-H, Huang J, Hsieh C-Y (2020) A survey on HHL algorithm: from theory to application in quantum machine learning. Phys Lett A 126595
4. Rebentrost P, Mohseni M, Lloyd S (2014) Quantum support vector machine for big data classification. Phys Rev Lett 113(13):130503
5. Wiebe N, Braun D, Lloyd S (2012) Quantum algorithm for data fitting. Phys Rev Lett 109(5):050505
6. Preskill J (2018) Quantum computing in the NISQ era and beyond. Quantum 2:79
7. Tang E (2019) A quantum-inspired classical algorithm for recommendation systems. In: Proceedings of the 51st annual ACM SIGACT symposium on theory of computing, pp 217–228
8. Duan B, Yuan J, Liu Y, Li D (2018) Efficient quantum circuit for singular-value thresholding. Phys Rev A 98(1):012308

Chapter 16
Quantum-Classical Hybrid Algorithms

16.1 Why Bother?

At the moment, quantum computers are extremely noisy . Yet, one would still like to harness the advantage of existing quantum computational device if possible. The conviction comes from our understanding of quantum mechanics. In theory, we should be able to get quantum advantage with quantum theory. John [1] has coined the word 'Noisy Intermediate-Scale Quantum' (NISQ) computers to describe all computational devices with limited number of qubits in which we have no good control over the qubits. In the NISQ era, it is likely that some form of hybrid algorithms could be useful [2] (Fig. 16.1).

Hybrid quantum algorithms use both classical and quantum resources to solve potentially difficult problems [3, 4]. The underlying idea is that one should assign appropriate tasks to appropriate devices. If quantum computers can solve certain task like finding the overlap of wave functions more efficiently, then we should let a quantum computer performs that task. Also with the current noisy quantum computers, it is unlikely that they can perform optimization efficiently. However, classical computers have a proven record of solving these problems. So, certain optimization should be carried out with a classical computer.

16.2 Variational Quantum Eigensolvers

The variational quantum eigensolver (VQE) is a hybrid classical-quantum algorithm that determines the ground state energy of a Hamiltonian [5]. The quantum computer determines the expectation value of the energy. This is typically computed by a quantum algorithm. The minimization of the energy is however performed with a classical optimization algorithm.

In VQE, we would like to find the minimum value of the expectation of a Hermitian matrix, H, given by

© The Author(s), under exclusive license to Springer Nature Singapore Pte Ltd. 2023   249
B. E. Baaquie and L.-C. Kwek, *Quantum Computers*,
https://doi.org/10.1007/978-981-19-7517-2_16

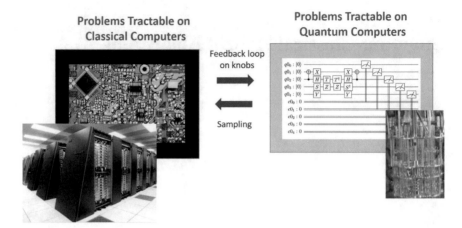

Fig. 16.1 Using the conventional (classical) computer to perform computations best suited for a classical computer and feeding the output to a quantum computer that works on the part best suited for a quantum computer. Published with permission of © Belal E. Baaquie and L. C. Kwek. All Rights Reserved

$$V = \frac{\langle \psi(\theta) | H | \psi(\theta) \rangle}{\langle \psi(\theta) | \psi(\theta) \rangle} \tag{16.1}$$

where $|\psi(\theta)\rangle$ is an unknown eigenstate of H (not necessarily normalized). If H is the Hamiltonian of a quantum system, $|\psi(\theta)\rangle$ and V are the ground state and the ground state energy respectively.

Here, we consider the case where we are interested in determining the lowest energy state (ground state) of a system Hamiltonian H. The ground state energy satisfies

$$H|\Psi_{GS}\rangle = E_{GS}|\Psi_{GS}\rangle.$$

It is known that if the Hamiltonian acts at most on k qubits, then the problem is known to be QMA-complete for $k \geq 2$.

To understand a little about the last statement, we first digress a little to explain quantum complexity theory. The Quantum Merlin–Arthur (QMA) class is an extension of the classical class non-deterministic polynomial (NP) in computational complexity theory . The class NP concerns problems that are hard to solve yet it is easy to verify that the solution is correct. One such problem is the prime number factorization, that we have seen earlier. It is hard to factorize a large number that is made up of two large prime numbers, but it is easy to verify the answers are right or wrong once we are given the factors. The probabilistic extension of NP problems is the class Merlin–Arthur (MA). In NP, the verification or proof can be done with absolute certainty. In MA, the verification is accepted with a certain probability, say greater than 2/3 (or one standard deviation as in a normal distribution test). MA is the class

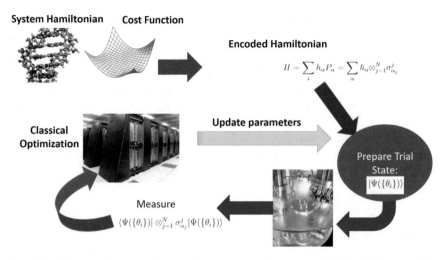

Fig. 16.2 Schematic diagram of the variational quantum eigensolver (VQE) in which part of the computations is done in a quantum computer and optimization is done in a classical computer. Published with permission of © Belal E. Baaquie and L. C. Kwek. All Rights Reserved

of problems (typically called languages in computational theory) where a computationally hard problem can be verified by an untrusted Merlin who can convince with high probability a verifier (Arthur), who has access only to polynomial-time computation.

In the original formulation [5], we first write the Hamiltonian for N qubits into the form

$$H = \sum_i h_\alpha P_\alpha = \sum_\alpha h_\alpha \otimes_{j=1}^N \sigma_{\alpha_j}^j \tag{16.2}$$

where h_α are coefficients and P_α are called Pauli strings. Pauli strings are tensor products of Pauli matrices including the identity matrix.

To perform VQE, one proceeds with the following steps [6]:

1. Map the problem for finding the ground state of a Hamiltonian to the form in Eq. 16.2.
2. Prepare a trial state $|\Psi(\{\theta_i\})\rangle$ that depends on some set of parameters $\{\theta_i\}$.
3. We use a quantum computer to evaluate the expectation values

$$\langle \Psi(\{\theta_i\})| \otimes_{j=1}^N \sigma_{\alpha_j}^j |\Psi(\{\theta_i\})\rangle$$

.

4. We feed the measured values from the quantum computer to find the energy $E = \sum_\alpha h_\alpha \langle \Psi(\{\theta_i\})| \otimes_{j=1}^N \sigma_{\alpha_j}^j |\Psi(\{\theta_i\})\rangle \geq E_{\text{exact}}$ and adjust the parameters θ in a classical computer. This is the optimization process.
5. We update the set of parameters $\{\theta_i\}$ and proceed with the iteration (Fig. 16.2).

16.3   Quantum Approximate Optimization Algorithm

The Quantum Approximate Optimization Algorithm (QAOA) is a canonical NISQ era algorithm that can provide approximate solutions in polynomial time for combinatorial optimization problems. Proposed originally by Farhi et al. [7], QAOA is regarded as a special case of VQE. It is also related to the quantum adiabatic algorithm [8, 9].

The cost function $\hat{C}$ depends on the bit strings that form the computational basis. The latter is designed to encode a combinatorial problem in those strings. Using the computational basis vectors,

$$|e_i\rangle = \{|00\cdots00\rangle, |00\cdots01\rangle \cdots |11\cdots11\rangle\},$$

one can define the problem Hamiltonian H_P as

$$H_P \equiv \sum_{i=1}^{n} C(e_i)|e_i\rangle, \tag{16.3}$$

and the mixing Hamiltonian H_M as

$$H_M \equiv \sum_{i=1}^{n} \hat{\sigma}_x^i, \tag{16.4}$$

The initial state in the QAOA algorithm is chosen to be an uniform superposition state

$$|+_n\rangle = H^{\otimes n}|0\rangle^{\otimes n} = \frac{1}{\sqrt{2^n}} \sum_{i=0}^{2^n-1} |i\rangle \tag{16.5}$$

The final quantum state is given by applying H_P and H_M alternately on the initial state p-times:

$$|\Psi(\boldsymbol{\gamma}, \boldsymbol{\beta})\rangle \equiv e^{-i\beta_p H_M} e^{-i\gamma_p H_P} \cdots e^{-i\beta_1 H_M} e^{-i\gamma_1 H_P} |D\rangle. \tag{16.6}$$

A quantum computer evaluates the objective function

$$\hat{C}(\boldsymbol{\gamma}, \boldsymbol{\beta}) \equiv \langle\Psi(\boldsymbol{\gamma}, \boldsymbol{\beta})|H_P(\boldsymbol{\gamma}, \boldsymbol{\beta})|\Psi(\boldsymbol{\gamma}, \boldsymbol{\beta})\rangle, \tag{16.7}$$

and a classical optimizer updates the $2p$ angles $\boldsymbol{\gamma} \equiv (\gamma_1, \gamma_2, \cdots, \gamma_p)$ and $\boldsymbol{\beta} \equiv (\beta_1, \beta_2, \cdots, \beta_p)$ until the objective function $\hat{C}$ is maximized, i.e., $\hat{C}(\boldsymbol{\gamma}^*, \boldsymbol{\beta}^*) \equiv \max_{\boldsymbol{\gamma}, \boldsymbol{\beta}} \hat{C}(\boldsymbol{\gamma}, \boldsymbol{\beta})$. Here, p is often referred to as the QAOA level or depth. Since the maximization at level $p - 1$ is a constrained version of the maximization at level p, the performance of the algorithm improves monotonically with p.

16.4 MaxCut Problem

To illustrate QAOA, we apply the algorithm to the MaxCut problem. Consider a graph with four edges (m vertices) and four vertices (n) as shown in Fig. 16.3a. The problem is to partition the four vertices into two sets A and B which will maximize the (cost) function

$$C(z) = \sum_{\alpha=1}^{n} C_\alpha(z) \tag{16.8}$$

where $C(z)$ enumerates the number of edges cut and αs are the labels for the edges. The term $C_\alpha(z) = 1$ if z places one vertex from the αth edge in set A and the other vertex in set B. If this is not possible, $C_\alpha(z) = 0$. It turns out that finding the cut that yields the maximum possible value of C is a hard problem (actually NP-complete in complexity theory).

We assign the vertices to set A and set B with a bit string z. In the example in Fig. 16.3a, $z = z_1 z_2 z_3 z_4$ since there are four vertices. If we assign the ith vertex to set A, then $z_i = 0$, otherwise, if the ith vertex is assign to set B, we write $z_i = 1$. In Fig. 16.3b, $z = 0101$ since the vertices 1 and 3 are in set A and vertices 2 and 4 are in set B.

To implement QAOA, we consider the objective (or cost) function

$$C_\alpha = \frac{1}{2}(1 - \sigma_i^z \sigma_j^z) \tag{16.9}$$

where the αth edge is between the vertices (i, j). Clearly, C_α has eigenvalue 1 if and only if ith and jth qubits yield different z-axis measurements corresponding to the different partitions.

The algorithm QAOA begins with a uniform superposition of all possible n-bit strings. In the case here, the number of vertices $n = 4$.

Fig. 16.3 a An example of a graph with four vertices and four edges: in the MaxCut Problem, one wishes to partition the vertices into two distinct sets A and B. One such solution is given in **b**. Published with permission of © Belal E. Baaquie and L. C. Kwek. All Rights Reserved

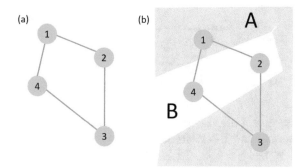

$$|+_4\rangle = \frac{1}{\sqrt{2^n}} \sum_{z\in\{0,1\}^n} |z\rangle \tag{16.10}$$

Our aim is to explore the bit strings z for a superposition that yields the largest value for the C operator. Using $2p$ angle parameters,

$$|\gamma, \beta\rangle = U_{B_p} U_{C_p} U_{B_{p-1}} U_{C_{p-1}} \cdots U_{B_1} U_{C_1}$$

where

$$U_{B_k} = \prod_{j=1}^{n} e^{-i\beta_k \sigma_j^z},$$

$$U_{C_k} = \prod_{all\ edges(m,n)} e^{-i\gamma_k(1-\sigma_m^z\sigma_n^z)/2},$$

where the matrices σ_i^α ($\alpha = x, y, z$) are Pauli matrices and the label i refers to the ith qubit.

These operators $e^{-i\gamma_k(1-\sigma_m^z\sigma_n^z)/2}$ (in U_{B_k}) and $e^{-i\beta_k\sigma_j^z}$ (in U_{C_k}) can be implemented on quantum circuit as shown in Fig. 16.4a, b. The single-qubit operation $e^{-i\beta_k\sigma_j^z}$ is straightforward. It is just a rotation matrix along the Pauli-x axis. For the two-qubit gate, we see that

$$e^{-i\gamma_k(1-\sigma_m^z\sigma_n^z)/2} = \text{CNOT}_{mn} \cdot e^{+i\gamma_k\sigma_m^z/2} \cdot \text{CNOT}_{mn}. \tag{16.11}$$

To see how Eq. (16.11) is obtained, we need to note that the CNOT gate can be cast into the form

$$\text{CNOT}_{mn} = |0\rangle_{mm}\langle 0| \otimes I_n + |1\rangle_{mm}\langle 1| \otimes \sigma_n^x.$$

Thus,

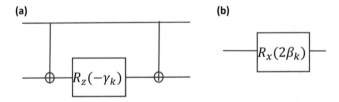

(a)   **(b)**

Fig. 16.4   a Quantum circuit for the entangling gates $e^{-i\gamma_k(1-\sigma_j^z\sigma_k^z)/2}$. **b** Circuit for the single-qubit operation: $e^{-i\beta_k\sigma_j^z}$. Published with permission of © Belal E. Baaquie and L. C. Kwek. All Rights Reserved

$$\text{CNOT}_{mn} \cdot e^{+i\gamma_k \sigma_n^z/2} \cdot \text{CNOT}_{mn}$$
$$= \left(|0\rangle_{mm}\langle 0| \otimes I_n + |1\rangle_{mm}\langle 1| \otimes \sigma_n^x\right) \cdot e^{+i\gamma_k \sigma_n^z/2} \cdot$$
$$\left(|0\rangle_{mm}\langle 0| \otimes I_n + |1\rangle_{mm}\langle 1| \otimes \sigma_n^x\right)$$
$$= |0\rangle_{mm}\langle 0| \otimes e^{+i\gamma_k \sigma_n^z/2} + |1\rangle_{jj}\langle 1| \otimes \sigma_n^x e^{+i\gamma_k \sigma_n^z/2} \sigma_n^x$$
$$= |0\rangle_{mm}\langle 0| \otimes e^{+i\gamma_k \sigma_n^z/2} + |1\rangle_{mm}\langle 1| \otimes e^{-i\gamma_k \sigma_n^z/2}, \tag{16.12}$$

where we have used $\sigma^x e^A \sigma^x = e^{\sigma^x A \sigma^x}$ and $\sigma_x \sigma^z \sigma^x = -\sigma^z$. But now we need to know a small trick,

$$|0\rangle_{mm}\langle 0| + |1\rangle_{mm}\langle 1| = I$$
$$|0\rangle_{mm}\langle 0| - |1\rangle_{mm}\langle 1| = \sigma_m^z, \tag{16.13}$$

so we can replace $|0\rangle_{mm}\langle 0|$ by $\frac{1}{2}(I + \sigma_m^z)$ and $|1\rangle_{mm}\langle 1|$ by $\frac{1}{2}(I - \sigma_m^z)$. Substituting back into Eq. (16.12) gives

$$\text{CNOT}_{mn} \cdot e^{+i\gamma_k \sigma_n^z/2} \cdot \text{CNOT}_{mn}$$
$$= \frac{1}{2}(I_m + \sigma_m^z) \otimes e^{+i\gamma_k \sigma_n^z/2} + \frac{1}{2}(I_j - \sigma_m^z) \otimes e^{-i\gamma_k \sigma_n^z/2}$$
$$= \frac{1}{2}\left(I_m \otimes \left(e^{+i\gamma_k \sigma_n^z/2} + e^{-i\gamma_k \sigma_n^z/2}\right)\right)$$
$$+ \frac{1}{2}\left(\sigma_m^z \otimes \left(e^{+i\gamma_k \sigma_n^z/2} + e^{-i\gamma_k \sigma_n^z/2}\right)\right)$$
$$= \frac{1}{2}\left(I_m \otimes (2\cos(\gamma_k/2))\right) + \frac{1}{2}\left(\sigma_m^z \otimes \left(2i\sin(\gamma_k/2)\sigma_n^z\right)\right)$$
$$= \left(I_m \otimes (\cos(\gamma_k/2))\right) + \left(\sigma_m^z \otimes \left(i\sin(\gamma_k/2)\sigma_n^z\right)\right)$$
$$= \left(I_m \otimes I_n \cos(\gamma_k/2) + i\sin(\gamma_k/2)\sigma_m^z \otimes \sigma_n^z\right)$$
$$= e^{i\gamma_k \sigma_m^z \otimes \sigma_n^z/2}$$
$$= e^{i\gamma_k/2} e^{-i\gamma_k(I - \sigma_m^z \otimes \sigma_n^z)/2}, \tag{16.14}$$

which is the required gate up to a global phase. These quantum gates allow us to estimate the expectation value $\langle \gamma, \beta | C | \gamma, \beta \rangle$.

We perform classical optimization over the circuit parameters (γ, β) to obtain the best state $|\gamma, \beta\rangle$ that is most likely to yield an approximate partition $|z\rangle$ on performing a measurement in the computational basis. In the case of the graph in Fig. 16.3b, we see that the optimal partitions correspond to the strings $|0101\rangle$ or $|1010\rangle$. The optimization process essentially corresponds to an evolution of the initial state to the plane spanned by the basis states $\{|0101\rangle, |1010\rangle\}$ (Fig. 16.5).

Fig. 16.5 Evolution of the
state $|+_4\rangle$ to the desired state
on the plane spanned by the
two optimal solutions.
Published with permission of
© Belal E. Baaquie and L. C.
Kwek. All Rights Reserved

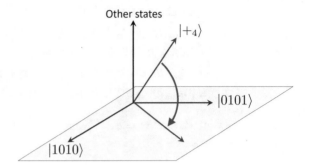

References

1. John P (2018) Quantum computing in the NISQ era and beyond. Quantum 2:79
2. Bharti K, Cervera-Lierta A, Kyaw TH, Haug T, Alperin-Lea S, Anand A, Degroote M, Heimonen H, Kottmann JS, Menke T, Mok W-K, Sim S, Kwek L-C, Aspuru-Guzik A (2022) Noisy intermediate-scale quantum algorithms. Rev Mod Phys 94:015004
3. McClean JR, Romero J, Babbush R, Aspuru-Guzik A (2016) The theory of variational hybrid quantum-classical algorithms. New J Phys 18(2):023023
4. Zhu D, Linke NM, Benedetti M, Landsman KA, Nguyen NH, Alderete CH, Perdomo-Ortiz A, Korda N, Garfoot A, Brecque C et al (2019) Training of quantum circuits on a hybrid quantum computer. Sci Adv 5(10):eaaw9918
5. Peruzzo A, McClean J, Shadbolt P, Yung M-H, Zhou X-Q, Love PJ, Aspuru-Guzik A, O'brien JL (2014) A variational eigenvalue solver on a photonic quantum processor. Nat Commun 5:4213
6. Moll N, Barkoutsos P, Bishop LS, Chow JM, Cross A, Egger DJ, Filipp S, Fuhrer A, Gambetta JM, Ganzhorn M et al (2018) Quantum optimization using variational algorithms on near-term quantum devices. Quantum Sci Technol 3(3):030503
7. Farhi E, Goldstone J, Gutmann S (2014) A quantum approximate optimization algorithm. *arXiv preprint* arXiv:1411.4028
8. Farhi E, Goldstone J, Gutmann S, Sipser M (2000) Quantum computation by adiabatic evolution. arXiv preprint quant-ph/0001106
9. Bengtsson A, Vikstål P, Warren C, Svensson M, Gu X, Kockum AF, Krantz P, Križan C, Shiri D, Svensson I-M et al (2020) Improved success probability with greater circuit depth for the quantum approximate optimization algorithm. Phys Rev Appl 14(3):03401

Chapter 17
Quantum Error Correction

17.1 Introduction

Fault-tolerant quantum computation essentially relies on quantum error-correcting codes. By repeated measurements of the error syndromes and the application of the corresponding correction operations, encoded states can be stored reliably for extended periods of time The field of quantum error correction is motivated almost entirely from classical error correction [1–4]. In classical error correction, coding essentially relies on extensive data copying. The design of many error-correcting codes depends on the concept of adding redundancy. However, this is not exactly possible in quantum theory since it is easily shown that it is impossible to copy an unknown state [5], see also Chap. 7 Sect. 4.7 and any measurement on a quantum state destroys the state itself.

17.2 Simple Quantum Errors

Classical computation works on two binary bits, 0 or 1. Errors can occur in the transmission of data from one point to another via bit-flip errors. Typically, such errors arise from stray electromagnetic fields in the transmission lines or from faulty computer chips causing the bit to change from 0 to 1 or vice verse. The analogous type of error in a quantum computer is one that affects the basis states $|0\rangle$ and $|1\rangle$: $|0\rangle \to |1\rangle$, or $|1\rangle \to |0\rangle$. Such error is described with the X (or σ_x) operator:

$$X = \begin{pmatrix} 0 & 1 \\ 1 & 0 \end{pmatrix}. \tag{17.1}$$

© The Author(s), under exclusive license to Springer Nature Singapore Pte Ltd. 2023
B. E. Baaquie and L.-C. Kwek, *Quantum Computers*,
https://doi.org/10.1007/978-981-19-7517-2_17

In general, X acts on an arbitrary qubit $\alpha|0\rangle + \beta|1\rangle = |\psi\rangle$:

$$X|\psi\rangle = \alpha|1\rangle + \beta|0\rangle \tag{17.2}$$

In quantum systems, aside from bit-flip errors, we expect another type of errors, namely phase-flip errors. A phase-flip error does not change the computational basis states. It simply adds a phase to one of the basis state, i.e., $|1\rangle \rightarrow -|1\rangle$. A phase-flip error is one that transforms the basis states as follows[1]:

$$|0\rangle \rightarrow |0\rangle$$
$$|1\rangle \rightarrow e^{i\pi}|1\rangle = -|1\rangle. \tag{17.3}$$

A phase-flip error is described by the Pauli Z operator. The operator Z acts on an arbitrary qubit $\alpha|0\rangle + \beta|1\rangle = |\psi\rangle$:

$$Z|\psi\rangle = \alpha|0\rangle - \beta|1\rangle. \tag{17.4}$$

Having described the actions of X and Z, one question that is often asked is what about the Y operator? The operator Y is really a combination of X and Z in concatenation since $Y = -iZX$.

17.3   Kraus Operators

In introductory quantum mechanics, one typically considers quantum states and its evolution within a closed system, i.e., the system is sufficiently isolated from its surrounding. In quantum information, one needs to be concerned not just with closed system, but also systems that interact with the environment. So suppose we prepare a system with state $|\psi\rangle = \alpha|0\rangle + \beta|1\rangle$. And the environment at time $t = 0$ is some pure state $|\xi\rangle$.

After a finite interval of time, let us assume that the quantum system undergoes a bit flip or a phase flip or both. Let the probabilities associated with these operations, X, Y and Z be p_1, p_2 and p_3 respectively.

The state of the system after this interval is described by:

$$|\psi\rangle|\xi\rangle \rightarrow \sqrt{p_1}\,(X|\psi\rangle)\,|\xi_1\rangle + \sqrt{p_2}\,(Y|\psi\rangle)\,|\xi_2\rangle + \sqrt{p_3}\,(Z|\psi\rangle)\,|\xi_3\rangle +$$
$$\sqrt{1 - p_1 - p_2 - p_3}\,(I|\psi\rangle)\,|\xi_4\rangle. \tag{17.5}$$

---

[1] More generally, in a phase-flip error, the phase need not be π, it could be an arbitrary angle: $|1\rangle \rightarrow e^{i\pi}|1\rangle$.

For simplicity, we assume that the states associated with the environment $\{|\xi_i\rangle\}$ are distinguishable from each other, so that $\langle\xi_i|\xi_j\rangle = \delta_{ij}$. Note that there is a probability $1 - p_1 - p_2 - p_3$ that the state remains untouched (identity operation).

If we traced out the environmental degree of freedom, then we see that the state $|\psi\rangle\langle\psi| = \rho$ becomes

$$\rho \to p_1 X|\psi\rangle\langle X + p_2 Y|\psi\rangle\langle Y p_3 Z|\psi\rangle\langle Z$$
$$(1 - p_1 - p_2 - p_3)|\psi\rangle\langle, \tag{17.6}$$

which we can rewrite as

$$\rho \to \sum_{i=1}^{4} K_i \rho K_i^\dagger \tag{17.7}$$

where the $K_i = \sqrt{p_i}\sigma_i (i = 1, 2, 3)$ and $K_4 = \sqrt{1 - p_1 - p_2 - p_3}I$. Here, $\sigma_1 = X$, $\sigma_2 = Y$ and $\sigma_3 = Z$ as in the usual notations. The operators K_i are called Kraus operators and they satisfy

$$\sum_{i=1}^{4} K_i^\dagger K_i = I \tag{17.8}$$

17.4 Nine-qubit Code

In classical error correction codes, repetition code is one way to overcome errors. Typically instead of sending a single classical bit b over a channel, one repeats the bit a number of times, typically three times, and decides the results based on majority votes.

Unlike classical repetition codes, quantum codes need to safeguard against not just bit-flip errors, i.e., $0 \to 1$ and vice verse, but also phase-flip errors. The state $1/\sqrt{2}(|0\rangle + |1\rangle)$ can become $1/\sqrt{2}(|0\rangle + e^{i\pi}|1\rangle) = 1/\sqrt{2}(|0\rangle - |1\rangle)$, a completely orthogonal state.

One of the easiest ways to overcome both bit-flip and phase-flip error is to encode $|0\rangle$ and $|1\rangle$ as

$$|\bar{0}\rangle = \frac{1}{2\sqrt{2}}(|000\rangle + |111\rangle)(|000\rangle + |111\rangle)(|000\rangle + |111\rangle) \tag{17.9}$$

$$|\bar{1}\rangle = \frac{1}{2\sqrt{2}}(|000\rangle - |111\rangle)(|000\rangle - |111\rangle)(|000\rangle - |111\rangle) \tag{17.10}$$

These codes do not violate the no-cloning theorem. An arbitrary code word is a linear superposition of these two states:

$$\alpha|\bar{0}\rangle + \beta|\bar{1}\rangle \tag{17.11}$$

The two logical states $|\bar{0}\rangle$ and $|\bar{1}\rangle$ can be regarded as two layers of onion peels. The inner layer, i.e., $|000\rangle \pm |111\rangle$ corrects bit-flip errors and any correction is done with majority votes. For example, if there is a bit flip in the second qubit,

$$|010\rangle \pm |101\rangle \rightarrow |000\rangle \pm |111\rangle. \tag{17.12}$$

The outer layer corrects for phase flips. In this case, we take the majority of the three signs.

$$|+\rangle|-\rangle|+\rangle \rightarrow |+\rangle|+\rangle|+\rangle \tag{17.13}$$

where we have used $|\pm\rangle = 1/\sqrt{2}\,(|000\rangle \pm |111\rangle)$. Naturally, one can check that the code works even if there is both a bit-flip and a phase-flip error.

Now that we know the code, we still need to know how to correct for errors. First, we need to determine if the first three qubits are the same, and if not which is different. To do this, we perform the following operations on the first three qubits:

$$Z \otimes Z \otimes I \text{ and } I \otimes Z \otimes Z. \tag{17.14}$$

The first operation provides information on an bit-flip error on qubit 1 or 2, while the second operation tells us that a bit-flip occurs on qubit 2 or 3. It is now a bit of simple detective work: the two pieces of information tells us precisely where the error occurs. We then repeat the procedure for the other two sets of three qubits. Note that it is important that we do not collapse the superposition used in the code. To do this, we attached an ancillary qubit and perform the controlled-Z operations to the first and second qubit as follows:

$$
\begin{aligned}
(|0\rangle + |1\rangle)) \sum_{ijk} \alpha_{ijk}|ijk\rangle &\rightarrow \sum_{ijk} \alpha_{ijk}(|0\rangle|ijk\rangle + (-1)^{i\oplus j}|1\rangle|ijk\rangle) \\
&= \sum_{ijk} c_{ijk}\left(|0\rangle + (-1)^{\mathrm{parity}(i,j)}\,|1\rangle\right)|ijk\rangle, \\
&= \left(|0\rangle + (-1)^{\mathrm{parity}(i,j)}|1\rangle)\right) \sum_{ijk} c_{ijk}|ijk\rangle
\end{aligned}
$$

since $i \oplus j = i + j$, Mod 2 and effectively detects the parity of (i, j). Thus, measuring the ancilla tells us the eigenvalue of $Z \otimes Z \otimes I$ and nothing about the encoded data.

Next, we need to correct for phase error. To do this, we perform the following measurements:

$$
\begin{aligned}
&X \otimes X \otimes X \otimes X \otimes X \otimes X \otimes I \otimes I \otimes I \\
&\text{and } I \otimes I \otimes I \otimes X \otimes X \otimes X \otimes X \otimes X \otimes X.
\end{aligned} \tag{17.15}
$$

These two measurements detect phase errors on the first six and last six qubits. As for bit-flip error, the error detected anti-commutes with the operator measured. The nine-qubit error correction codes is also known as the Shor code.

17.5 General Properties of Quantum Error-Correcting Codes

In this section, we follow closely Sect. 17.2 of the excellent review by Daniel Gottesman [1]. There are many other useful references as well, see for example [6–8].

An important feature of quantum error correction is linearity of quantum mechanics. Because of this feature, we only need to correct a subset of errors, for instance in the one-qubit case, $\{I, X, Y, Z\}$ where X, Y, Z are the Pauli operators. If a quantum code corrects errors A and B, it also connects a linear combination of A and B.

In the presence of noise, any two orthogonal logical states, $|\bar{0}\rangle$ and $|\bar{1}\rangle$ remain orthogonal with errors, i.e., $E_1|\bar{0}\rangle$ and $E_2|\bar{1}\rangle$ are orthogonal:

$$\langle \bar{0}|E_1^\dagger E_2|\bar{1}\rangle = 0. \tag{17.16}$$

As noted in Ref. [3], if the condition in Eq. (17.16) was not satisfied, then the errors would destroy the perfect distinguishability of orthogonal codewords and the encoded quantum information would be damaged. Therefore, it is sufficient to distinguish error E_1 from error E_2 when they act on $|\bar{0}\rangle$ and $|\bar{1}\rangle$ for $E_1 \neq E_2$:

$$\langle \bar{0}|E_1^\dagger E_2|\bar{0}\rangle = \langle \bar{1}|E_1^\dagger E_2|\bar{1}\rangle = 0 \tag{17.17}$$

When we cannot distinguish between the two errors, it is still sometimes perfectly all right if one constructs new operators $F_1 = (E_1 + E_2)/2$ and $F_2 = (E_1 - E_2)/2$, assuming $\langle \bar{0}|E_1^\dagger E_2|\bar{0}\rangle$ is real so that

$$\langle \bar{0}|F_1^\dagger F_2|\bar{0}\rangle = \langle \bar{1}|F_1^\dagger F_2|\bar{1}\rangle = 0 \tag{17.18}$$

There is an additional requirement that the errors can be inverted:

$$\langle \bar{0}|E_i^\dagger E_i|\bar{0}\rangle = \langle \bar{1}|E_i^\dagger E_i|\bar{1}\rangle \tag{17.19}$$

Moreover, all errors acting on the Hilbert space form a linear subspace, $\mathcal{E}$. Let $\{E_i\}$ be a basis of $\mathcal{E}$. Let $\mathcal{C}$ be a subspace of the Hilbert space that constitutes the space of error-correcting codes and $\{\psi_i\}$ be a basis for $\mathcal{C}$. Consider logical qubits $|0\rangle_L$ and $|1\rangle_L$ on a two-qubit subspace subject to the following interaction operators [9]:

$$E_0 = \begin{pmatrix} \sqrt{1-2q} & 0 & 0 & 0 \\ 0 & 1 & 0 & 0 \\ 0 & 0 & 1 & 0 \\ 0 & 0 & 0 & \sqrt{1-2q} \end{pmatrix};$$

$$E_1 = \begin{pmatrix} \sqrt{q/2} & 0 & 0 & 0 \\ 0 & 0 & 0 & \sqrt{q/2} \\ \sqrt{q/2} & 0 & 0 & 0 \\ 0 & 0 & 0 & \sqrt{q/2} \end{pmatrix};$$

$$E_2 = \begin{pmatrix} \sqrt{q/2} & 0 & 0 & 0 \\ 0 & 0 & 0 & \sqrt{q/2} \\ -\sqrt{q/2} & 0 & 0 & 0 \\ 0 & 0 & 0 & -\sqrt{q/2} \end{pmatrix}. \tag{17.20}$$

You can easily check that $\sum_i E_i^\dagger E_i = 1$. Also, it is not hard to verify that

$$\langle i_L | E_i^\dagger E_j | i_L \rangle = \langle j_L | E_i^\dagger E_j | j_L \rangle \tag{17.21}$$

$$\langle i_L | E_i^\dagger E_j | j_L \rangle = 0 \tag{17.22}$$

where $|k_L\rangle = |0\rangle_L$ or $|1\rangle_L$ for $k = i, j$.

17.6   Classical Linear Codes

Classical error correction has been developed over many years. A linear code is an error-correcting code in which any linear combination of codewords is also a codeword. The repetition code in which we repeat a classical bit a number of times and correct through majority vote is an example of a linear code. Nonlinear codes outperform linear codes in terms of error detection and correction. However, these codes are more complicating. The human visual system uses codes as well. However, human vision does not operate linearly, and the code used in vision is typically nonlinear.

One of the key notions in the theory of linear codes is Hamming distance. Suppose two codewords $x = x_1 x_2 x_3 \ldots x_n$ and $y = y_1 y_2 y_3 \ldots y_n$, we define for every i, the distance

$$d(x_i, y_i) = \begin{cases} 1, & x_i \neq y_i \\ 0 & x_i = y_i \end{cases}, \tag{17.23}$$

then the Hamming distance between the two codewords is $d(x, y) = \sum_{i=1}^{n} d(x_i, y_i)$.

For instance, the Hamming distance between the code word 0101 and 1001 is 3. It is easily verified that $d(x, y)$ is a metric in the strict sense:

$$d(x, y) = 0 \Leftrightarrow x = y \tag{17.24}$$

$$d(x, y) = d(y, x) \text{ symmetry} \tag{17.25}$$

$$d(x, y) \leq d(x, z) + d(z, y) \text{ triangle inequality.} \tag{17.26}$$

Let C be a code of length n over a set of alphabets A. In the case of binary bits, the alphabets are $\{0, 1\}$ and the set of codes of length 2 are $|A|^2 \equiv \{00, 01, 10, 11\}$ where $|A|$ denotes the size of the set of the alphabets. The nearest neighbor decoding rule explicitly says that every $x \in A^n$ is decoded to the codeword $c_x \in C$ closest to x such that $d(x, c_x) = \min_{c \in C}(d(x, c))$. For any code C, the distance of the code is defined as $d(C) = \min\{d(c_1, c_2) | c_1, c_2 \in C, c_1 \neq c_2\}$.

With these definitions, we can now embark on some descriptions of linear codes. A linear code of length n and dimension k with Hamming distance d is denoted as an $[n, k, d]$ code. Another way of looking at an $[n, k, d]$ code is: with a binary string of length n, we have a space of 2^n strings. We designate a subset of k-bit string as the codewords such that the Hamming distance between any two codewords is at least d.

The space C of the code is spanned by a basis of k vectors $v_1, v_2, \ldots v_k$ so that an arbitrary codeword is expressed as a linear combination of these basis vectors:

$$v(\alpha_1, \alpha_2, \ldots, \alpha_k) = \sum_i \alpha_i v_i, \tag{17.27}$$

where $\alpha_i \in \{0, 1\}$ and addition is just modulo 2.

The k basis vectors may be assembled into a $k \times n$ matrix:

$$G = \begin{pmatrix} v_1 \\ v_2 \\ \ldots \\ v_k \end{pmatrix}. \tag{17.28}$$

The matrix G is called the generator matrix of the code. Equation (17.27) can then be written succinctly as:

$$v(\boldsymbol{\alpha}) = \boldsymbol{\alpha} G \tag{17.29}$$

We can also define the parity check matrix H of size $(n - k) \times n$ whose rows are $(n - k)$ linearly independent vectors that are orthogonal to the code space. Orthogonality is defined with respect to the bit-wise inner product modulo 2. Clearly we have:

$$H G^T = 0 \tag{17.30}$$

where G^T is the transpose of matrix G.

For a classical bit, the only kind of error is a bit flip. A string v under a bit-flip error e is expressed as:

$$v \rightarrow v + e. \tag{17.31}$$

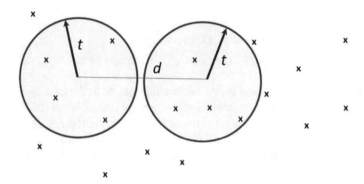

Fig. 17.1 Figure showing the minimum distance and error correction capability of a linear code. Published with permission of © Belal E. Baaquie and L. C. Kwek. All Rights Reserved

Errors can be detected by applying the parity check matrix:

$$H(v + e) = hv + He = He, \qquad (17.32)$$

and He is called the syndrome of the error e.If we denote the set of all possible errors $\{e_i\}$ as $\mathcal{E}$, error recovery is possible if and only if all errors possess distinct syndromes. We can unambiguously diagnose the error for a given syndrome He by flipping the bits specified by e as:

$$v + e \rightarrow (v + e) + e = v \qquad (17.33)$$

For any two errors $e_1 \neq e_2$, if $He_1 = He_2$, we may mistake e_1 error as e_2 error, then our attempt at recovery will yield a damaged code:

$$v + e_1 \rightarrow v + (e_2 + e_2) \neq v \qquad (17.34)$$

The distance of a code C, d, is the minimum weight of any vector in the code, where the weight is the number of 1's in the string v. Classical coding theory says that a linear code with distance $d = 2t + 1$ can correct t errors, see Fig. 17.1. For codes with $n = 7, k = 4$ and a minimum distance $d_{\min} = 7, t = \left\lfloor \dfrac{d_{\min} - 1}{2} \right\rfloor = 1$.

In short, a linear code with a distance $d = 2t + 1$ can correct t errors and the code assigns a distinct syndrome to each $e \in \mathcal{E}$, where $\mathcal{E}$ contains all the vectors of weight t or less. If $He_1 = He_2$, then

$$0 = He_1 + He_2 = H(e_1 + e_2)$$

and then $e_1 + e_2 \in C$. But if e_2 and e_2 and each codeword has weight no larger than t, then the weight of $e_1 + e_2$ should not be greater than zero and no larger than $2t$. Since $d = 2t + 1$, no such vector exists and hence $He_1 \neq He_2$.

For any code C, one can also construct a dual code. For a $k \times n$ generator matrix G, and the $(n - k) \times n$ parity check matrix H of a code C, we see that $HG^T = 0$. By taking the transpose, it follows that $GH^T = 0$, and we may now regard H^T as a generator and G as the parity check of an $(n - k)$-dimensional code, denoted by $C^\perp$ and known as the dual of the code C.

For any code C and its dual code $C^\perp$, it can be shown that

$$\sum_{v \in C} (-1)^{v.u} = \begin{cases} 2^k, \ u \in C^\perp \\ 0, \ u \notin C^\perp \end{cases} \tag{17.35}$$

17.7   An Example of a Linear Code: Hamming Code

We digress a little here to discuss an example of a linear code: a Hamming Code with seven bits. Hamming codes are codes that can correct a single error. A good reference is Ref. [10]. As an example, let us consider the following four-bit codes (x_1, x_2, x_3, x_4), augmented by a three-bit parity check (p_1, p_2, p_3):

x_1	x_2	x_3	x_4	p_1	p_2	p_3
0	0	0	0	0	0	0
0	0	0	1	0	1	1
0	0	1	0	1	1	0
0	0	1	1	1	0	1
0	1	0	0	1	1	1
0	1	0	1	1	0	0
0	1	1	0	0	0	1
0	1	1	1	0	1	0
1	0	0	0	1	0	1
1	0	0	1	1	1	0
1	0	1	0	0	1	1
1	0	1	1	0	0	0
1	1	0	0	0	1	0
1	1	0	1	0	0	1
1	1	1	0	1	0	0
1	1	1	1	1	1	1

where the three check bits are given by

$$p_1 = x_1 \oplus x_2 \oplus x_3, \tag{17.36}$$
$$p_2 = x_2 \oplus x_3 \oplus x_4, \tag{17.37}$$
$$p_3 = x_1 \oplus x_2 \oplus x_4. \tag{17.38}$$

It is not difficult to check that the Hamming distance between these code words is at least 3. Thus these codes are called [7, 4, 3] codes, meaning that it is a seven-bit text that encodes a four-bit text, whose Hamming distance is at least 3. Moreover, we see from previous section that we can correct at least one error.

Another way to describe the $[7, 4, 3]$ code is

$$
\begin{pmatrix} x_1 \\ x_2 \\ x_3 \\ x_4 \\ p_1 \\ p_2 \\ p_3 \end{pmatrix} = \begin{pmatrix} 1\,0\,0\,0 \\ 0\,1\,0\,0 \\ 0\,0\,1\,0 \\ 0\,0\,0\,1 \\ 1\,1\,1\,0 \\ 0\,1\,1\,1 \\ 1\,1\,0\,1 \end{pmatrix} \begin{pmatrix} x_1 \\ x_2 \\ x_3 \\ x_4 \end{pmatrix} \tag{17.39}
$$

which follows from Eqs. (17.36)–(17.38).

When the transmitted message arrives at the intended recipient, the recipient subjects the received message $(x_1', x_2', x_3', x_4', p_1', p_2', p_3')$ to a decoder by computing

$$
s_1 = p_1' \oplus x_1' \oplus x_2' \oplus x_3', \tag{17.40}
$$
$$
s_2 = p_2' \oplus x_2' \oplus x_3' \oplus x_4', \tag{17.41}
$$
$$
s_3 = p_3' \oplus x_1' \oplus x_2' \oplus x_4'. \tag{17.42}
$$

Now suppose the message sent is $(0, 0, 1, 0, 1, 1, 0)$. Suppose also that after going through a noisy channel, the received message is $(0, 0, 1, 0, 0, 1, 0)$. Clearly in this case, we can compute $(s_1, s_2, s_3) = (1, 0, 0)$, telling us that an error (see table below) has occurred in the parity bit p_1. This three-bit vector (s_1, s_2, s_3) is called a syndrome. Depending on the value of the syndrome, we can correctly identify the error and correct it. More generally, we can compare the result of the syndrome with the following table:

Syndrome			Error						
0	0	0	0	0	0	0	0	0	0
0	0	1	0	0	0	0	0	0	1
0	1	0	0	0	0	0	0	1	0
0	1	1	0	0	0	1	0	0	0
1	0	0	0	0	0	0	1	0	0
1	0	1	1	0	0	0	0	0	0
1	1	0	0	0	1	0	0	0	0
1	1	1	0	1	0	0	0	0	0

Note that the syndrome is determined by another matrix multiplication on the received message:

$$
\begin{pmatrix} s_1 \\ s_2 \\ s_3 \end{pmatrix} = \begin{pmatrix} 1\,1\,1\,0\,1\,0\,0 \\ 0\,1\,1\,1\,0\,1\,0 \\ 1\,1\,0\,1\,0\,0\,0 \end{pmatrix} \begin{pmatrix} x_1' \\ x_2' \\ x_3' \\ x_4' \\ p_1' \\ p_2' \\ p_3' \end{pmatrix} \tag{17.43}
$$

17.8 Quantum Linear Codes: CSS Codes

Finally we conclude with a short description of a family of quantum error-correcting linear codes known as the Calderbank-Shor-Steane (or CSS) codes.

Suppose C_1 be a classical linear code with $(n - k_1) \times n$ parity check matrix H_1 and C_2 is a subcode of C_1 with $(n - k_2) \times n$ parity check matrix H_2, where $k_2 < k_1$. The first $(n - k_2)$ rows of H_2 are the same as the rows of H_1 but there are still $(k_1 - k_2)$ rows in H_1 that are linearly independent in C_1 but not in C_2. The subcode C_2 then defines an equivalence relation in C_1. For any $u_1, U_2 \in C_1$, we say that u_1 and u_2 are equivalent ($u_1 \equiv u_2$) if then there exist a $\tilde{u} \in C_2$ such that $u_1 = u_2 + \tilde{u}$. The equivalence classes form the cosets of C_2 in C_1.

The Calderbank-Shor-Steane (CSS) code is a family of linear codes $[2^r - 1, 2^r - 1 - 2r, 3]$ for $r \geq 3$ that makes full use of classical theory of linear (Hamming) codes to correct bit-flip errors (X errors) and the dual code to correct phase-flip errors (Z errors) [11, 12]. Steane investigates the case of $r = 3$ corresponding to $[7, 1, 3]$ in his 1996 paper.

The parity check matrix for the $[7, 4, 3]$ classical code is

$$H = \begin{pmatrix} 1\,0\,1\,0\,1\,0\,1 \\ 0\,1\,1\,0\,0\,1\,1 \\ 0\,0\,0\,1\,1\,1\,1 \end{pmatrix} \tag{17.44}$$

It can check easily that the Hamming distance of the code is $d = 3$. The generator matrix G of the code is

$$G = \begin{pmatrix} 1\,0\,1\,0\,1\,0\,1 \\ 0\,1\,1\,0\,0\,1\,1 \\ 0\,0\,0\,1\,1\,1\,1 \\ 1\,1\,1\,0\,0\,0\,0 \end{pmatrix}. \tag{17.45}$$

The dual of the Hamming code is the $[7, 3, 4]$ code generated by H. For the CSS code, we choose C_1 to be the Hamming code and C_2 as the the dual code. Note that C_2 is an even subcode of C_1.

We define two bases: the F and P bases. In the F basis, the two orthogonal codewords are:

$$|0\rangle_F = \frac{1}{\sqrt{8}} \sum_{\text{even} v \in \text{Hamming}} |v\rangle$$

$$|1\rangle_F = \frac{1}{\sqrt{8}} \sum_{\text{odd} v \in \text{Hamming}} |v\rangle. \tag{17.46}$$

Since both $|0\rangle_F$ and $|1\rangle_F$ are superpositions of Hamming codewords, bit flips can be diagnosed within the basis by performing the parity check with the H matrix. In the Hadamard rotated basis, these codes words become the P basis:

$$|0\rangle_F \rightarrow |0\rangle_P \equiv \frac{1}{4} \sum_{v \in \text{Hamming}} |v\rangle = \frac{1}{\sqrt{2}}(|0\rangle_F + |1\rangle_F), \tag{17.47}$$

$$|1\rangle_F \rightarrow |1\rangle_P \equiv \frac{1}{4} \sum_{v \in \text{Hamming}} (-1)^{Wt(v)}|v\rangle = \frac{1}{\sqrt{2}}(|0\rangle_F - |1\rangle_F). \tag{17.48}$$

In this basis the states are still superposition of (hamming) codewords and so bit flips in the P basis are phase flips in the original basis and can still be corrected with the H parity check.

References

1. Daniel G (2010) An introduction to quantum error correction and fault-tolerant quantum computation. In: Quantum information science and its contributions to mathematics, Proceedings of symposia in applied mathematics, vol 68, pp 13–58
2. McMahon D (2007) Quantum computing explained. Wiley
3. Preskill J (1998) Lecture notes for physics 229: quantum information and computation. California Institute of Technology, vol 16
4. Nielsen MA, Chuang I (2002) Quantum computation and quantum information
5. Wooters WK, Zurek WK (1982) Quantum no-cloning theorem. Nature 299:802
6. Steane AM (1998) Introduction to quantum error correction. Philosophical transactions of the royal society of London. Series A: Math Phys Eng Sci 356(1743):1739–1758
7. Devitt SJ, Munro WJ, Nemoto K (2013) Quantum error correction for beginners. Rep Progress Phys 76(7):076001
8. Lidar DA, Brun TA (2013) Quantum error correction. Cambridge University Press
9. Emanuel K, Raymond L, Lorenza V (2000) Theory of quantum error correction for general noise. Phys Rev Lett 84(11):2525
10. Blahut RE (2003) Algebraic codes for data transmission. Cambridge University Press
11. Steane AM (1996) Error correcting codes in quantum theory. Phys Rev Lett 77(5):793
12. Calderbank AR, Shor PW (54) Good quantum error-correcting codes exist. Phys Rev A 54(2):1098

Chapter 18
One-Way Quantum Computer

So far, we have described a quantum computer technology that relies on gates and connecting wires to implement any algorithm. Such a paradigm is sometimes called the quantum circuit model. In this model, every quantum algorithm is implemented through a quantum circuit, and the basic building blocks of quantum circuits are quantum gates. Each quantum gate operates on one or two qubits, which are called single-qubit gates and two-qubit (possibly entangling) gates respectively. An example of entangling two-qubit gates is the controlled-NOT gate (or controlled-Z), which is given in the computational basis by the matrix

$$\begin{pmatrix} 1 & 0 & 0 & 0 \\ 0 & 1 & 0 & 0 \\ 0 & 0 & 0 & 1 \\ 0 & 0 & 1 & 0 \end{pmatrix}. \tag{18.1}$$

If a set of quantum gates is sufficient to construct any arbitrary quantum circuit, we say that this set is universal for quantum computing. An important universal set of quantum gates is composed by the controlled-NOT gate and arbitrary single-qubit quantum gates. It is also interesting to note that in this model, most computations proceed with unitary, and therefore reversible, evolutions until measurements are made.

As we have seen in earlier chapters, in circuit model, a typical circuit diagram given in Fig. 18.1 comprises of the inputs, and time flows from the left to the right. In this specific example, there are a total of three qubits as inputs. The left end of the circuit shows the initial state of the qubits, which is $|\psi\rangle \otimes |0\rangle \otimes |0\rangle$, where $|\psi\rangle$ is a single-qubit state. The very right side gives the final state of the qubits, which is $Z |m_1\rangle \otimes H |m_2\rangle \otimes |\psi'\rangle$, where $m_1, m_2 \in \{0, 1\}$ and $|\psi'\rangle$ is some single-qubit state whose actual value depends on $|\psi\rangle$ and θ. Each quantum gate in Fig. 18.1 is explained in Fig. 18.2.

B. E. Baaquie and L.-C. Kwek, *Quantum Computers*,
https://doi.org/10.1007/978-981-19-7517-2_18

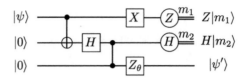

Fig. 18.1 A quantum circuit. Published with permission of © Belal E. Baaquie and L. C. Kwek. All Rights Reserved

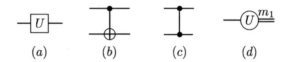

Fig. 18.2 **a** A single-qubit gate U; **b** A controlled-NOT gate with the top qubit as the control qubit and the bottom qubit as the target qubit; **c** A controlled-Z gate; **d** A single-qubit measurement in the basis $\{U\,|0\rangle,\,U\,|1\rangle\}$, and the measurement result m_1 is obtained which will give an output state $U\,|m_1\rangle$. Published with permission of © Belal E. Baaquie and L. C. Kwek. All Rights Reserved

In order to design and build efficient quantum circuits, it is useful to exploit special features of quantum systems that will optimize the performance. It is believed that entanglement, as one of the main pillars of quantum information, provides an important key for the speedup of quantum algorithms overs their classical analogs [1, 2]. It is also likely that efficient quantum algorithms require a minimal amount of the entanglement, and otherwise the quantum algorithm should be simulable with classical computers. The inputs to a typical quantum circuit usually take the form $|0\rangle \otimes |0\rangle \cdots \otimes |0\rangle$, which can be written concisely as $|00\cdots 0\rangle$. Such inputs are not entangled. Thus, the entanglement needed for the computation must be generated within the circuit. This means that some quantum gates used in the quantum circuit model should be capable of generating entanglement for quantum circuits. These gates are called quantum entangling gates. One important example is the controlled-NOT gate, which we have just mentioned earlier.

The realization of quantum circuits in various physical systems has been extensively studied. These approaches include implementation through nuclear magnetic resonance (NMR) [3], the manipulations of atoms or ions in ion traps [4], the manipulation of neutral atom [5], implementation with cavity QED [6], the optical platform with linear or nonlinear optical devices [7], manipulation of electrons or atoms in the solid state [8], superconducting devices [9], and a 'unique' approach to quantum information proposed in the quantum information roadmap [10]. Regardless of the approach to the realization of large-scale quantum computing, DiVincenzo has elegantly summarized five important criteria for the physical implementation of future quantum computers [11], and each one of these criteria has been carefully examined [4–11]. These criteria are:

- A scalable physical system equipped with well-characterized qubit;
- An ability to initialize the state of the qubits to a simple fiducial state;
- Long relevant decoherence times;

- A 'universal' set of quantum gates; and
- A capability to perform relevant measurement.

To date, none of these systems is fully capable of realizing a large-scale fully programmable quantum computer in the foreseeable future without any glitches: each system presents its own unique set of challenges and problems, including noises. One important stumbling block in many of these systems is that all quantum entangling gates cannot be implemented with high fidelity [12]. Thus, it is desirable to reduce the number of quantum entangling gates needed for quantum computation to the bare minimum.

To overcome some of these limitations, researchers have actively sought other paradigms of quantum computation. Although these alternative paradigms are equivalent to the quantum circuit model in terms of computational power, they are very different in terms of real physical realization. Among these alternative paradigms, some of more promising ones are measurement-based quantum computing [13], topological quantum computing [14] and adiabatic quantum computing [15]. In this chapter, we focus on the measurement-based quantum computing (MBQC) model as introduced by Raussendorf and Briegel in 2001 [13]. In MBQC, we start with highly entangled states and only single-qubit measurements are performed to realize the quantum algorithms. These measurements 'collapse' the wave functions of the initially highly entangled resource, and the scheme is therefore no longer reversible.

18.1 Measurement-Based Quantum Computation

In this section, we briefly describe how MBQC works. Figure 18.3 provides a schematic illustration of MBQC. In this figure, each circle represents a qubit in a two-dimensional square lattice. Qubits that are connected with a solid line are neighbors. A system of qubits is first prepared as a cluster state $|\Psi_C\rangle$. Each qubit is then measured sequentially, from the left columns to the right columns, so the information flows from left to right. The arrow in each circle illustrates the direction of the spin that is measured, and the choice of directions depends on the results of earlier measurements. At the end of the procedure, all the spins are measured, and the measurement results together give the result of the computation. As mentioned before, since measurement constitutes an irreversible process, MBQC is also known in the literature, as the 'one-way quantum computing'.

In Sect. 18.2, we take a closer look at the cluster state and in Sect. 18.3, we show how MBQC can simulate any quantum circuit efficiently, hence universal quantum computing can also be implemented by MBQC.

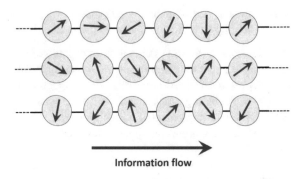

Information flow

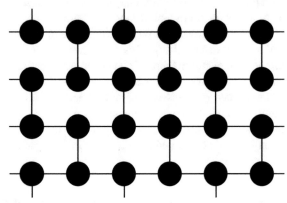

18.2   The Cluster State

The cluster state was first introduced by Briegel and Raussendorf shortly before they introduced the model of MBQC [16]. The term 'cluster state' actually refers to a family of highly entangled quantum states. These states are quantum states of qubits associated with certain lattices (or just simply arbitrary graphs). For any arbitrary graph of n vertices, one can always define a cluster state. For example, the graph can be a 2D square lattice discussed in Fig. 18.3, or the 2D honeycomb lattice given in Fig. 18.4, or a simple graph of three vertices given in Fig. 18.5a.

The creation of the cluster state associated with a given graph is described in three easy steps.

- Initialize every qubit to the state $|0\rangle$.
- Apply a Hadamard gate on each of the qubits so that each node becomes $1/\sqrt{2}(|0\rangle + |1\rangle)$.
- Apply a controlled-Z gate to each pair of qubits whose corresponding graph vertices are connected by a solid line.

As an example, the cluster state associated with the graph of three vertices given in Fig. 18.5a can be realized in the circuit model with the circuit given in Fig. 18.5(b).

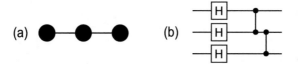

- First, we initialize the three qubits as $\psi_1 = |0\rangle \otimes |0\rangle \otimes |0\rangle$.
- Next, we apply Hadamard to each qubit so that we get

$$\psi_2 = \frac{1}{\sqrt{2}}(|0\rangle + |1\rangle) \otimes \frac{1}{\sqrt{2}}(|0\rangle + |1\rangle) \otimes \frac{1}{\sqrt{2}}(|0\rangle + |1\rangle)$$

$$= \frac{1}{2\sqrt{2}}(|0\rangle + |1\rangle) \otimes (|0\rangle + |1\rangle) \otimes (|0\rangle + |1\rangle)$$

$$= \frac{1}{2\sqrt{2}}(|000\rangle + |001\rangle + |010\rangle + |011\rangle + |100\rangle + |101\rangle$$
$$+ |110\rangle + |111\rangle)$$

- Apply CZ gate between qubit 1 and 2 and then between qubit 2 and 3. The order is immaterial.

$$\psi_3 \xrightarrow{CZ_{12}} \frac{1}{2\sqrt{2}}(|000\rangle + |001\rangle + |010\rangle + |011\rangle + |100\rangle + |101\rangle$$
$$- |110\rangle - |111\rangle)$$

$$\xrightarrow{CZ_{23}} \frac{1}{2\sqrt{2}}(|000\rangle + |001\rangle + |010\rangle - |011\rangle + |100\rangle + |101\rangle$$
$$- |110\rangle + |111\rangle)$$

$$\frac{1}{2}(|0+0\rangle + |0-1\rangle + |1-0\rangle + |1+1\rangle)$$

where ψ_i are the intermediate states and $|+\rangle = \frac{1}{\sqrt{2}}(|0\rangle + |1\rangle)$ and $|-\rangle = \frac{1}{\sqrt{2}}(|0\rangle - |1\rangle)$. Of course, the final graph state, that is generated, is given by ψ_3.

Consider ψ_3. Suppose we now apply the Pauli-X operator (i.e., σ_x) to the first site, which we denote as X_1, and the Paul-Z operator to the second site, which we denote as Z_2. What happens?

Recall that

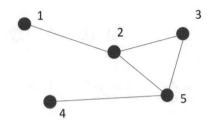

$$X\,|0\rangle = |1\rangle\,,\quad X\,|+\rangle = |+\rangle\,,$$
$$X\,|1\rangle = |0\rangle\,,\quad X\,|-\rangle = -\,|-\rangle\,,$$
$$Z\,|0\rangle = |0\rangle\,,\quad Z\,|+\rangle = |-\rangle\,,$$
$$Z\,|1\rangle = -\,|1\rangle\,,\quad Z\,|-\rangle = |+\rangle$$

We see that

$$X_1 Z_2 |\psi_3\rangle = \frac{1}{2\sqrt{2}} \big(|100\rangle + |101\rangle + (-1)\,|110\rangle - (-1)\,|111\rangle$$
$$+ |000\rangle + |001\rangle - (-1)\,|010\rangle + (-1)\,|011\rangle\big)$$
$$= |\psi_3\rangle.$$

Similarly, one checks also that $Z_1 X_2 Z_3$ acting on $|\psi_3\rangle$ does not change the state; i.e., the state is invariant under the operator $Z_1 X_2 Z_3$. Moreover, the state is also invariant under $Z_2 X_3$.

Thus, we see that the state is unchanged by the operators

$$\{X_1 Z_2,\ Z_1 X_2 Z_3,\ Z_2 X_3\}.$$

How do we get these operators that allow the state ψ_3 to stay unchanged? For each node (or vertex) i in Fig. 18.5a, we apply an X_i operator and, at the same time, we apply a Z_j operator to any other nodes j that are connected to the node i by an edge. These operators that leave the graph state unchanged are called stabilizers. Thus, for the graph with five nodes shown in Fig. 18.6, the stabilizers are

$$\{X_1 Z_2,\ Z_1 X_2 Z_3 Z_5,\ Z_2 X_3 Z_5,\ X_4 Z_5,\ Z_2 Z_3 Z_4 X_5\}$$

More generally, for each vertex j in a given graph, we denote the neighboring qubits of j by $\mathrm{nb}(j)$. It is easily shown that the graph or cluster state $|\Psi_C\rangle$ associated with the graph is an eigenstate of the operator $X_j \bigotimes_{k \in \mathrm{nb}(j)} Z_k$ with eigenvalue 1, where X_j (Z_k) is the Pauli X (Z) operator acting on the jth (kth) qubit. That is,

$$X_j \bigotimes_{k \in \mathrm{nb}(j)} Z_k |\Psi_C\rangle = |\Psi_C\rangle \tag{18.2}$$

We can glean the same insight from a slightly different angle. We again take the simple example given in Fig. 18.5. We denote the qubits from left to right as 1, 2, 3, and the controlled-Z gate acting on the jth and kth qubits by S_{jk}. Then the cluster state associated with the graph is given by (according to the circuit given by 18.5b)

$$|\Psi_C\rangle = S_{23} S_{12} H_1 H_2 H_3 |000\rangle,\qquad(18.3)$$

where H_j is the Hadamard operator acting on the j-th qubit. Now we show that Eq. (18.2) holds. That is,

$$X_1 Z_2 |\Psi_C\rangle = |\Psi_C\rangle,$$
$$Z_1 X_2 Z_3 |\Psi_C\rangle = |\Psi_C\rangle,$$
$$Z_2 X_3 |\Psi_C\rangle = |\Psi_C\rangle.\qquad(18.4)$$

To show this, observe the following identities:

$$Z_j S_{jk} = S_{jk} Z_j \quad \text{and} \quad X_j S_{jk} = S_{jk} X_j Z_k,$$
$$Z_j H_j = H_j X_j \quad \text{and} \quad X_j H_j = H_j Z_j.\qquad(18.5)$$

Substituting Eq. (18.3) into Eq. (18.5) immediately result in Eq. (18.4). In general, it is straightforward to show that Eq. (18.2) holds for any cluster state associated with any graph based on a similar argument.

One sees that the operator $X_j \bigotimes_{k \in \text{nb}(j)} Z_k$ has eigenvalues $+1$, and the operators corresponding to different vertices commute. For instance, $X_1 Z_2$ and $Z_1 X_2 Z_3$ and $Z_2 X_3$ commute with each other. Therefore, if we choose a Hamiltonian H_C as

$$H_C = -\sum_j X_j \bigotimes_{k \in \text{nb}(j)} Z_k,\qquad(18.6)$$

where the summation is running over all vertices of the graph, then $|\Psi_C\rangle$ is obviously the unique ground state of the Hamiltonian H_C. Moreover, this Hamiltonian is obviously gapped; i.e., there is a gap between the ground state and the next excited state, and frustration-free (as $|\Psi_C\rangle$ is the ground state of each term in the summation). So it would be great if we could find such Hamiltonian in Nature. But alas, such a Hamiltonian is in general not a two-body nearest-neighbor Hamiltonian and most Hamiltonians in Nature, relying on say electromagnetic (Coulomb for instance) or gravitational forces, are just two-body terms.

18.3 Simulation of Basic Quantum Gates

In this subsection, we show that MBQC based on certain cluster states can be used to perform universal computation. As discussed before, in order to implement univer-

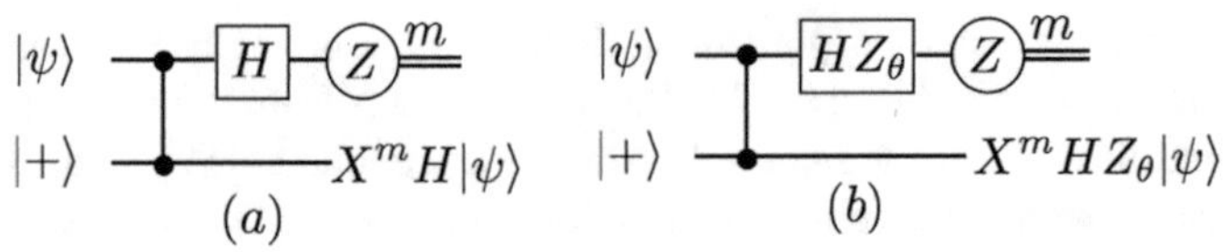

Fig. 18.7 (This figure is redrawn from Eqs. (18.11) and (18.12) in [23]). **a** A circuit for one-bit teleportation; **b** Generalized one-bit teleportation. Published with permission of © Belal E. Baaquie and L. C. Kwek. All Rights Reserved

sal quantum computation, one needs to realize *any* arbitrary single-qubit gate and a suitable entangling gate, for instance, the controlled-NOT gate. Apart from the original discussion on how MBQC simulates these gates [13], there are other alternative discussions [17–23]. Here we follow the explanation given by Nielsen in [23], based on the primitive circuit given in Fig. 18.7a that is proposed in [24].

The circuit in Fig. 18.7a is sometimes known as one-bit teleportation scheme, where $|+\rangle = \frac{1}{\sqrt{2}}(|0\rangle + |1\rangle)$. To show how it works, let $|\psi\rangle = \alpha |0\rangle + \beta |1\rangle$. Then note

$$H_1 S_{12}(\alpha |0\rangle + \beta |1\rangle) \otimes \frac{1}{\sqrt{2}}(|0\rangle + |1\rangle)$$

$$= \frac{1}{\sqrt{2}}(|0\rangle \otimes H |\psi\rangle + |1\rangle \otimes X H |\psi\rangle), \qquad (18.7)$$

where S_{12} is the controlled phase gate acting on the two qubits. Measuring the first qubit in the $\{0, 1\}$ basis (or Z measurement) gives the desired result. This circuit is then directly generalized to the one given in Fig. 18.7b, since Z_θ commutes with S_{12}. We then measure the first qubit and depending on the outcome of measurement, we get $H |\psi\rangle$ if the outcome is $|0\rangle$ (i.e., $m = 0$) or $X H |\psi\rangle$ if the outcome is $|1\rangle$ (i.e., $m = 1$). In short, we have $X^m H |\psi\rangle$.

In Fig. 18.7b, we added a Z_θ gate where $Z_\theta = e^{i\frac{\theta}{2} Z}$, with Z as the Pauli-Z operator. Note that $H Z_\theta$ acting on the control-phase operator means that Z_θ acts on S_{12} first and then followed by the Hadamard operator, H. However, Z_θ commutes with S_{12}, so in accordance with Eq. (18.7), the result is just

$$\frac{1}{\sqrt{2}}(|0\rangle \otimes H Z_\theta |\psi\rangle + |1\rangle \otimes X H Z_\theta |\psi\rangle).$$

In this case, we have

$$X^m H Z_\theta |\psi\rangle$$

depending on the outcome m of the first qubit.

We are now ready to show that any arbitrary single-qubit gate can be performed on a quantum state. To do so, we simply concatenate the 'generalized' teleportation circuit in Fig. 18.7b as in Fig. 18.8.

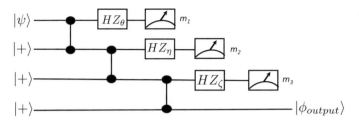

Fig. 18.8 Concatenation of the 'generalized' teleportation circuit to form the arbitrary single-qubit gate. Published with permission of © Belal E. Baaquie and L. C. Kwek. All Rights Reserved

In Fig. 18.8, the output states in the concatenated circuit are

$$\left|\phi_{\text{output}}\right\rangle = X^{m3} H Z_\zeta X^{m2} H Z_\eta X^{m1} H Z_\theta \left|\psi\right\rangle . \tag{18.8}$$

We then note that $XZ = -ZX$ and $HX = ZH$ to get

$$
\begin{aligned}
\left|\phi_{\text{output}}\right\rangle &= X^{m3} H Z_\zeta X^{m2} H Z_\eta X^{m1} H Z_\theta \left|\psi\right\rangle \\
&= (-1)^{m2} X^{m3} Z^{m2} H Z_\zeta H Z_\eta X^{m1} H Z_\theta \left|\psi\right\rangle \\
&= (-1)^{m2} (-1)^{m1} X^{m3} Z^{m2} X^{m1} H Z_\zeta H Z_\eta H Z_\theta \left|\psi\right\rangle \\
&= (-1)^{m2} (-1)^{m1} X^{m3} Z^{m2} X^{m1} X_\zeta H H Z_\eta X_\theta H \left|\psi\right\rangle \\
&= (-1)^{m1+m2} X^{m3} Z^{m2} X^{m1} X_\zeta Z_\eta X_\theta H \left|\psi\right\rangle
\end{aligned} \tag{18.9}
$$

where we note that in the last equality, we have used H^2 as the identity matrix. Depending on the outcomes of the measurement, we see that we have performed the following transformation

$$U(\zeta, \eta, \theta) = X_\zeta Z_\eta X_\theta = e^{i\frac{\zeta}{2}X} e^{i\frac{\eta}{2}Z} e^{i\frac{\theta}{2}X} \tag{18.10}$$

on the input state $\left|\psi\right\rangle$. Since, up to a global phase, $U(\zeta, \eta, \theta)$ is just the most general form for an arbitrary single-qubit rotation, we have shown that we can perform arbitrary unitary operation on a single qubit.

For universal quantum computation, we need to demonstrate that we can also achieve a two-qubit entangling gate. It can be shown that no one-dimensional cluster state array can simulate a two-qubit entangling gate. One needs to have a two-dimensional array. To see how we can do this, let us consider the two-dimensional cluster state in Fig. 18.9. This cluster state is a four-qubit state. Qubits 1 and 4 are the input states, and we assume that they are separable with the states $\alpha|0\rangle + \beta|1\rangle$ and $\alpha'|0\rangle + \beta'|1\rangle$, respectively.

It is not hard to show that the resulting state after applying the control-phase gates is

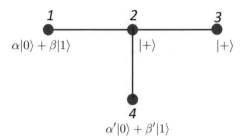

Fig. 18.9 A four-qubit cluster or graph state can be used to simulate a Controlled-NOT gate. The state $|+\rangle$ is $\dfrac{1}{\sqrt{2}}(|0\rangle + |-\rangle)$. Published with permission of © Belal E. Baaquie and L. C. Kwek. All Rights Reserved

$$|\text{2D state}\rangle = \frac{1}{2}(|00\rangle(\alpha\alpha'|+\rangle|0\rangle + \alpha\beta'|+\rangle|1\rangle)$$
$$|01\rangle(\alpha\alpha'|-\rangle|0\rangle - \alpha\beta'|-\rangle|1\rangle)$$
$$|10\rangle(\beta\alpha'|+\rangle|0\rangle + \beta\beta'|+\rangle|1\rangle)$$
$$-|11\rangle(\beta\alpha'|-\rangle|0\rangle - \beta\beta'|-\rangle|1\rangle)). \qquad (18.11)$$

We then perform measurements on qubit 1 and qubit 2 in the X basis. Suppose we get $|++\rangle = |+\rangle_1 \otimes |+\rangle_2$. We see that the state of qubit 3 and 4 becomes

$$\alpha\alpha'|00\rangle + \alpha\beta'|11\rangle + \beta\alpha'|10\rangle + \beta\beta'|01\rangle$$
$$= \alpha'(\alpha|0\rangle + \beta|1\rangle)_3|0\rangle_4 + \beta'(\alpha|1\rangle + \beta|0\rangle)_3|1\rangle_4 \qquad (18.12)$$

Clearly, this is a CNOT gate acting the original separable state

$$(\alpha|0\rangle + \beta|1\rangle)_3 \otimes (\alpha'|0\rangle + \beta'|1\rangle)_4$$

on qubit 3 and 4 with qubit 4 as the control bit which we denote as CNOT_{43}. We see that in Eq. (18.12), a controlled NOT operation is applied to the state $(\alpha|0\rangle + \beta|1\rangle)_3$ whenever the state of the second qubit is $|1\rangle_4$.

What happens if the outcome is not $|++\rangle_{12}$? Suppose the measurement outcomes of qubit 1 and 2 is $|+-\rangle_{12}$. In this case, we see that the state of qubit 3 and 4 becomes

$$\alpha\alpha'|10\rangle + \alpha\beta'|01\rangle + \beta\alpha'|00\rangle + \beta\beta'|11\rangle$$
$$= (X \otimes X)\,\text{CNOT}_{43}\,(X \otimes X)\,(\alpha|0\rangle + \beta|1\rangle)_3(\alpha'|0\rangle + \beta'|1\rangle)_4 \qquad (18.13)$$

For the outcome $|-+\rangle_{12}$, qubits 3 and 4 become

$$\alpha\alpha'|00\rangle + \alpha\beta'|11\rangle - \beta\alpha'|10\rangle - \beta\beta'|01\rangle$$
$$= (Z \otimes Z)\,\text{CNOT}_{43}(\alpha|0\rangle + \beta|1\rangle)_3(\alpha'|0\rangle + \beta'|1\rangle)_4, \qquad (18.14)$$

Finally, for the outcome $|--\rangle_{12}$, qubits 3 and 4 become

$$\alpha\alpha'|10\rangle + \alpha\beta'|01\rangle - \beta\alpha'|00\rangle - \beta\beta'|11\rangle$$
$$= -(Z \otimes Z)\,\mathrm{CNOT}_{43}\,(X \otimes Id)\,(\alpha|0\rangle + \beta|1\rangle)_3(\alpha'|0\rangle + \beta'|1\rangle)_4. \quad (18.15)$$

So regardless of the measurement outcome, we see that we have performed a controlled-NOT operation on qubit 3 and qubit 4 with the latter (qubit 4) as the control bit.

18.4 Resource States for MBQC

It is shown in Sect. 18.3 that arbitrary single-qubit gates and the controlled NOT gate can be simulated by MBQC with the appropriate cluster states or graph state. Universal quantum computing can then be implemented by concatenating these graphs together. It should be noted that not all graph states are appropriate as resource states for universal MBQC, for instance graph states associated with the tree graphs (i.e., graph without cycles) are not necessary resource states for MBQC [25–28]. Yet, many of these graphs, for instance, the cluster states associated with a square lattice or the honeycomb lattice [29], are demonstrated to be useful.

The advantage of MBQC for the implementation of quantum computing in the real world is that once the highly entangled resource state is prepared, only single-qubit measurements are needed throughout the entire process of computation, and quantum entangling gates are unnecessary. The challenge in implementing quantum entangling gates in a circuit model is the high noise and low fidelity. In MBQC, one avoids entangling gates but the computation requires the preparation of a highly entangled resource. Such a resource, e.g., cluster state, typically uses a three-step procedure outlined in Sect. 18.2. The third step requires the application of a controlled-Z gates on pairs of qubits connected by edges in the graph. So, a blind adherence to this approach provides no advantage over the quantum circuit model.

An appealing idea that overcomes the need for entangling operations completely is to seek for a physical system in which one could obtain the required entangled resource states as the unique ground state. By cooling some quantum many-body, hopefully just two-body, systems, one hopes to generate such entangled resource. Yet, interactions in Nature are described solely by two-body process. So we hope to achieve these ground states with systems whose Hamiltonians are equipped only with two-body terms. We already know that the cluster state is a unique ground state of the Hamiltonian H_C given in Eq. (18.6), and that H_C is gapped and frustration-free. However, H_C involves many-body interactions, not just two-body interactions and such Hamiltonians are generally difficult to find in nature. For instances, the H_C associated with a 2D square lattice involves five-body interactions and the H_C associated with the 2D honeycomb lattice involves four-body interactions. Even the H_C associated with a 1D chain involves at least three-body interactions.

One therefore hopes that there exists a Hamiltonian with just two-body interactions that gives a particular useful cluster state as the unique ground state. Unfortunately, this turns out to be impossible as observed by Nielsen [23]. The idea behind the proof is the following: if the Hamiltonian involves at most two-body interactions, then its ground state energy is totally determined by the two-particle reduced density matrices of the states. So, given any cluster state, $|\Psi_C\rangle$, which acts as a resource state for MBQC, some of the eigenstates of H_C have the same two-particle reduced density matrices as $|\Psi_C\rangle$. Therefore, if the cluster state $|\Psi_C\rangle$ is the ground state, then some other eigenstates of H_C are also ground states, meaning the cluster state cannot be the unique ground state of any two-body Hamiltonian. Given this no-go theorem, what can do next?

In the original proposal by Raussendorf and Briegel [30], cluster states are created by unitary evolution of Ising-type spin–spin interactions. There has been a proposal to generate macroscopic cluster state in 3D lattice arrays of ultracold bosonic and fermionic atoms [31–34]. Experimental efforts in this direction for sufficiently large-scale cluster states are still rather challenging. Moreover, we have seen that an alternative method of generating the resource state as the unique ground state of a short-ranged interacting Hamiltonian with a finite spectral gap by cooling the system to low-enough temperature is not feasible.

If we relax the condition for generating exact ground states, one can generate an approximate cluster state as ground state of a two-body interacting Hamiltonian through perturbation [35]. An alternative idea toward this goal is to exploit higher dimensional states called tri-cluster state [36]. The tri-cluster state is a universal resource state with a nearest-neighbor interacting parent Hamiltonian equipped with a nonzero spectral gap on a hexagonal lattice. This quantum state has a local Hilbert space of dimension six and contains the cluster state in three different bases, hence the name tri-cluster state. Due to the extra dimensions, the tri-cluster state can be further converted to a cluster state of qubit local Hilbert space (i.e., of two levels) by the so-called quantum state reduction [28]. The technique has been extended to three-dimensional graph states [37] and finite temperature states [26].

In recent years, continuous variable large-scale photonic cluster states have been constructed [38, 39]. To demonstrate that these resources can be used for measurement-based quantum computation, these states however need the preparation of highly squeezed states [40], which are experimentally challenging. For discrete variable quantum computation, i.e., using 0s and 1s, experimental progress is more difficult [41]. Yet overall, this paradigm in which quantum entanglement is established right at the start provides a viable and interesting alternative to the unitary circuit approach where quantum entanglement is created on demand through entangling gates throughout the algorithm.

References

1. Vidal G (2003) Efficient classical simulation of slightly entangled quantum computations. Phys Rev Lett 91(14):147902
2. Noah Linden, Sandu Popescu (2001) Good dynamics versus bad kinematics: is entanglement needed for quantum computation? Phys Rev Lett 87(4):047901
3. Cory D, Heinrichs T (2004) Nuclear magnetic resonance approaches to quantum information processing and quantum computing. Quantum Inf Sci Technol Roadmap v2.0. http://quist.lanl.gov
4. Wineland D, Heinrichs T (2004) Ion trap approaches to quantum information processing and quantum computing. Quantum Inf Sci Technol Road v2.0. http://quist.lanl.gov
5. Caves C, Heinrichs T (2004) Neutral atom approaches to quantum information processing and quantum computing. Quantum Inf Sci Technol Roadmap v2.0. http://quist.lanl.gov
6. Chapman M, Heinrichs T (2004) Cavity qed approaches to quantum information processing and quantum computing. Quantum Inf Sci Technol Roadmap v2.0. http://quist.lanl.gov
7. Kwiat P, Milburn G, Heinrichs T (2004) Optical approaches to quantum information processing and quantum computing. Quantum Inf Sci Technol Roadmap v2.0. http://quist.lanl.gov
8. Clark R, Awschalom D, DiVincenzo D, Hammel PC, Steel D, Birgitta Whaley K, Heinrichs T (2004) Solid state approaches to quantum information processing and quantum computing. Quantum Inf Sci Technol Roadmap v2.0. http://quist.lanl.gov
9. Orlando T, Heinrichs T (2004) Superconducting approaches to quantum information processing and quantum computing. Quantum Inf Sci Technol Roadmap v2.0. http://quist.lanl.gov
10. Lloyd S, Hammel PC, Heinrichs T (2004) "unique" qubit approaches to quantum information processing and quantum computing. Quantum Inf Sci Technol Roadmap v2.0. http://quist.lanl.gov
11. Divincenzo DP (2000) The physical implementation of quantum computation. Fortschritte der Physik 48:771–783
12. Alamos L (2004) National security. Quantum Inf Sci Technol Roadmap v2.0. http://quist.lanl.gov
13. Robert R, Briegel Hans J (2001) A one-way quantum computer. Phys Rev Lett 86(22):5188–5191
14. Kitaev A, Laumann C (2009) Topological phases and quantum computation. arXiv:0904.2771
15. Farhi E, Goldstone J, Gutmann S, Sipser M (2000) Quantum computation by adiabatic evolution. arXiv:quant-ph/0001106
16. Briegel HJ, Raussendorf R (2001) Persistent entanglement in arrays of interacting particles. Phys Rev Lett 86:910–913
17. Nielsen MA (2003) Quantum computation by measurement and quantum memory. Phys Lett A 308:96–100
18. Leung DW (2001) Two-qubit projective measurements are universal for quantum computation. arXiv:quant-ph/0111122
19. Leung DW (2003) Quantum computation by measurements. arXiv:quant-ph/0310189
20. Aliferis P, Leung DW (2004) Computation by measurements: a unifying picture. Phys Rev A 70(6):062314
21. Childs AM, Leung DW, Nielsen MA (2005) Unified derivations of measurement-based schemes for quantum computation. Phys Rev A 71(3):032318
22. Jorrand P, Perdrix S (2005) Unifying quantum computation with projective measurements only and one-way quantum computation. In: Ozhigov YI (ed) Society of photo-optical instrumentation engineers (SPIE) conference series, volume 5833 of society of photo-optical instrumentation engineers (SPIE) conference series, pp 44–51
23. Nielsen Michael A (2006) Cluster-state quantum computation. Rep Math Phys 57(1):147–161
24. Zhou X, Leung DW, Chuang IL (2000) Methodology for quantum logic gate construction. Phys Rev A 62(5):052316
25. Shi Y-Y, Duan L-M, Vidal G (2006) Classical simulation of quantum many-body systems with a tree tensor network. Phys Rev A 74(2):022320

26. Li Y, Browne DE, Kwek LC, Raussendorf R, Wei T-C (2011) Thermal states as universal resources for quantum computation with always-on interactions. Phys Rev Lett 107(6):060501
27. Kyaw TH, Li Y, Kwek L-C (2014) Measurement-based quantum computation on two-body interacting qubits with adiabatic evolution. Phys Rev Lett 113(18):180501
28. Kwek LC, Wei Z, Zeng B (2012) Measurement-based quantum computing with valence-bond-solids. Int J Mod Phys B 26(02):1230002
29. van den Nest M, Miyake A, Dür W, Briegel HJ (2006) Universal resources for measurement-based quantum computation. Phys Rev Lett 97(15):150504
30. Raussendorf R, Briegel HJ (2001) A one-way quantum computer. Phys Rev Lett 86(22):5188
31. Mandel O, Greiner M, Widera A, Rom T, Hänsch TW, Bloch I (2003) Controlled collisions for multi-particle entanglement of optically trapped atoms. Nature 425(6961):937–940
32. Gao X, Wang S-T, Duan L-M (2017) Quantum supremacy for simulating a translation-invariant Ising spin model. Phys Rev Lett 118(4):040502
33. Mamaev M, Blatt R, Ye J, Rey AM (2019) Cluster state generation with spin-orbit coupled fermionic atoms in optical lattices. Phys Rev Lett 122(16):160402
34. Amico L, Boshier M, Birkl G, Minguzzi A, Miniatura C, Kwek L-C, Aghamalyan D, Ahufinger V, Anderson D, Andrei N (2021) Roadmap on atomtronics: state of the art and perspective. AVS Quantum Sci 3(3):039201
35. Bartlett SD, Rudolph T (2006) Simple nearest-neighbor two-body hamiltonian system for which the ground state is a universal resource for quantum computation. Phys Rev A 74(4):040302
36. Chen X, Zeng B, Gu Z-C, Yoshida B, Chuang IL (2009) Gapped two-body hamiltonian whose unique ground state is universal for one-way quantum computation. Phys Rev Lett 102(22):220501
37. Wei T-C, Li Y, Kwek LC (2014) Transitions in the quantum computational power. Phys Rev A 89(5):052315
38. Asavanant W, Shiozawa Y, Yokoyama S, Charoensombutamon B, Emura H, Alexander RN, Takeda S, Yoshikawa J, Menicucci NC, Yonezawa H et al (2019) Generation of time-domain-multiplexed two-dimensional cluster state. Science 366(6463):373–376
39. Yokoyama S, Ukai R, Armstrong SC, Sornphiphatphong C, Kaji T, Suzuki S, Yoshikawa J, Yonezawa H, Menicucci NC, Furusawa A (2013) Ultra-large-scale continuous-variable cluster states multiplexed in the time domain. Nat Photon 7(12):982–986
40. Menicucci NC, Van Loock P, Gu M, Weedbrook C, Ralph TC, Nielsen MA (2006) Universal quantum computation with continuous-variable cluster states. Phys Rev Lett 97(11):110501
41. Larsen MV, Guo X, Breum CR, Neergaard-Nielsen JS, Andersen UL (2019) Deterministic generation of a two-dimensional cluster state. Science 366(6463):369–372

Part IV
Summary

Chapter 19
Efficiency of a Quantum Computer

To understand why the workings of a quantum computer is radically different from a classical computer—a difference that makes it far more powerful than any possible classical computer—some of the results obtained are summarized.

19.1 Quantum Algorithms

Some of the quantum algorithms discussed are reviewed to see what general lessons can be drawn from them. Two cardinal points should be kept in mind: (a) quantum superposition, called quantum parallelism in the popular literature, and quantum entanglement play a crucial role in the advantage of a quantum over a classical algorithm, and (b) quantum measurement theory plays an essential and counterintuitive role in obtaining results that are impossible for a classical computer. In particular, unlike a classical computer each step in the execution of a quantum computation cannot, in principle, be observed: one can only observe the initial and final quantum states of a quantum algorithm. A key to the superiority of the quantum computer over a classical one arises from the use of the qubit as a basic unit of computation. The qubit allows for the superposition of two classical bits, and which leads to many advantages over a classical algorithm.

- **Deutsch algorithm**. The qubit in the Deutsch allows for transformations of the quantum algorithm, based on superposition, that gives a result no classical algorithm can provide.
- **Deutsch–Jozsa algorithm**. This algorithm demonstrates the power of quantum computing arising from superposition. Note that, from Eq. 10.5, the final amplitude for the output qubit state of Deutsch–Jozsa algorithm is given by

© The Author(s), under exclusive license to Springer Nature Singapore Pte Ltd. 2023
B. E. Baaquie and L.-C. Kwek, *Quantum Computers*,
https://doi.org/10.1007/978-981-19-7517-2_19

$$\left| \frac{1}{N} \sum_x (-1)^{f(x)} \right|^2 \tag{19.1}$$

The fact that the $f(x)$ is simultaneously evaluated for *all* x is directly employed in obtaining the result above. In contrast, a classical algorithm can only evaluate $f(x)$ for x one at a a time; or it requires parallel capabilities for evaluating f for all x, which soon becomes impractical when the number of configurations exponentially increase. As n gets larger, the resources required to evaluate $f(x)$ by classical computer scale as a power of n.

Constructive and **destructive** interference that takes place in the summation given in Eq. 19.1 is a direct result of quantum parallelism, which is impossible in a classical computer.

- **Grover algorithm**. The Hadamard gate $H^{\otimes n}$ is applied to the input n-qubit state $|0\rangle^{\otimes n}$ and maps the input qubit into a linear combination, with equal amplitude, of all the computational basis states. Recall from Eq. 11.10

$$|0\rangle \;\rightarrow\; \frac{1}{\sqrt{N}} \sum_{x=1}^{N} |x\rangle$$

The ability of Hadamard gates to map the input qubit into this linear combination is due to the superposition principle of quantum mechanics.

The n-qubits evolve in the Hilbert state space of the qubits. The advantage of the Grover algorithm over the classical case lies in the fact that it is the state vector $|\psi\rangle$ that evolves in Hilbert space; due to the superposition principle, one has the *simultaneous* updating of all the basis states, as given in Eq. 11.7

$$|\psi\rangle = \frac{1}{\sqrt{N}} \sum_{x=0}^{N-1} (-1)^{f(x)} |x\rangle H |1\rangle \tag{19.2}$$

What Eq. 19.2 tells us is that the state vector $|\psi\rangle$ is being evolved; the right hand side of Eq. 19.2 are the components of the state vector in the computational basis states, and these components are all evolved simultaneously. See Sect. 4.12 and Eq. 4.42 for a more detailed discussion. It should be noted that using all the values of $f(x)$ in Eq. 19.2 does not mean that we know the value of the function $f(x)$: for this one would need to make separate measurements that would evaluate the value of f for each x.

For $n > 100$, *simultaneously* evaluating $f(x)$ for 2^n values x of the argument is impossible for a classical computer: it would need to evaluate, separately, the value of f for each of the 2^n values of x.[1]

The key to the Grover algorithm is that the amplification (rotation) gate G acts on *all* the n-qubits at the *same time* and yields

---

[1] 2^{100} is more than the atoms in the Universe.

$$G|\phi\rangle = G|\xi\rangle + G|\eta\rangle$$

and which is only possible due to quantum superposition.

Repeated applications of G on all the n-qubits bit is possible only if the intermediate states are *not measured*: if any measurement is performed before the final output state is ready, the quantum superposition would be destroyed. The repeated application of G allows for the amplification of the amplitude of the sought for qubit $|\xi\rangle$.

• **The Shor algorithm**. This algorithm uses the subtle properties of quantum measurement and entangled states to pick out the numerical value of A, which is required for factorization into primes. From Eq. 13.4, we have the superposed and entangled state given by

$$|\psi_f\rangle = \frac{1}{\sqrt{Q}} \sum_{i=0}^{r-1} \left[\sum_{k=0}^{A-1} |x_i + kr\rangle \right] \otimes |f(x_i)\rangle \tag{19.3}$$

We can extract the quantity A from Eq. 19.3 using the fact the right hand side is an entangled state of qubits of the first and second registers. A measurement on the qubits of the second register, namely $|f(x_i)\rangle$, which does not need knowledge of the quantity A, yields the value of A on performing the measurement.

The generalized Born rule shows that, as given in Eq. 13.4, the process of measurement of the qubits of the second register, due to the collapse of the state vector, picks out a single state vector from Eq. 19.3, say $|f(x_0)\rangle$, from the summation $\sum_{i=0}^{r-1}$ and yields

$$|\psi_f\rangle \rightarrow \text{Measurement} \rightarrow |\psi_3(x_0)\rangle |f(x_0)\rangle \tag{19.4}$$

The state vector $|\psi_3(x_0)\rangle$ is picked out by the generalized Born measurement since it is entangled with the state vector $|f(x_0)\rangle$. The value of x_0 is arbitrary and $|f(x_0)\rangle$ has some likelihood of being detected. This likelihood does not concern us since the analysis for obtaining A is valid for all x_0 and $|f(x_0)\rangle$ is ignored henceforth. The post-measurement state vector for the k degree of freedom is given by

$$|\psi_3(x_0)\rangle = \frac{1}{\sqrt{A}} \sum_{k=0}^{A-1} |x_0 + kr\rangle; \quad \langle\psi_3(x_0)|\psi_3(x_0)\rangle = 1 \tag{19.5}$$

Note the important fact that the pre-factor of $1/\sqrt{A}$ is a result of the generalized Born measurement that requires the state $|\psi_3(x_0)\rangle$ to be normalized. Since Q is known and A can be extracted from Eq. 19.5, the equation $Q = rA$ allows one to obtain the value of r, and which in turn leads to the factorization of N into its primes. The Shor algorithm shows that a quantum algorithm, due to the non-classical process of a quantum measurement, superposition and entanglement, yields a result that is not possible for a classical algorithm.

Equation 19.3 is the quantum superposition of entangled states of a large number of qubits ($Q > N^2$). It is this quantum superposition and entanglement that lies at the root of the exponential speeding up of the Shor algorithm compared to any classical algorithm. Furthermore, as shown by the Shor algorithm, quantum theory of measurement plays a key role in all quantum algorithms.

19.2   Memory and Speed of Quantum Computations

Memory of a computer, which depends on the amount of information that can be stored in the computer, is one factor in the speed of an algorithm. The other factor is how fast are the computer circuits updated. In 2022, the largest memory of a classical computer was 1024×10^{15} bytes, where one byte is eight bits. For an algorithm using n-bits, the classical algorithm requires a memory of 2^n bits, as discussed in Sect. 4.12, and all the 2^n bits for a classical computer have to be updated individually—whether it be sequential or in parallel. The world's fastest supercomputer, as of June 2021, was Japan's Fugaku at a speed of 442 petaflops, where one petaflop is 10^{15} floating-point operations[2] per second.

The quantum computer has the following three advantages over a classical computer that are a consequence of quantum superposition, entanglement, and quantum measurement theory. The are the following three sources of the great speed of quantum computations.

- The storage for memory of a quantum computer is not 2^n bits but n-bits due to quantum superposition and is discussed in Sect. 4.12.
- A classical algorithm has only one register (computational basis state) at every stage of the calculation, whereas the quantum algorithm has a superposition of many qubits, which in particular can be elements of the computational basis states.
- Born's theory of quantum measurement, discussed in detail in Sect. 4.5, is the key element that separates classical from quantum superposition. Quantum measurement causes the projection of one of the component states from a superposed state in Hilbert space to the observed state in the experimental apparatus. This is one of the reasons why quantum parallelism is possible.
- The updating of the quantum system evolves the entire state vector. In particular, it does not involve the updating of the individual components in its representation by a superposition in terms of the qubits (such as the 2^n computational basis states). See Sect. 4.12 and Eq. 4.42.
- The quantum superposition of many qubits as well as the simultaneous updating of all the superposed states are two sources of quantum parallelism: this greatly decreases the memory required and increases the speed of quantum algorithms in comparison with the classical case.

---

[2] Fugaku has been superseded as the fastest supercomputer in the world by Hewlett Packard Enterprise Frontier, or OLCF-5, in May 2022.

The points above are discussed in some detail below.

1. As mentioned earlier, it is misleading to think that the quantum computer is simultaneously evaluating f for all possible value of the degree of freedom in Eqs. 19.1 and 19.2: to obtain this information about f in the laboratory, we need to make many measurements. However, it is also true is that the quantum algorithm employs the values of f for every x in the superposed state. For example, in the Deutsch–Jozsa algorithm in Eq. 19.1 and the Grover algorithm in Eq. 19.2 $f(x)$ for all values of x are *simultaneously* employed in obtaining the output qubit (state vector). What is remarkable is that even with without knowing the value of $f(x)$ in the laboratory, the quantum algorithm can use this information. This is because the value of f for every x exists in Hilbert space.

2. The allowed states for the n-qubits are all elements of Hilbert space, which is $S^2 \otimes \cdots \otimes S^2$. The information and memory of a quantum computer is stored in Hilbert space, in the form of state vectors. All the steps in the execution of the quantum algorithm take place in Hilbert space. Quantum superposition and entanglement—which are the key features of a quantum computer that makes it superior to a classical computer—are both properties of state vectors that exist in Hilbert space. A more detailed discussion of this point is given in Sect. 19.3.

   For a quantum computer, all the n qubits—when they are in a superposed state—evolve simultaneously in Hilbert space. To understand quantum algorithms, one needs to locate the arena of quantum computations and quantum algorithms in Hilbert space. This point is discussed more elaborately in the following Sect. 19.3.

3. We review the results discussed in Sect. 4.12. A major resource of the quantum computer is the way information is stored. A classical computer has to physically store the value of every bit. For n binary degrees of freedom, the number of allowed configurations of a classical computer is 2^n: the memory required for the all possible configurations of the degrees of freedom is 2^n deterministic bits.

   A classical computer carrying out computations with n binary bits needs a physical memory device with 2^n units of memory. For large n, the physical memory device needed becomes huge. In contrast, the quantum computer needs only n-qubits, even though all the 2^n configurations for the degrees of freedom are employed in the quantum algorithm.

   The reason that the quantum computer needs only n-qubits is because, as mentioned earlier, only the initial and output states of the quantum computation are observable: the execution of the quantum algorithm takes place in Hilbert space. For example, for a n-qubit calculation, one needs to record, for example, only the following two computational basis states:

$$|\text{input}\rangle = |00110011\ldots11001\rangle; \quad |\text{output}\rangle = |10100001\ldots1100\rangle$$

Every time the algorithm is run, the same output register can be used for recording the new outcome, which is uncertain. The remaining 2^n configurations required for the algorithm are employed in Hilbert space, which is not observable.

4. The speed of the quantum computer is determined by the speed of the updating, as well as the fact that there is parallel updating of all the components of the state vector. As mentioned above, and given in Eq. 4.42, the updating of the quantum system evolves the entire state vector: hence the simultaneous evolution of all the components of state vector in terms of the computational basis states—which can run into millions for an algorithm like the Shor algorithm—results in an exponential speeding up compared to a classical computer. For example, as mentioned in Chap. 1, quantum computing system of 'Jiuzhang' is 100 trillion times faster than the world's fastest existing supercomputer.
5. The upper limit on the speed of quantum computations is imposed by how fast the quantum state can be updated. One line of reasoning is that the time-energy uncertainty principle limits the speed with which one can evolve a quantum state to another orthogonal state. Let the minimum time be T_M; similar to the uncertainty principle, T_M is given by the following Margolus–Levitin theorem

$$T_M \geq \frac{h}{4\langle E \rangle}$$

where $\langle E \rangle$ is the average energy of the quantum state being evolved. However, it has been shown that the speed of computation is not limited by the Margolus–Levitin theorem but by other considerations such as information density and information transmission speed, which is less than the speed of light [1].

19.3   Where Does Quantum Computation Take Place?

The question and answer below is a continuation of the interview in Sect. 1.1. Many of the points covered in earlier discussions are revisited for greater clarity, and we review the earlier discussion in Sects. 4.12 and 19.1. The answer below is based on the Copenhagen interpretation of quantum mechanics: open questions regarding this interpretation are discussed in Sect. 4.8.

Hobson: What are quantum computers teaching us about our quantum mechanical universe?

Baaquie: Quantum computers bring to the forefront the paradoxes, enigmas and conundrums of quantum mechanics that have been, and continue to be, encased in a thick shell of formalism [2].

To illustrate these paradoxes, consider how information is stored for a classical computation, which we can then compare with a quantum computation. To make things explicit and clear, consider a computation involving three binary bits. The 2^3-eight configurations, as enumerated in Sect. 4.1, are

$$\{000, 001, 010, 011, 100, 101, 110, 111\}$$

The input and output for both the classical and quantum computation is a determinate 3-bits string, which could, for example, be

$$Input : 011; \quad Output : 100$$

The classical computer's algorithm, in principle, needs to store all the eight configurations given above, since any of the eight configurations may be generated as the algorithm is executed, and each intermediate step of the classical computation is a configuration of the classical computer that can be directly observed. In contrast, none of the intermediates steps of a quantum algorithm can, in principle, be observed: **only** the **input** and **output** states of the quantum algorithm can be observed—this is discussed in Sect. 5.2. Hence, the quantum computer only needs the input and output basis states, which are two 3-bits strings.

In summary, as discussed in Sect. 4.12, for a computation requiring say 300 bits, a classical computer needs to have a storage device that can store 2^{300}-binary string configurations, and which requires a physical device that would have more atoms than the entire Universe. In contrast, this very same computation can be carried out by a quantum computer with just 300 qubits.

So the question arises: where is the information of the quantum computer being stored and where does the process of quantum computation take place? To answer this, one needs to start with the nature of a quantum degree of freedom. Recall from Sect. 4.1 that the quantum degree of freedom is an indeterminate entity that cannot, as such, be observed in physical spacetime: what can be observed are only the particular and determinate values of the degree of freedom. Every time its state vector is observed, the underlying quantum degree of freedom—with some likelihood—is always observed to be in a determinate state.

The Copenhagen interpretation of quantum mechanics and the one used in quantum computers, which was pioneered by Niels Bohr, Max Born and Werner Heisenberg, provides the following explanation: the indeterminate quantum degrees freedom of a quantum computer is described by qubits (state vector) that 'exists' in Hilbert space—a mathematical space that is a superstructure of quantum mechanics. Both the non-trivial phenomena of quantum superposition and quantum entanglement are properties of elements (vectors) of Hilbert space.

It is a remarkable fact that, as discussed by Baaquie [2], the Copenhagen interpretation of quantum mechanics implies that **Hilbert space exists outside of space and time**.

The hardware (quantum device) subjects the state vector of the degree of freedom to various process using electric and magnetic fields, photons and so on, so as to evolve the quantum state vector in Hilbert space, as shown in Fig. 4.6. In particular, the hardware evolves the state vector coherently in Hilbert space—as dictated by the quantum algorithm—and in doing so executes the quantum algorithm, as shown in Fig. 4.6.

As discussed in Sect. 5.2, before the completion of the quantum algorithm no measurement should be performed on the degree of freedom—since doing so will pre-maturely terminate the algorithm. On the completion of the computation, a measurement is performed on the quantum device that causes the qubit (state vector) in Hilbert space to 'collapse'—with different likelihoods—to one of the possible determinate output states determined by the measuring device. For a quantum computer, as mentioned earlier, the possible output states can be the computational basis states—which are the particular and determinate values of the degrees of freedom.[3] The process of measurement projects the degree of freedom—with different likelihood—from Hilbert space to one of the possible computational basis states, which is observed output state; see Fig. 4.6.

In sum, the process of quantum computation takes place in Hilbert space that exists outside spacetime and the results are observed by measuring devices that are in spacetime.

The degrees of freedom and its corresponding Hilbert space are fundamental ingredients of quantum mechanics, and quantum computers bring out this aspect of quantum mechanics with full force. It is as if, while the different steps of the quantum algorithm are being carried out by the evolution of the quantum device in physical spacetime, we are 'looking' at Hilbert space—and the indeterminate quantum degree of freedom that generates it.

19.4 Conclusions

Quantum computing represents a paradigm shift in theoretical computer science. Since all classical algorithms can run on a quantum computer, classical algorithms are a special case of quantum algorithms. Information science can now be seen as an application of the underlying principles of quantum mechanics.

As our discussion on quantum mechanics shows, quantum computations provide an example, in the setting of information science, that once again brings out the mysteries, paradoxes and conundrums of quantum mechanics.

Quantum superposition plays a major role in the efficiency of a quantum computer and quantum entanglement has played a minor role. Entanglement usually came into play when the input state is subjected to the oracle's unitary transformation—and which results in the input state becoming entangled with its ancillary qubit. The ancillary qubits are sometimes used to store the computation's output. At present, most quantum computers do not rely on quantum entanglement, which is currently used for error corrections and in building hardware. Moreover, one needs to prove that, even without error corrections, the quantum algorithms yield an improvement over classical computations.

Both software and hardware need to be developed for quantum computing to make progress. For programmers, the 'hardware description language' using gates and circuits is cumbersome and error prone and requires that each and every step be clearly defined in terms of quantum gates to implement a quantum algorithm.

---

[3] Recall the degree of freedom can have many determinate representations depending on the measuring device.

High-level languages do not refer to the hardware required for carrying out logical operations, and such a high-level programming language has yet to be developed for quantum computers.

Building hardware for quantum computers poses formidable—but apparently not insurmountable—technological challenges. For any realistic problem, one would require about 100 °C of freedom. In 2019, Google had the most powerful quantum computer with 72 qubits.[4] IonQ claims to have the most powerful quantum computer in 2022, based on application-oriented industry benchmarks.[5]

Quantum mechanics started with Schördingers equation in 1926 and it has taken almost a century to understand its implications for information and computer science. What is remarkable is that the hardware of a quantum computer is a macroscopic physical object that obeys the laws of quantum mechanics, bringing forth the paradoxes and enigmas of quantum theory—which explains the microscopic behavior of nature—to a new and currently unexplored regime.

Quantum computers have the potential for an epoch-making transformation in the science of information processing. Quantum algorithms hold the promise of solving problems that have so far been considered insoluble. It is hoped that quantum computers will usher in a new wave of high technology and open up new vistas for the fundamental sciences as well.

References

1. Jordan SP (2017) Fast quantum computation at arbitrarily low energy. Phys Rev A 95:032305
2. Baaquie BE (2013) The theoretical foundations of quantum mechanics. Springer, UK

---

[4] https://www.technologyreview.com/2019/09/18/132956/ibms-new-53-qubit-quantum-computer-is-the-most-powerful-machine-you-can-use/.

[5] https://www.hpcwire.com/off-the-wire/ion-qs-aria-quantum-computer-achieves-20-algorithmic-qubits/.

Index

Printed in the United States
by Baker & Taylor Publisher Services